G. H. Liang, PhD
D. Z. Skinner, PhD
Editors

Genetically Modified Crops
Their Development, Uses, and Risks

Pre-publication
REVIEWS,
COMMENTARIES,
EVALUATIONS . . .

"This book is a useful reference for all researchers working on the improvement of crops through plant transformation. Breeders of all field crops (corn, wheat, rice, soybean, cotton, sorghum, and alfalfa), vegetable crops, horticultural crops, and turfgrasses can get an overview of the progress of genetic engineering efforts in a particular crop. Molecular biologists and cell/tissue culture specialists will find Chapter 1 through Chapter 3 very informative for understanding the methods, strategies, and mechanisms of transgene integration. Both scientists and the general public need to carefully evaluate the benefits and risks of transgenic crop plants by first understanding all facets of each one. This book covers those facets, from the source of the gene, compositions of a gene construct, method of gene delivery, and result of gene integration and expression, to effects of the transgene on plants and the ecology."

Richard R.-C. Wang, PhD
Research Geneticist,
USDA-ARS Forage & Range
Research Laboratory,
Logan, Utah

"This book will help the public gain a better understanding of agricultural biotechnology's merits and risks, thus reducing the public's fear and anxiety. In addition, it will also benefit plant breeders by helping them to decide where to focus their research efforts and to gain more efficiency in breeding methodology."

Paul Sun, PhD
Vice President, Research,
Dairyland Seed Co., Inc.,
Clinton, Wisconsin

More pre-publication
REVIEWS, COMMENTARIES, EVALUATIONS . . .

"This book is useful to graduate and advanced undergraduate students, as well as the interested public for a general understanding of plant transformation and for experienced scientists in specific areas of interest.

It is excellent—full of high-quality information, and illustrated with clear figures and tables. It has an application-oriented explanation and up-to-date references. One exciting part of the book is the review of some of the plant transformation protocols that include some validated modifications to be considered as resources for advanced undergraduate and beginning graduate students who desire orientation on relevant topics and methods used in plant transformation research. It also is appropriate as a textbook for advanced courses in plant transformation and as supplemental references for undergraduate courses in plant breeding or other related courses. *Genetically Modified Crops* is useful as a tool to help students understand research results that are published in refereed journal articles. For students who desire to apply a method, the book provides many examples of pitfalls associated with a particular technique. Such information is quite useful and often is difficult to pry from the detailed literature. The information listed in this book would enable the federal regulatory agencies to contribute technical expertise to the realm of national discussions and the making of science-based decisions on guidelines for commercial release of other genetically modified crops."

Suleiman S. Bughrara, PhD
Assistant Professor,
Department of Crop and Soil Sciences,
Michigan State University,
East Lansing, Michigan

Food Products Press®
An Imprint of The Haworth Press, Inc.
New York • London • Oxford

NOTES FOR PROFESSIONAL LIBRARIANS
AND LIBRARY USERS

This is an original book title published by Food Products Press®, an imprint of The Haworth Press, Inc. Unless otherwise noted in specific chapters with attribution, materials in this book have not been previously published elsewhere in any format or language.

CONSERVATION AND PRESERVATION NOTES

All books published by The Haworth Press, Inc. and its imprints are printed on certified pH neutral, acid-free book grade paper. This paper meets the minimum requirements of American National Standard for Information Sciences-Permanence of Paper for Printed Material, ANSI Z39.48-1984.

Genetically Modified Crops
Their Development, Uses, and Risks

FOOD PRODUCTS PRESS®
Crop Science
Amarjit S. Basra, PhD
Senior Editor

Heterosis and Hybrid Seed Production in Agronomic Crops edited by Amarjit S. Basra

Intensive Cropping: Efficient Use of Water, Nutrients, and Tillage by S. S. Prihar, P. R. Gajri, D. K. Benbi, and V. K. Arora

Physiological Bases for Maize Improvement edited by María E. Otegui and Gustavo A. Slafer

Plant Growth Regulators in Agriculture and Horticulture: Their Role and Commercial Uses edited by Amarjit S. Basra

Crop Responses and Adaptations to Temperature Stress edited by Amarjit S. Basra

Plant Viruses As Molecular Pathogens by Jawaid A. Khan and Jeanne Dijkstra

In Vitro Plant Breeding by Acram Taji, Prakash P. Kumar, and Prakash Lakshmanan

Crop Improvement: Challenges in the Twenty-First Century edited by Manjit S. Kang

Barley Science: Recent Advances from Molecular Biology to Agronomy of Yield and Quality edited by Gustavo A. Slafer, José Luis Molina-Cano, Roxana Savin, José Luis Araus, and Ignacio Romagosa

Tillage for Sustainable Cropping by P. R. Gajri, V. K. Arora, and S. S. Prihar

Bacterial Disease Resistance in Plants: Molecular Biology and Biotechnological Applications by P. Vidhyasekaran

Handbook of Formulas and Software for Plant Geneticists and Breeders edited by Manjit S. Kang

Postharvest Oxidative Stress in Horticultural Crops edited by D. M. Hodges

Encyclopedic Dictionary of Plant Breeding and Related Subjects by Rolf H. G. Schlegel

Handbook of Processes and Modeling in the Soil-Plant System edited by D. K. Benbi and R. Nieder

The Lowland Maya Area: Three Millennia at the Human-Wildland Interface edited by A. Gómez-Pompa, M. F. Allen, S. Fedick, and J. J. Jiménez-Osornio

Biodiversity and Pest Management in Agroecosystems, Second Edition by Miguel A. Altieri and Clara I. Nicholls

Plant-Derived Antimycotics: Current Trends and Future Prospects edited by Mahendra Rai and Donatella Mares

Concise Encyclopedia of Temperate Tree Fruit edited by Tara Auxt Baugher and Suman Singha

Landscape Agroecology by Paul A Wojkowski

Concise Encyclopedia of Plant Pathology by P. Vidhyasekaran

Molecular Genetics and Breeding of Forest Trees edited by Sandeep Kumar and Matthias Fladung

Testing of Genetically Modified Organisms in Foods edited by Farid E. Ahmed

Agrometeorology: Principles and Applications of Climate Studies in Agriculture by Harpal S. Mavi and Graeme J. Tupper

Concise Encyclopedia of Bioresource Technology edited by Ashok Pandey

Genetically Modified Crops: Their Development, Uses, and Risks edited by G. H. Liang and D. Z. Skinner

Genetically Modified Crops
Their Development, Uses, and Risks

G. H. Liang, PhD
D. Z. Skinner, PhD
Editors

Food Products Press®
An Imprint of The Haworth Press, Inc.
New York • London • Oxford

Published by

Food Products Press®, an imprint of The Haworth Press, Inc., 10 Alice Street, Binghamton, NY 13904-1580.

Cover design by Jennifer M. Gaska.

Library of Congress Cataloging-in-Publication Data

Genetically modified crops : their development, uses, and risks / G. H. Liang, D. Z. Skinner, editors.
p. cm.
Includes bibliographical references and index.
ISBN 1-56022-280-8 (case : alk. paper)—ISBN 1-56022-281-6 (soft : alk. paper)
1. Transgenic plants. 2. Crops—Genetic engineering. I. Liang, G. H. (George H.) II. Skinner, D. Z. (Daniel Z.)
SB123.57.G479 2004
631.5'233—dc22

2003022328

CONTENTS

ABOUT THE EDITORS

George H. Liang, PhD, is Professor of Agronomy and Chairman of Genetics at Kansas State University in Manhattan, Kansas. He has been teaching Plant Genetics and Quantitative Genetics at the graduate level for the past 34 years. As a plant geneticist, he has been working on crop transformation using various means since 1993, including pioneer work with *Agrobacterium*-mediated transformation in bentgrass and sorghum. Together with his colleagues, he has published 12 book chapters, 120 refereed journal articles, and two books: *Plant Genetics,* Second Edition and *Experiment Principles and Methods*.

Dr. Liang is recipient of the American Society of Agronomy Fellow award in 1997, the Crop Science Fellow in 2002, and the distinguished faculty award by Gamma Sigma Delta in 2002. He was consultant for the World Bank and the United Nations Development Programme in China. He was elected honorary professor by China Agricultural University, Chinese Academy of Agricultural Sciences, Henan Agricultural University, and Shandong Academy of Agricultural Sciences.

Daniel Z. Skinner, PhD, currently works for the USDA-ARS, and the Wheat Genetics, Quality, Physiology and Disease Research Unit and Crop and Soil Sciences Department at Washington State University as a research leader and supervisory research geneticist (plants). His adjunct professorships include Associate Professor at the Department of Microbiology, Faculty of Science, at Kasetsart University in Bangkok, Thailand; Associate Professor of Agronomy and Associate Professor of Genetics at Kansas State University; and Associate Professor in the Department of Crop and Soil Sciences at Washington State University. His work has been published in many journals, including *Weed Science, Journal of Industrial Microbiology and Biotechnology, Electronic Journal of Biotechnology, Theoretic and Ap-*

plied Genetics, Plant Science, Molecular Breeding, and *Trends in Agronomy.*

Dr. Skinner has worked with plant and fungal genetics and biotechnology for more than 20 years. He has conducted studies designed to assess the potential spread of insect-borne alfalfa pollen from fields of transgenic plants using rare, naturally occurring molecular markers to simulate transgenes. He also investigated the formation of interspecific hybrids among plant species, and found that mobile genetic elements are a factor in genetic rearrangements that occur in the hybrid genome.

CONTRIBUTORS

Tom Clemente is associated with the Plant Science Initiative, Department of Agronomy and Horticulture, University of Nebraska, Lincoln, Nebraska.

Hui Duan is associated with the Department of Plant Science, University of Connecticut, Storrs, Connecticut.

Mistianne Feeney is Graduate Research Assistant, Centre for Environmental Biology, Department of Biological Sciences, Simon Fraser University, Burnaby, British Columbia, Canada.

J. Jayaraj is affiliated with the Department of Agronomy, Kansas State University, Manhattan, Kansas.

J. M. Jeoung is affiliated with the Department of Agronomy, Kansas State University, Manhattan, Kansas.

Junda Jiang is associated with the University of Arkansas Rice Research and Extension Center, Stuttgart, Arkansas.

Karl J. Kramer is Research Chemist, Grain Marketing and Production Research Center, Agricultural Research Service, U.S. Department of Agriculture, Manhattan, Kansas.

Qi Li is associated with the Department of Plant Science, University of Connecticut, Storrs, Connecticut.

Yi Li is associated with the Department of Plant Science, University of Connecticut, Storrs, Connecticut.

Keming Luo is associated with the Biotechnology Center, Southwest Agricultural University, Beipei, Chongqing, People's Republic of China.

Irina F. Makarevitch is associated with the Department of Agronomy and Plant Genetics, University of Minnesota, St. Paul, Minnesota.

Richard J. McAvoy is associated with the Department of Plant Science, University of Connecticut, Storrs, Connecticut.

Amitava Mitra is associated with the Department of Plant Pathology, University of Nebraska, Lincoln, Nebraska.

S. Muthukrishnan is Professor, Department of Biochemistry, Kansas State University, Manhattan, Kansas.

James Oard is associated with the Department of Agronomy, Louisiana State University, Baton Rouge, Louisiana.

Paula M. Olhoft is associated with the Department of Agronomy and Plant Genetics, University of Minnesota, St. Paul, Minnesota.

David W. Ow is associated with the Plant Gene Expression Center, Agricultural Research Service, U.S. Department of Agriculture, Albany, California, and the Department of Plant and Microbial Biology, University of California, Berkeley, California.

Sung Hun Park is associated with the Vegetable and Fruit Improvement Center and the Department of Entomology, Texas A&M University, College Station, Texas.

Yan Pei is associated with the Biotechnology Center, Southwest Agricultural University, Beipei, Chongqing, People's Republic of China.

Zamir K. Punja is Professor and Director, Centre for Environmental Biology, Department of Biological Sciences, Simon Fraser University, Burnaby, British Columbia, Canada.

Deborah A. Samac is associated with the U.S. Department of Agriculture, Agricultural Research Service, and the University of Minnesota, St. Paul, Minnesota.

James W. Smith is associated with the Vegetable and Fruit Improvement Center and the Department of Entomology, Texas A&M University, College Station, Texas.

Roberta H. Smith is associated with the Vegetable and Fruit Improvement Center and the Department of Entomology, Texas A&M University, College Station, Texas.

David A. Somers is associated with the Department of Agronomy and Plant Genetics, University of Minnesota, St. Paul, Minnesota.

Paul St. Amand is affiliated with the Agronomy Department, Kansas State University, Manhattan, Kansas.

Sergei K. Svitashev is associated with the Department of Agronomy and Plant Genetics, University of Minnesota, St. Paul, Minnesota.

Stephen J. Temple is associated with the U.S. Department of Agriculture, Agricultural Research Service, and the University of Minnesota, St. Paul, Minnesota.

M. R. Tuinstra is affiliated with the Department of Agronomy, Kansas State University, Manhattan, Kansas.

Xiaoling Wang is associated with the Department of Plant Biology and Pathology at Rutgers University, New Brunswick, New Jersey.

J. T. Weeks is affiliated with the Simplot Plant Sciences, J. R. Simplot Company, Boise, Idaho.

Jack M. Widholm is Professor of Plant Physiology, Department of Crop Sciences, Edward R. Madigan Laboratory, University of Illinois, Urbana, Illinois.

Yan H. Wu is associated with the Department of Plant Science, University of Connecticut, Storrs, Connecticut.

John Wurst is associated with the Department of Plant Science, University of Connecticut, Storrs, Connecticut.

Degang Zhao is associated with the Department of Plant Science, University of Connecticut, Storrs, Connecticut.

Barbara A. Zilinskas is associated with the Department of Plant Biology and Pathology at Rutgers University, New Brunswick, New Jersey.

Preface

Agricultural biotechnology is a major issue today, facing not only scientists but consumers and producers as well. A large number of people in the United States and Canada consume food and beverages produced from crops and enzymes that are modified through biotechnology processes. Canola oil, soybean oil, corn oil, tofu, potato and potato chips, cheese, milk, fruits, and even beer are examples of genetically modified products. In addition, fiber crops, such as cotton, and vegetables, such as tomato, produced from genetically modified plants also are on the market. Biotechnology-modified foods, a success of the scientific community, clearly are a reality. On the one hand, these crop plants do provide advantages and are beneficial to society. On the other hand, the production and marketing of genetically modified food has created fear and anxiety among the public. There is a definite need for this fear and anxiety to be addressed.

Many meetings and workshops have been organized to discuss the methodologies, advantages, and disadvantages of biotechnologically processed food products. Chapters in this book are a continuation of these efforts. Topics under discussion range from safety to evolution, ethics to ecology, and environment to legality. All contributors are genuinely concerned with the processes and the products. Regardless of the differences in personal viewpoints, the basic tenet of using biotechnology in food production is to resolve the problem of ever-increasing world population versus food production with limited crop land. Food insecurity and malnutrition still exist, even in the twenty-first century when humans have already landed on the moon, communication-satellite technology is in use by every household where there is a phone, human genome mapping is nearly completed, cloning of livestock and even humans is no longer a dream, and a computer has been installed in nearly every home, classroom, or office in schools, governmental agencies, and private companies in many countries. Yet one of the most basic requirements of human life—food—still is not in adequate supply for every human being on earth. Do those who are hungry or suffer from malnutrition have the right to express their desire and need? Should we hear from them? Conversely, in order to fulfill such needs, should we not attempt to provide some answers?

The processes of genetically modifying crops or livestock have been in existence since antiquity, although the techniques used now are revolutionary. The production of a mule by crossing a horse and donkey is an example of an old technique. Triticale, a product of crossing wheat and rye, is another. Many scientists working on livestock, crops, fish, poultry, forest trees, enzymes, and even insects such as silkworm have changed the genetic constitution of certain species by adding genetic material from other species to produce a more adapted form for human consumption and utilization. Transformation, a branch of agricultural biotechnology, simply expands the existing knowledge and skill from our ancestors to produce more adapted forms of crops that are environmentally friendly and beneficial to both producers and consumers.

This book is intended for use not only by scientists but also by the general public who are interested in being informed of the methods used and risks of introducing genetically modified organisms into the environment. To this end, the following questions are posed and answers are attempted:

- What are the techniques and goals of producing genetically modified crops?
- What genetic materials have been, or are going to be, used in transformation?
- What are the results of such endeavors?

Those who have contributed to the chapters in this book are all researchers who have hands-on experience in transformation and are willing to provide and share their firsthand information with their colleagues and the public.

Chapter 1

Transformation: A Powerful Tool for Crop Improvement

D. Z. Skinner
S. Muthukrishnan
G. H. Liang

INTRODUCTION

The demand for producing transgenic crops is the direct result of an ever-increasing world population. The World Bank reported that enough staple food is produced to feed everyone in the world, and yet 800 million people are food insecure and 180 million of them are preschool children (Weeks, 1999). Clearly, equitable food distribution presents a challenge. In addition, production could become limiting in the future, with a projected addition of 1 billion people in Asian countries and an 80 percent increase in population in sub-Saharan Africa by 2020. Distribution of people throughout the world is highly uneven, and the countries with the largest gains in population are often the ones without the resources to accommodate greater masses.

To meet the food demand by the year 2020, it is predicted that a 60 percent increase in the food supply will be necessary. Also, a 200 percent increase was projected for meat and dairy products, both of which must be converted from grain and forage, driven in part by increasing wealth in Asia and Southeast Asia. To compound the problem, the increase in food supply will not come from an increase in cropland. Rather, it must come from increases in productivity per unit area of land and unit volume of water (Pinstrup-Andersen and Pandya-Lorch, 1999). Increases in cultivated area will contribute less than 20 percent to the increase in global cereal production between

1993 and 2020. Most of the growth in land for cereal production will be concentrated on the land with relatively low production potential in sub-Saharan Africa; some expansion of cultivated land will occur in South America, but cultivated area for cereals in Asia will remain stagnant.

If increase in cropland area does not contribute much to increased cereal production, then meeting the demand for cereals will depend on yield improvement. However, rate of increase in yield is declining because prices of cereals are low, forcing some farmers to abandon cereal crops for more profitable crops, which in turn reduces continuous investment in research, infrastructure, and irrigation facilities. Also, sustaining the same rate of yield gain will be difficult in certain regions where input of water and fertilizer is already high and optimum profitable economic return has been approached. The forecasted yield increase will drop to an annual rate of 1.5 percent from 1993 to 2020 in contrast to an increase of 2.3 percent from 1982 to 1994 (Pinstrup-Andersen and Pandya-Lorch, 1999).

Much of the yield increase in cereals will likely go toward meeting projected increases in demand for meats. Due to population growth, coupled with increased income and changes in lifestyle in developing countries, demand for meat is projected to increase by 2.8 percent per year from 1993 to 2020. While per capita demand for cereals is projected to increase by 8 percent, demand for meat will increase by 43 percent (Pinstrup-Andersen and Pandya-Lorch, 1999). The increase in demand for meat will make it necessary to produce more feed from already limited land resources.

Without proper management, availability of fresh water will emerge as the most important constraint to global food production. Furthermore, water often is poorly distributed among countries, within countries, and between seasons. The second longest river in China, the Yellow River, for example, was without water in downstream areas for more than 200 days in 1997, and many of the water springs in the northern provinces are out of water due to heavy use of underground water. Today, 28 countries with a total population exceeding 300 million people face water stress. By 2025, the number of people facing a water shortage could increase to 3 billion across 50 countries (Rosegrant et al., 1997).

Agriculture is by far the largest user of water, accounting for 72 percent of global water withdrawals and 87 percent of withdrawals in developing countries. Establishing sound policies or reforming the current ones to reduce wasteful use of water will provide improved efficiency in water use and increase crop production per unit of land. Nevertheless, global demand for irrigation water will continue to grow as demand for cereal production increases. The cost of developing new sources of water will be high, and special sources of water supplies, such as desalination, reuse of wastewater, and water harvesting, are unlikely to provide universal availability of water, although these methods may be established and helpful for certain limited localities.

Crops also need protection from pests. Naturally occurring pests inflict tremendous yield and storage losses. For example, stalk rot of sorghum incited by *Fusarium* species could cause an $80 million loss each year in Kansas alone. Other pathogen threats also exist. There are scab, rust, and virus disease in cereals; leaf and stem diseases in forage crops; and root diseases in almost all the crops. All of these are common occurrences in crop fields. Likewise, insect damage is equally serious. In addition to pests, abiotic stresses, such as drought and extreme temperatures, could cause more damage not just to field crops but also to turfgrasses. For example, there are 23 million acres of turfgrass or lawn in the United States, and abiotic stress is known to lower the yield and reduce turfgrass quality.

Meeting the demand of a rapidly growing population, reducing the widening social and economic disparities both within and among countries, and at the same time safeguarding the earth's natural resources, the quality of the atmosphere, and other critical components of the environment—all represent tremendous challenges facing us today. Genetically modified crops have the potential to contribute to solving these challenges.

INCORPORATION OF TRANSGENES INTO CROPS

Most of the methods currently used for plant transformation employ a technique for delivering the DNA into the cell without regard to its ultimate intracellular location. Once inside the appropriate cellular location by a chance process, the DNA is integrated into the chromosomal (or organellar) DNA, usually by a nonspecific recombination process. The exception is *Agrobacterium*-mediated transformation, which delivers the DNA specifically into the nuclear compartment with high efficiency and also provides a mechanism for its integration.

In plant transformation, two physical barriers prevent the entry of DNA into a nucleus, namely the cell wall and the plasma membrane. Most plant transformation methods overcome these barriers using physical and/or chemical approaches. In protoplast-mediated transformation, the cell wall is digested with a mixture of enzymes which attack cell wall components, yielding individual protoplasts that can be maintained intact using appropriate osmoticum in the medium. Entry of DNA is then facilitated by the addition of permeabilizing agents, such as polyethylene glycol, that allow DNA uptake, presumably by coating the negatively charged DNA inside the cell and various cellular compartments in a random process.

A second method, often referred to as biolistic transformation, utilizes fine metal particles (typically tungsten or gold) coated with DNA that are usually accelerated with helium gas under pressure. Other methods, such as microinjection, silicon carbide fiber treatment, vacuum infiltration, dipping of floral parts, sonication, and electroporation, cause transient microwounds in the cell wall and the plasma membrane, allowing the DNA in the medium to enter the cytoplasm before repair or fusion of the damaged cellular structures. However, many of these methods are tedious and result in variable transformation efficiencies. In the following sections, we provide details and relative merits of several, but not all, available protocols.

In all transformation techniques, the desired transgene is placed under the control of a promoter which produces high-level constitutive or inducible expression of the gene in specific or all tissues. In addition, another gene that allows detection or selection of the transformed cells is introduced in the same or different vector DNA. The presence of a screenable or selectable marker gene greatly simplifies the identification of transformed plants and increases the efficiency of recovery of transgenic plants. Typical screenable markers are the *gus* gene, encoding a β-glucuronidase, and the *gfp* gene encoding a green fluorescent protein. Selectable markers that have been most useful are those conferring resistance to antibiotics such as kanamycin, paromomycin, and hygromycin, or to herbicides such as bialaphos, glyphosate, or cyanamide. In addition, selection also can be achieved by including metabolites that require special enzymes for their utilization as a carbon source, such as phosphomannose isomerase and xylose isomerase.

METHODOLOGY OF PRODUCING TRANSGENIC PLANTS

Most of the important crop species have been successfully transformed, at least in the laboratory. Two major approaches have been widely used to produce transgenic crop plants, both monocots and dicots. One is biolistic bombardment and the other is *Agrobacterium*-mediated transformation. In the biolistic protocol, the primary delivering systems are the BioRad PDS 1000/He helium-powered gun and similar designs, or the particle inflow gun. The second approach involves various strains of *Agrobacterium tumefaciens,* such as LBA4404, EHA105, and C58, carrying a T-DNA vector. The transgene and the selectable marker are inserted in the vector between two unique sequences, called the left border and the right border, which are utilized in insertion of the T-DNA into the host chromosomal DNA (Kado, 1998; Zupan and Zambryski, 1995).

Parameters involved in the biolistic gun include pressure (ranging from 900 to 1300 psi), particle size (0.6 to 1.1 μm), and type of material (gold and tungsten), target distance (7.5 to 10 cm), and target material (cell suspension, callus, meristem, protoplast, immature embryo). These parameters vary somewhat with the crop involved. Disadvantages of using the biolistic gun are low success frequency, high copy numbers that often are correlated with gene silencing, patent issues, and cost.

Agrobacterium-mediated transformation may correct some of the weaknesses encountered with the biolistic approach and has been successful in rice, maize, sorghum, bentgrass, and other crops; its effectiveness with other crops, especially wheat, remains questionable. Further, some cultivar versus *Agrobacterium* strain specificity may limit the range of cultivars that can be successfully transformed. Development of reliable and efficient protocols are needed to improve the efficiency and range of both approaches in transforming crop plants.

For *Agrobacterium*-mediated transformation, calli are induced from the explants, such as immature embryos from wheat and sorghum and mature seed from creeping bentgrass. To illustrate, the procedure used for sorghum is briefly described. Calli are induced from zygotic immature embryos (1 to 2 mm in diameter). *Agrobacterium* culture carrying the desired transgene/selectable marker is grown overnight and resuspended to reach an $OD_{600} = 0.5$ in I_6 liquid medium (a modification of I_3 medium [Casas et al., 1993]) containing 100 μM acetosyringone. The calli are cocultured with the *Agrobacterium* for 30 min and blotted dry for a few seconds. Then the calli are placed on semisolid I_6 medium containing 2.5 $g{\cdot}L^{-1}$ phytagel and 100 μM acetosyringone and cultured for three to four days at 25°C in darkness. The calli should be washed three times with cefotaxime (300 to 500 $mg{\cdot}L^{-1}$) solution and cultured in I_6 medium without selection for one week in darkness. The calli will then be transferred to fresh medium containing a selection agent, e.g., 3 $mg{\cdot}L^{-1}$ bialaphos or 12.5 $mg{\cdot}L^{-1}$ cyanamide hydratase. They will be cultured at 25 to 27°C for six weeks with subculture to fresh medium once every two weeks. Next, shoots formed from bialaphos-resistant calli will be selected and transferred to rooting medium (half-strength MS medium with 0.2 $mg{\cdot}L^{-1}$ IAA, 0.2 $mg{\cdot}L^{-1}$ NAA) containing 3 $mg{\cdot}L^{-1}$ bialaphos or 12.5 $mg{\cdot}L^{-1}$ cyanamide hydratase. Finally shoots with well-developed root systems may be transferred to soil.

The efficiency of tissue culture methods is affected by species, surfactant use, and presence of acetosyringone. DNA delivery efficiency was significantly decreased in the absence of acetosyringone when embryo-derived calli were used in transformation of rice (Hiei et al., 1994) and maize (Ishida et al., 1996). However, when wheat cell suspension culture was used in transformation with *Agrobacterium,* acetosyringone was not essential (Cheng et al., 1997), suggesting that different tissues may have different competence for *Agrobacterium* infection. Use of a surfactant, such as Silwet, proved to be essential to DNA delivery, possibly indicating that surface-tension-free cells favored *Agrobacterium* attachment.

Transformation techniques not involving tissue culture are desirable for crop species in which many cultivars or inbred lines do not respond to tissue culture. These techniques include soaking and vacuum infiltration transformation of *Arabidopsis* inflorescence with *Agrobacterium* (Bechtold, 1994; Bent et al., 1994), transformation via the pollen tube pathway (Luo and Wu, 1988; Zhang, personal communication, 1999), and pollen transformation via biolistics (Ramaiah and Skinner, 1997). The dipping method adopted in *Arabidopsis* must be modified for cereal crops by altering the plant stage of infiltration, the concentration of bacterial culture, and the duration of treatment. Further, each tiller (for wheat, barley, etc.) of the same plant must be kept separate. If this technique works, selection for transformants is made directly from seed-derived seedlings and tissue culture is avoided. Thus, genotypes that respond negatively to callus induction and plant regeneration will not present a problem; however, it is likely some genotypes will be more amenable to infiltration transformation than others.

CONFIRMATION OF PUTATIVE TRANSGENIC PLANTS AND TRANSFORMATION EFFICIENCY

Commonly used methods to confirm the putative transgenic plants are polymerase chain reaction (PCR), Southern blotting, Western blotting, Northern blotting, functional assay (testing the presence of the selectable marker and the target gene), in situ hybridization, and progeny analysis (segregation of the target gene). Not all transgenic plants produce the same amount of protein from the target gene and selection based on the Western blot is necessary. This is because a positive correlation usually exists between the effectiveness of the gene in the bioassay and the amount of protein it produces. For example, the level of rice chitinase accumulating in transgenic sorghum plants with the *chi11* gene was positively correlated with resistance to sorghum stalk rot (Krishnaveni et al., 2001).

Polymerase Chain Reaction

PCR, a simple and rapid procedure, is utilized to confirm whether a putatively transgenic plant that has survived selection is indeed transgenic. Usually, two primers (one forward and one reverse) specific for the selectable marker (*bar* or *cah* gene, for example) are used in a PCR reaction with genomic DNA extracted from the transgenic plants. A thermostable DNA polymerase amplifies the region between the two primers during the multiple amplification cycles of the PCR, which yields a DNA fragment of predicted size (the length equal to the number of base pairs between the two primers in the transgene). This fragment is easily detected on an agarose gel by staining with ethidium bromide. PCR is a very sensitive and rapid method for identification of transgenic plants in the seedling stage and requires only a small amount of plant tissue. The presence of additional transgenes in the same plants carrying the selectable marker can be detected with a different set of primers using the same template DNA.

Southern Blotting/Hybridization Analysis

In Southern blotting, DNA fragments from transgenic plants generated by digestion with restriction enzyme(s) are first separated according to fragment size by electrophoresis through an agarose gel. The DNA fragments then are transferred to a solid support, such as a nylon membrane or a nitrocellulose sheet. The transfer is effected by simple capillary action, sometimes assisted by suction or electric current. The DNA binds to the solid support, usually because the support has been treated to carry a net positive charge, or some other means of binding, such as inducing covalent binding of the DNA to the support. DNA fragments maintain their original positions in the gel after transfer to the membrane. Hence, larger fragments will be localized toward the top of the membrane and smaller fragments toward the bottom. The positions of specific fragments can then be determined by "probing" the membrane. The probe consists of the DNA fragments of interest, such as a cloned gene, which has been labeled with a radioactive isotope or some other compound that allows its visual detection. Under the proper set of conditions, the denatured single-stranded probe will hybridize to its complementary single strands of genomic DNA affixed to the membrane. In this way, the size of the fragment on which the probe resides in the genomic DNA can be determined. In transgenic plant experiments, the Southern blot often is used to determine whether an introduced gene is indeed present in the plant DNA and whether multiple transgenic plants carry the introduced gene on the same size of DNA fragment (suggesting a single transformation event) or on different sized fragments (suggesting independent transformation events). The results of Southern blots also indicate whether a single copy of the gene has been inserted or if multiple copies are likely to be present. The Southern blot derives its name from its inventor, E. M. Southern (Southern, 1975).

Northern Blotting

Northern blotting—the name was derived as a play on words from the Southern blot—is very similar to the Southern blot, except that instead of restriction enzyme-digested DNA, native RNA is separated according to size by electrophoresis through an agarose gel and then transferred to a solid support. The rest of the Northern blot procedure is very similar to that of the Southern blot and it is used to determine whether the introduced gene has been transcribed into messenger RNA and accumulates in the transgenic plant.

Western Blotting

The Western blotting procedure detects the protein of the transgene in an extract of proteins prepared from various parts of the transgenic plants and is, therefore, an assay for a functional transgene. In this technique the proteins are first electrophoresed in an SDS-polyacylamide gel and the proteins are then transferred to nitrocellulose of polyvinylidene difluoride (PVDF) membrane by electrophoretic transfer. The membrane is then treated with an antibody specific for the protein encoded by the transgene followed by a second antibody coupled to an enzyme, which can act on a chromogenic (or fluorogenic) substrate leading to visualization of the transgene protein with increased sensitivity. The expression level of the protein can be quantified using known amounts of the transgene-encoded protein.

Functional Assay

When the selectable markers used are antibiotic-resistant or herbicide-resistant genes, a function assay can be made by spraying antibiotics or smearing herbicide on the leaves of those putative transgenic seedlings or plants in later segregating populations. Sensitive plants typically will turn brown and shrivel up whereas resistant (transgenic) plants stay healthy and green. Such an assay provides the initial screening of large numbers of putative transgenic plants and reduces the work load by eliminating escapes during selection.

Progeny Test

With stable transformed genes, progeny testing should show the presence and activity of the selectable marker and target genes, such as the gene *gfp* encoding green fluorescence protein, or *bar* and disease resistance. However, segregation does not always follow the typical Mendelian fashion. For example, among the progeny in the *Agrobacterium*-mediated wheat transformation experiments reported by Cheng and colleagues (1997), segregation in the T_1 generation had ratios of 32:0, 1:34, 0:40, and 74:0 in addition to the expected 1:1, 3:1, or 15:1 ratios. This variability indicates aberrant segregation. However, in other cases, segregation follows the normal Mendelian pattern. For example, among six sorghum T_0 transgenic plants produced by biolistic bombardment, all showed typical 3:1 segregation ratios in the T_1 generation (Zhu et al., 1998).

CROP SPECIES AMENABLE TO TRANSFORMATION

Any crop that is able to produce calli from explants and is capable of callus regeneration into plants with high efficiency is amenable to transformation using biolistic bombardment and *Agrobacterium tumefaciens*. However, it is known that response to tissue culture is highly genotype dependent. Furthermore, somaclonal variation, spontaneous genetic variation occurring in cells growing in vitro (Larkin and Scowcroft, 1981), could occur during tissue culture processes. Thus, to confirm that the improved phenotype of the transgenic plants is due to the transgene, two controls should be included—one from seed-derived plants and the other from nontransformed tissue culture-derived plants.

For those crops or genotypes that show a negative response to tissue culture, a transformation procedure independent from tissue culture should be considered and tested. It will be a great accomplishment to perfect a procedure bypassing tissue culture, because many cultivars of wheat and rice and inbred lines of maize and sorghum do not respond to tissue culture operations.

CURRENT AND FUTURE TRANSGENIC CROPS

The great majority of transgenic crops currently grown in the field are derived from the "magic bullet" type of transformation. That is, a single gene has been inserted into the crop species; the product of that gene causes a demonstrable change in phenotype of the plants. For example, many crop species express an endotoxin from various *Bacillus thuringiensis* (Bt) strains, conferring resistance to various insect pests. Numerous examples have been reported in the literature (Arpaia et al., 2000; Leroy et al., 2000; Breitler et al., 2000; Mandaokar et al., 2000; Acciarri et al., 2000; Giles et al., 2000; Walker et al., 2000; Harcourt et al., 2000; and Zhao et al., 2000).

Padgette and colleagues (1995) determined that a cloned 5-enolpyruvylshikimate-3-phosphate synthase gene from *Agrobacterium tumefaciens* strain CP4, engineered to express in plants, substituted for the endogenous gene that is inhibited by the nonspecific herbicide glyphosate (*N*-phosphonomethyl-glycine). The bacterial gene was not inhibited by the herbicide, resulting in transgenic plants, now commonly referred to as Roundup Ready, able to withstand applications of glyphosate that kill all surrounding plants without this gene. Other herbicide resistance genes have been introduced into crop plants as well, including resistance to L-phosphinothricin due to the phosphinothricin-acetyl transferase *(pat)* gene from *Streptomyces viridochromogenes,* or resistance to the closely related bialaphos (phosphinothricyl-L-alanyl-L-alanine) due to the *bar* (bialaphos resistance) gene isolated from *Streptomyces hygroscopicus*. These genes are used routinely as selectable markers in plant transformation experiments, as well as the functional genes in the herbicide-resistant Liberty-Link series of field crops.

Other experiments of single transgenes conferring a trait of interest include constitutive expression of a bacterial citrate synthetase gene, which reportedly conferred tolerance to acidic soil in tobacco and papaya (de la Fuente et al., 1997). Attempts have been made to interfere with normal growth and nutrient assimilation by pests and parasites by expressing in plants enzymes such as chitinase, or enzyme inhibitors such as protease inhibitors. Some success has been reported with this approach. For example, several plant species have shown enhanced resistance to fungal diseases when transformed to constitutively express a chitinase gene isolated from rice (Lin et al., 1995; Tabei et al., 1998; Nishizawa et al., 1999; Yamamoto et al., 2000; Takatsu et al., 1999).

Recently, combinations of genes have been expressed in plants to confer multiple resistance. Examples include 'Herculex I' maize, engineered to express resistance to glufosinate herbicides and to express a Bt toxin effective against lepidopterans, and 'NewLeaf Plus' potatoes, engineered to express a Bt toxin and genes that confer viral resistance to viruses. It is likely that many more "magic bullet" combinations will be constructed and marketed in the future.

Potential and existing transgene combinations necessitate the study of the interactions of the transgenes to elucidate any effect one gene has on the efficacy of the others. Some investigation has been conducted into the expression of multiple transgenes in the same plant (e.g., Maqbool and Christou, 1999; Maqbool et al., 2001). However, these combinations of genes are not interdependent. That is, the Bt gene in 'Herculex I' maize does not rely on the glufosinate herbicide resistance gene for expression, nor vice versa. In the future, entire metabolic pathways will be engineered to express several intermediate gene products to gain expression of a particular desired metabolite, increasing the potential for transgene interactions.

The first successful attempt to engineer a system into plants involved expressing a form of the mammalian antiviral 2-5A system in tobacco plants (Mitra et al., 1996). Two genes were built into the same vector and transformed into tobacco plants. Some of the resulting plants expressed both genes, resulting in a resistance response to three different plant pathogenic viruses (Mitra et al., 1996). A more complex system was developed in the construction of a novel form of rice (*Oryza sativa* L.). Through careful characterization of metabolic products naturally occurring in rice, it was deduced that if four enzymes—phytoene synthase, phytoene desaturase, zeta-carotene desaturase, and lycopene cyclase—were expressed in rice, provitamin A would result. After a relatively short eight years of work, "Golden Rice" was developed and found to express provitamin A in quantities theoretically sufficient to be pharmacologically therapeutic (Potrykus, 2001). Genes cloned from ornamental flowers were used in the construction of "Golden Rice." As the biochemical pathways of increasingly complex traits are elucidated, we can expect to see more engineered pathways to gain expression of nutritionally beneficial metabolites in high-yielding crops, with the genes cloned from model systems that lend themselves to the isolation of the necessary genes. The knowledge and techniques gained from the comparatively simple transformation studies done to date are just the beginning of what may develop into a second "green revolution."

FOOD SAFETY AND RISK ANALYSIS

It is only natural for humans to be concerned with food safety. This concern is increasing, at least in industrialized countries, as evidenced by the growing demand for organic foods, public objections to producing genetically modified organisms, requiring labeling of food to show the origin and mode of producing and processing, and establishment of regulations for producing, processing, storing, and transporting foods. In spite of regulation and precaution, recalls of manufactured food products and outbreaks of food poisoning are not uncommon. The extensive recall of food products containing Star-Link corn in 2001 is an example. In addition, the proposed use of terminator seed technology to protect the transgenic crops from illegal and unauthorized use of patented seed has met with strong objection. Chapter 14 discusses the risk analysis.

Clearly, transgenic crops will play a major role in modern agriculture. The efficient and safe use of this technology will depend on the free and open sharing of the developing technology and the rigorous examination of the productivity and safety of the resulting crops. In the chapters that follow, authors at the forefront of this developing technology present reviews of the current state of the art for field and horticultural crops. It is hoped that this collection will serve as a platform for the continued exchange of results and ideas for improving the productivity of the world's supply of food, feed, and fiber.

REFERENCES

Acciarri, N., G. Vitelli, G. Mennella, F. Sunseri, and G.L. Rotino (2000). Transgenic resistance to the Colorado potato beetle in Bt-expressing eggplant fields. *HortScience* 35:722-725.

Arpaia, S., L. Demarzo, G.M. Di-Leo, M.E. Santora, G. Mennella, and J.J.A. van Loon (2000). Feeding behaviour and reproductive biology of Colorado beetle adults fed transgenic potatoes expressing the *Bacillus thuringiensis* Cry3B endotoxin. *Entomologia Experimentalis et Applicata* 95:31-37.

Bechtold, N., J. Ellis, and G. Pelletier (1994). *Agrobacterium* mediates gene transfer by infiltration of adult *Arabidopsis thalina. Plant Journal* 5:421-427.

Bent, A., B.N. Kunkel, D. Kahlbeck, K.L. Brown, R. Schmidt, J. Giraudat, J. Leung, and B.J. Staskawicz (1994). RPS2 of *Arabidopsis thalina:* A leucine-rich repeat class of plant diseases resistance genes. *Science* 265:1856-1860.

Breitler, J.C., V. Marfa, M. Royer, D. Meynard, J.M. Vassal, B. Vercambre, R. Frutos, J. Messeguer, R. Gabarra, and E. Guiderdoni (2000). Expression of a *Bacillus thuringiensis* cry1B synthetic gene products Mediterranean rice against the striped stem borer. *Plant Cell Reports* 19:1195-1202.

Casas, A.M., A.K. Kononowicz, U.B. Zehr, D.T. Tomes, J.D. Axtell, R.A. Bresan, and P.M. Hasegawa (1993). Transgenic sorghum plants via microprojectile bombardment. *Proceedings of the National Academy of Sciences of the United States of America* 90:11212-11216.

Cheng, M., J.E. Fey, S. Pang, H. Zhou, C.M. Hironaka, D.R. Duncan, T.W. Conner, and Y. Wan (1997). Genetic transformation of wheat mediated by *Agrobacterium tumefaciens. Plant Physiology* 115:971-980.

de la Fuente, J.M., V. Ramirez-Rodriguez, J.L. Cabrera-Ponce, and L. Herrera-Estrella (1997). Alluminum tolerance in transgenic plants by alteration of citrate synthesis. *Science* 276:1566-1568.

Giles, K.L., R.L. Hellmich, C.T. Iverson, and L.C. Lewis (2000). Effects of transgenic *Bacillus thuringiensis* maize grain on *B. thuringiensis*-susceptible *Ploida interpunctella* (Lepidoptera: Pyralidae). *Journal of Economic Entomology* 93:1011-1016.

Harcourt, R.L., J. Kyozuka, R.B. Floyd, K.S. Bateman, H. Tanaka, V. Decroocq, D.L. Llewellyn, X. Zhu, W.J. Peacock, and E.S. Dennis (2000). Insect- and herbicide-resistant transgenic eucalypts. *Molecular Breeding* 6:307-315.

Hiei, Y.S., S. Ohta, T. Komari, and T. Kumasho (1994). Efficient transformation of rice (*Oryza sativa* L.) mediated by agrobacterium and sequence analysis of the boundaries of the T-DNA. *Plant Journal* 6:271-282.

Ishida, Y., H. Saito, S. Ohta, Y. Hiei, T. Komari, and T. Kumashiro (1996). High efficiency transformation of maize (*Zea mays* L.) mediated by *Agrobacterium tumefaciens. Nature Biotechnology* 14:745-750.

Kado, C.I. (1998). *Agrobacterium*-mediated horizontal gene transfer. In *Genetic Engineering,* Volume 20 (1-24), Ed. Setlow, J.K. New York: Plenum Press.

Krishnaveni, S., J.M. Jeoung, S. Muthukrishnan, and G.H. Liang (2001). Transgenic sorghum plants constitutively expressing a rice chitinase gene show improved resistance to stalk rot. *Journal of Genetics and Breeding* 55:151-158.

Larkin, P.J. and W.R. Scowcroft (1981). Somaclonal variation—A novel source of variability from cell cultures for plant improvement. *Theoretical and Applied Genetics* 60:197-214.

Leroy, T., A.M. Henry, M. Royer, I. Altosaar, R. Frutos, D. Duris, and R. Philippe (2000). Genetically modified coffee plants expressing the *Bacillus thuringiensis Cry1Ac* gene for resistance to leaf miner. *Plant Cell Reports* 19:382-389.

Lin, W., C.S. Anuratha, K. Datta, I. Potrykus, S. Muthukrishnan, and S.K. Katta (1995). Genetic engineering of rice for resistance to sheath blight. *Biotechnology* 13:686-691.

Luo, Z.X. and R. Wu (1988). A simple method for the transformation of rice via the pollen-tube pathway. *Plant Molecular Biology Reporter* 6:165-174.

Mandaokar, A.D., R.K. Goya, A. Shukla, S. Bisaria, R. Bhalla, V.S. Reddy, A. Chaurasia, R.P. Sharma, I. Altosaar, and P.A. Kumar (2000). Transgenic tomato plants resistant to fruit borer (*Helicoverpa armigera* Hubner). *Crop Protection* 19:307-312.

Maqbool, S.B. and P. Christou (1999). Multiple traits of agronomic importance in transgenic indica rice plants: Analysis of transgene integration patterns, expression levels and stability. *Molecular Breeding* 5:471-480.

Maqbool, S.B., S. Riazuddin, N.T.L. Angharad, M.R. Gatehouse, J.A. Gatehouse, and P. Christou (2001). Expression of multiple insecticidal genes confers broad resistance against a range of different rice pests. *Molecular Breeding* 7:85-93.

Mitra, A., D.W. Higgins, W.G. Langenberg, H. Nie, D.N. Sengupta, and R.H. Silverman (1996). A mammalian 2-5A system functions as an antiviral pathway in transgenic plants. *Proceedings of the National Academy of Sciences of the United States of America* 93:6780-6785.

Nishizawa, Y., Z. Nishio, K. Nakazono, M. Soma, E. Nakajima, M. Ugaki, and T. Hibi (1999). Enhanced resistance to blast *(Magnaporthe grisea)* in transgenic Japonica rice by constitutive expression of rice chitinase. *Theoretical and Applied Genetics* 99:383-390.

Padgette, S.R., K.H. Kolacz, X. Delannay, D.B. Re, B.J. LaVallee, C.N. Tinius, W.K. Rhodes, Y.I. Otero, G.F. Barry, and D.A.W. Eichholtz (1995). Development, identification, and characterization of a gluphosate-tolerant soybean line. *Crop Science* 35:1451-1461.

Pinstrup-Andersen, P. and R. Pandya-Lorch (1999). Securing and sustaining adequate world food production for the third millennium. In *World Food Security and Sustainability: The Impacts of Biotechnology and Industrial Consolidation* (pp. 27- 53), Eds. Weeks, D.P., J.B.Segelken, and R.W.F. Hardy. Ithaca, NY: National Agricultural Biotechnology Council.

Potrykus, I. (2001). Golden rice and beyond. *Plant Physiology* 125:1157-1161.

Ramaiah, S.M. and D.Z. Skinner (1997). Particle bombardment: A simple and efficient method of alfalfa [*Medicago sativa* (L.)] pollen transformation. *Current Science* 73:674-682.

Rosegrant, M.W., C. Ringler, and R.V. Gerpacio (1997). Water and land resources and global food supply. Twenty-Third International Conference on Agriculture Economy, Sacramento, California, August 10-16.

Southern, E.M. (1975). Detection of specific sequences among DNA fragments separated by gel electrophoresis. *Journal of Molecular Biology* 98:503-517.

Tabei, Y., S. Kitade, Y. Nishizawa, N. Kikuchi, T. Kayano, T. Hibi, and K. Akutsu (1998). Transgenic cucumber plants harboring a rice chitinase exhibit enhanced resistance to gray mold *(Botrytis cinerea). Plant Cell Reports* 17:159-164.

Takatsu, Y., Y. Nishizawa, T. Hibi, and K. Akutsu (1999). Transgenic chrysanthemum (*Dendranthema grandiflorum* [Ramat] Kitamura) expressing a rice chitinase gene shows enhanced resistance to gray mold *(Botrytis cinerea). Scientia Horticulturae* 82:113-123.

Walker, D.R., J.N. All, R.M. McPherson, H.R. Boerma, and W.A. Parrott (2000). Field evaluation of soybean engineered with a synthetic cry1Ac transgene for resistance to corn earworm, soybean looper, velvetbean caterpillar (Lepidoptera:Noctuidae), and lesser corn stalk borer (Lepidoptera:Pyralidae). *Journal of Economic Entomology* 93:613-622.

Weeks, D.P. (1999). An overview. In *World Food Security and Sustainability: The Impacts of Biotechnology and Industrial Consolidation* (pp. 3-12), Eds. Weeks, D.P., J.B. Segelken, and R.W.F. Hardy. Ithaca, NY: National Agricultural Biotechnology Council.

Yamamoto, T., H. Iketani, H. Ieki, Y. Nishizawa, K. Notsuka, T. Hibi, T. Hayashi, and N. Matsuta (2000). Transgenic grapevine plants expressing a rice chitinase with enhanced resistance to fungal pathogens. *Plant Cell Reports* 19:639-646.

Zhao, J.Z., H.L. Collins, J.D. Tang, J. Cao, E.D. Earle, R.T. Roush, S. Herrero, B. Escriche, J. Ferre, and A.M. Shelton (2000). Development and characterization of diamondback moth resistance to transgenic broccoli expressing high levels of Cry1C. *Applied and Environmental Microbiology* 66:3784-3789.

Zhu, H., S. Muthukrishnan, S. Krishnaveni, G. Wilde, J.M. Jeoung, and G.H. Liang (1998). Biolistic transformation of sorghum using a rice chitinase gene. *Journal of Genetics and Breeding* 52:243-252.

Zupan, J.R. and P.C. Zambryski (1995). Transfer of T-DNA from *Agrobacterium* to the plant cell. *Plant Physiology* 107:1041-1047.

Chapter 2

Mechanism(s) of Transgene Locus Formation

David A. Somers
Paula M. Olhoft
Irina F. Makarevitch
Sergei K. Svitashev

INTRODUCTION

The production of genetically engineered plants was first achieved in 1983. A large number of plant species have since been genetically engineered, primarily using two different strategies for DNA delivery into totipotent cells. The first strategy employed T-DNA delivery mediated by disarmed strains of the soil bacterium *Agrobacterium tumefaciens* (De Block et al., 1984; Horsch et al., 1984). *Agrobacterium tumefaciens* was initially used in transformation of a broad range of dicotyledonous crops (Horsch et al., 1987). At that time, it appeared that *Agrobacterium* would not be useful for transformation of the monocotyledonous cereal grains (Potrykus, 1990; Vain ct al., 1995). This led to the concurrent development of direct DNA delivery methods as the second strategy for transformation of the recalcitrant monocot species. Fertile, transgenic plants produced via direct DNA delivery methods were reported soon after the success with *A. tumefaciens* (Paszkowski et al., 1984). Direct DNA delivery methods were initially based on protoplasts as the target cells for DNA delivery and employed polyethylene glycol (PEG)- and electroporation-mediated delivery of DNA (Paszkowski et al., 1984; Potrykus et al., 1987; Vain et al., 1995). In 1987, Klein and colleagues reported the development of microprojectile bombardment technology for delivery of DNA into intact plant cells (Klein et al., 1987, 1989; Klein and Jones, 1999; Sanford, 1988, 1990; Sanford et al., 1987, 1991, 1993). This development resulted in production of the first fertile transgenic plants in soybean (McCabe et al., 1988), maize (Fromm et al., 1990; Gordon-Kamm et al., 1990), wheat (Vasil et al., 1992, 1993), oat (Somers et al., 1992), and barley (Wan and Lemaux, 1994). Combined with improvements in plant tissue culture procedures, new selection systems for isolation of transgenic tissue cultures and plants, and optimization of DNA delivery parameters, microprojectile bombardment protocols have led to efficient transformation systems for a number of important crops (Christou et al., 1991; Christou, 1992; Walters et al., 1992; Weeks et al., 1993; Russell et al., 1993; Barcelo et al., 1994; Becker et al., 1994; Brar et al., 1994; Nehra et al., 1994; Koprek et al., 1996; Torbert et al., 1998). The development of transformation systems for cereal crops has been extensively reviewed in a book edited by Vasil (1999).

Agrobacterium-mediated genetic engineering is currently the most widely used plant genetic engineering strategy. *Agrobacterium*-mediated transformation of monocots was first demonstrated in rice in 1994 (Hiei et al., 1994) and is now routinely used to transform cereals, including rice, corn (Ishida et al., 1996), barley (Tingay et al., 1997), and wheat (Cheng et al., 1997). Researcher familiarity with *Agrobacterium*-mediated transformation systems and their simplicity and independence from DNA delivery devices are important rationale for the major shift toward using *Agrobacterium* for crop genetic engineering (Songstad et al., 1995). The greatest motivation to use *Agrobacterium* appears to be driven by molecular and cytogenetic analyses showing that higher frequencies of transgene loci produced via microprojectile bombardment and other direct DNA delivery methods exhibit complex structures composed of multiple copies of whole and rearranged delivered DNA compared to transgene loci produced via *Agrobacterium* (Pawlowski and Somers, 1996). Complex transgene loci are associated with problems of transgene expression (Finnegan and McElroy, 1994; Matzke and Matzke, 1995; Pawlowski et al., 1998; Kohli, Gahakwa, et al., 1999) and inheritance (Christou et al., 1989; Choffnes et al., 2001) and pose considerable difficulties in transgene locus characterization for regulatory approval before commercialization of transgenic crops. Thus, transformation methods that are simple and inexpensive and produce the simplest transgene locus structures, such as *Agrobacterium*-based methods, are readily adopted for both basic and applied uses of plant genetic engineering.

Impressive progress has been achieved since 1983 in developing methods for genetically engineering a diverse range of plant species, in commercialization of transgenic traits in a number of crops that are planted on substantial acreages throughout the world, and in characterization of transgene locus structure. However, much remains to be learned about the mechanisms that result in the formation of transgene loci, why the main transformation systems differ in the proportions of complex transgene loci produced, and how to routinely produce simple transgene loci.

A large body of literature describes the structure of transgene loci produced using both *Agrobacterium*-mediated and direct DNA delivery transformation; however, a paucity of studies prospectively investigate factors that affect transgene locus formation mechanism(s) and the resulting structure of transgene loci. The primary analytical method used to characterize transgene loci involves various restriction digestions of genomic DNA from transgenic plants followed by Southern blot analysis (Southern, 1975). Although these studies have been useful in documenting several features of transgene loci, Southern analyses are unable to provide detailed characterization of transgene loci or even good estimates of their size. More recently, fluorescent in situ hybridization (FISH) has been applied to studying transgene loci and has resulted in a number of new insights into their structure and size (Ambros et al., 1986; Moscone et al., 1996; Mouras et al., 1987; Mouras and Negrutiu, 1989; for review see Svitashev and Somers, 2001a). Polymerase chain reaction (PCR) has been used to isolate small portions of transgene loci (Kohli, Griffiths, et al., 1999) and genomic sequences flanking integrated delivered DNA; however, little direct DNA sequence data have been reported for complete or partial transgene loci. As transgene sequence data are reported, the realization emerging is that many features of transgene loci are more complex than previously thought. Moreover, these data are providing new insights into mechanism(s) of transgene locus formation. The aim of this chapter is to review the literature concerning the structure of nuclear transgene loci and relate these features to mechanism(s) leading to their formation in plants transformed via *Agrobacterium*-mediated T-DNA and direct DNA delivery.

The following definitions will be used in describing nuclear transgene loci and mechanisms of transgene locus formation. All DNA, including nonfunctional components such as plasmid backbone sequences, delivered into cells using either *Agrobacterium* or direct delivery will be referred to as *delivered DNA*. Upon integration into the recipient genome, the delivered DNA will be referred to as *integrated DNA* to avoid confusion created by the term *transgene DNA*. Although often intended to refer to all DNA integrated into a transgene locus, in some reports transgene DNA is used to describe the functional transgenes carried on the delivered DNA. The term *transgene locus* will be used to describe the heritable unit of the integrated DNA. Transgene loci frequently exhibit rearrangements in the delivered DNA, which are generally detected as deviations from the unit length and sequence of the delivered DNA. In very simple transgene loci produced from T-DNA, the delivered DNA is not rearranged; thus, transgene locus formation results from only genomic integration. In complex loci exhibiting multiple rearrangements, for example those produced via microprojectile bombardment, transgene locus rearrangements may occur before, during, or even after transgene integration. We refer to the process(es) leading to integration of delivered DNA into the nuclear genome and possible rearrangements of the delivered DNA in the resultant locus as *transgene locus formation*. Our rationale for this stems from evidence (to be discussed) supporting the simultaneous rearrangement and genomic integration of the delivered DNA via the same mechanism(s). Two mechanistic definitions are necessary. Homologous recombination (HR) occurs between homologous DNA sequences. Illegitimate recombination (IR) results in end-joining between noncontiguous and nonhomologous DNA sequences. The term *transgenic event* will be used to describe tissue cultures and plants derived from a single transformed or transgenic cell. Regenerated plants are referred to as the T_0 generation. Subsequent self-pollination of T_0 plants will be referred to as T_1 through T_n generations.

PLANT GENETIC ENGINEERING SYSTEMS

Plant genetic engineering systems are comprised of three major components: (1) a source of target cells that are totipotent or are germline cells which give rise to gametes, (2) a means of delivering DNA into the target cells, and (3) methods for selecting or identifying transgenic cells or plants. Success in developing genetic engineering systems for any plant species largely depends on the availability of technologies within these three components. Some aspects of each component also are likely to affect transgene locus structure. Although the role of these factors in transgene locus formation will be discussed in detail in later sections, it is important to describe these components to set the stage for those discussions.

Sources of Totipotent Target Cells

Currently, all crop transformation systems use tissue cultures that regenerate plants as the main source of totipotent target cells. Even when tissue eplants are used as sources of totipotent cells, such as the tobacco leaf disk system (Horsch et al., 1984), the differentiated cells are passaged through a tissue culture step for selection and plant regeneration. Thus, the vast majority of studies investigating transgene locus formation and structure in plants are inseparably confounded with the effects of tissue culture. Tissue culture causes cytogenetic, genetic, and epigenetic variation collectively referred to as tissue culture-induced or *somaclonal* variation. Exceptionally high frequencies of cytologically detected chromosomal changes such as polyploidy and genome alterations primarily associated with chromosome breakage and mutations in regenerated plants or their progeny have been documented (for reviews see Phillips et al., 1988; Olhoft and Phillips, 1999). Tissue culture-induced chromosome breakage and repair likely confound our ability to directly investigate mechanism(s) underlying transgene locus formation because double-strand break (DSB) repair processes are likely essential components of both processes. The cellular processes involved in integration of delivered DNA and transgene locus formation may be substantially modified by these tissue culture effects compared to the normal sporophytic or gametophytic states. Recently, Choi and colleagues (2000) reported an elevated frequency of chromosomal changes in transgenic barley plants compared to nontransgenic plants regenerated from tissue culture, indicating that tissue culture and transformation lead to increased mutation frequency. Explant source for tissue culture initiation has been implicated in affecting transgene locus complexity in transgenic *Arabidopsis* plants (Grevelding et al., 1993). It is difficult to determine if these effects are due to the tissue culture process per se, transgene integration, aborted transgene integration, or a combination of these processes.

Production of transgenic plants without employing a tissue culture step would provide a system for directly investigating transgene locus formation without the confounding effects of tissue culture. However, the only widely reported and corroborated non-tissue culture-based transformation systems exist for *Arabidopsis*. Seed transformation (Feldman and Marks, 1987) and later floral dip (Bechtold et al., 1993; Chang et al., 1994; Clough and Bent, 1998) methods for *Arabidopsis* have been widely used for several years but have not been extended to crop plants. It is expected that more non-tissue culture transformation systems will be developed for crop plants in the near future, simplifying investigation of mechanism(s) involved in transgene locus formation. The array of transgene rearrangements in *Arabidopsis* plants transformed via floral dip appear to be similar to those in plants transformed via a tissue culture-dependent process. However, comprehensive analyses of transgene loci in these *Arabidopsis* plants will be required to provide better determination of this observation.

DNA Delivery Methods

DNA delivery methods and the structure of the delivered DNA undoubtedly have major impacts on the structure of transgene loci. *Agrobacterium tumefaciens* causes crown gall disease by transferring genes from a tumor-inducing (Ti) plasmid into the plant genome. Deletion of the tumor-inducing genes from the Ti plasmid and replacing them with transgenes of interest allows the transfer of the transgenes into the plant genome. *Agrobacterium*-mediated DNA delivery from Ti or binary plasmids is initiated in the bacterium via excision of single-stranded (ss) DNA molecules delineated by 25 base pair (bp) sequences that constitute the left and right borders of the transferred (T) DNA (Hooykaas and Schilperoort, 1992; Zambryski et al., 1989; Zambryski, 1992). Export of the T-DNA from *Agrobacterium* into the plant cell is similar to both bacterial and viral mechanisms, while

T-DNA nuclear transport and integration into the plant genome is likely mediated by plant cell processes. Many excellent reviews detail the *Agrobacterium* and T-DNA transfer mechanisms (Sheng and Citovsky, 1996; Tinland, 1996; Gelvin, 2000; Zupan et al., 2000).

Two main characteristics of T-DNA likely influence transgene integration and transgene locus structure. First, T-DNA is a linear molecule, which greatly reduces the randomness of breakpoints compared to direct delivery of a closed-circular plasmid DNA. Second, several proteins encoded by the *Agrobacterium vir* genes are involved in T-DNA transport, its protection from plant cell nucleases, and nuclear targeting (Gelvin, 2000). Single-stranded nicks at both the left and right borders of the T-DNA in the Ti plasmid are initiated by the VirD1 and VirD2 proteins to release the ss-T-DNA with the VirD2 protein covalently bound to its 5' terminus. Both the T-strand and the ss-binding proteins, VirE2, are exported out of the *Agrobacterium* and eventually assembled into a T-strand coated with VirE2 proteins in the plant cell. These VirE2 proteins both target the DNA to the nucleus and protect the T-DNA from nucleases, which ultimately increases the efficiency of T-DNA integration (Zupan et al., 2000). Besides being a border-specific endonuclease, the VirD2 protein guides the T-strand from *Agrobacterium* into the plant cell. In addition, it has nuclear localization signals and ligase activity in vitro, and is responsible for the minimal sequence variation seen on the 5' end of integrated T-DNA.

Hansen and colleagues (1997) used direct gene transfer to introduce the *Agrobacterium* virulence genes *virD1* and *virD2,* with or without *virE2* together with a gene of interest flanked by right and left border repeats into maize protoplasts. Their results suggest that once target and helper DNA are in the plant cell, independent of the DNA delivery mechanism, the Vir proteins can interact with the T- DNA, resulting in simpler transgene loci in the plant genome. Transient expression of reporter genes encoded in the complement of the T-strand (Kapila et al., 1997) and extrachromosomal HR between T-DNAs (Offringa et al., 1990) demonstrate that some portion of the T-DNA becomes double-stranded (ds) after importation into the nucleus. Whether ss or ds forms of T-DNA are involved in genomic integration has not been determined (Gelvin, 2000).

Plant proteins involved in intranuclear trafficking of T-DNA have recently been identified (reviewed in Tzfira and Citovsky, 2002; Ward et al., 2002). These proteins interact with VirE2 and VirD2 on the T-strand and target it to the genome, presumably at actively transcribing regions. These recent findings explain the mechanisms underlying the observation that T-DNA is integrated primarily in transcribed regions of the genome (Koncz et al., 1989).

Because of the highly evolved nature of the mechanisms employed by *Agrobacterium* for synthesis of T-DNA, its transfer from the bacterium into the plant cell nucleus, and integration into the host genome, it is intuitive that a higher proportion of transformation events would exhibit a single copy of the intact T-DNA integrated into the recipient genome compared to the requisite breakage and integration of circular plasmid DNA delivered via direct methods (Pawlowski and Somers, 1996; Tinland, 1996; Tzfira and Citovsky, 2002). Little information is available on the amount of T-DNA delivered to the plant cell via *Agrobacterium* infection. However, it has been shown in numerous experiments that a single bacterium can deliver multiple T-strands originating from either two different binary plasmids or from dual T-DNA containing single plasmids (Ebinuma et al., 2001).

Direct DNA delivery is generally mediated either by chemical- or electroporation-induced uptake into protoplasts (Potrykus et al., 1987; Bates, 1994) or cells (D'Halluin et al., 1992) or by microprojectile bombardment of cells (Sanford, 1990). In microprojectile bombardment, it has been shown that DNA is carried into the nucleus on the microprojectile (Hunold et al., 1994). The mechanism for DNA transfer to the nucleus following direct DNA uptake and electroporation, which presumably delivers DNA into the cytoplasm, has not been described. In most studies, closed circular plasmid DNA is delivered via these approaches.

The amount of DNA delivered has been shown to influence transgene locus structure, with more DNA delivered resulting in a more complex transgene locus (Vain et al., 2002). Theoretically, a 1.1 μm microprojectile can carry from a few to at least several hundred copies of the transgene (for review see Klein and Jones, 1999). Hadi and colleagues (1996) reported delivery of 10 to 15 copies each of 12 different plasmids into soybean cells on the surface of one microprojectile and demonstrated that all 12 transgenes were integrated into the genome. Likewise, Chen and colleagues (1998) produced rice plants that were transformed with 11 different transgenes co-delivered on separate plasmids. These data suggest that as many as several hundred copies of the delivered DNA may be involved in transgene locus formation.

Some studies have reported delivery of linear (Fu et al., 2000; Loc et al., 2002) and single-stranded DNA (Uze et al., 1999). In some cases the state of the delivered DNA has been shown to have an impact on transgene locus structure. For example, Fu and colleagues (2000) reported that delivery of linearized DNA carrying only the transgene genes reduced the frequency of transgene locus rearrangements compared to delivery of the circular plasmid. In other studies linearization of delivered DNA had little effect in transgene locus structure (Uze et al., 1999; Breitler et al., 2002).

Selection of Transgenic Cells and Regenerated Plants

Selection of transformed cells or plants is a requisite component of most plant transformation systems because only a small frequency of the total number of target cells receive and integrate the delivered DNA. For example, it is estimated that one gram of 'Black Mexican Sweet' (BMS) maize cell suspension culture contains 1.8×10^7 cells. In a typical microprojectile bombardment experiment approximately 10^3 BMS cells may transiently express a reporter gene, but only 50 to 100 stably transformed colonies are selected per bombardment (Klein et al., 1989). Thus, only 10^{-4} to 10^{-5} of the original cell population gives rise to transgenic cell colonies in this system, indicating that stringent selection is required for isolation of transgenic events. Some studies have shown that when the proportion of target cells receiving DNA is very high, transgenic colonies and regenerated plants may be isolated in the absence of direct selection (Shillito et al., 1985). However, the efficiency of this approach in most plants is very low, as exemplified in microprojectile bombardment transformation of proliferating soybean meristems (McCabe et al., 1988) and barley (Ritala et al., 1994).

Selection of transgenic cells and tissues usually relies on positive selection for the expression of a transgene that confers gain-of-function resistance to a selective agent (for review see Birch, 1997). Selective agents are biochemical inhibitors that block cell metabolism, and combinations of selective agents and resistance genes have been reviewed (Weising et al., 1988; Wilmink and Dons, 1993). Other selection strategies use visual selection of cells expressing a reporter gene such as the green fluorescence protein (GFP) (for review see Stewart, 2001). An impressive example of visual selection was described by Kaeppler and colleagues (2000) who used *gfp* for selection of transgenic oat cells followed by regeneration and production of transgenic plants. It is likely that the chemical properties of the selective agent and the stringency with which selection is imposed are factors that may affect transgene locus structure. This is probably best exemplified in studies in which gradually increasing the concentration of selective agent over the course of long-term selection of transgenic cells resulted in selection of transgenic colonies that exhibited amplification of the selectable marker transgene (Jones et al., 1994).

Selection of transgenic cells, tissues, and organs for production of transgenic plants may impose a bias on the transgene loci recovered. For selection to yield transgenic plants, the transgene locus must express the selectable marker gene throughout the tissue culture phase and, in some cases, throughout plant regeneration. Thus, integration events leading to transgene locus structures that do not encode a complete selectable marker gene are integrated into regions of the genome which are not transcribed, or are prone to transgene silencing due to structural rearrangements or other factors, will be lost during the selection process. This potential narrowing of the variation in transgene locus structures to only those that express the selectable marker may have led to misconceptions about the locations of transgene integration, possible numbers of independent insertions in a single cell, and the range of structures which may be formed.

Other Factors That May Influence Transgene Integration

Integration of delivered DNA into plant genomes likely involves DNA replication and DSB repair via illegitimate recombination. Integration of DNA introduced by PEG treatment of protoplasts isolated from cell cultures synchronized in S phase revealed that transforming DNA was integrated into more loci, was present at greater copy numbers, and exhibited more rearrangements than the transgenes integrated during unsynchronized growth (Kartzke et al., 1990). Ultraviolet (UV) and X-ray irradiation also increase the frequency of transgene integration (Kohler et al., 1989; Gharti-Chhetri et al., 1990). Increased integration may result from increased numbers of nicks or breaks in the genomic DNA present during DNA replication and DNA damage, suggesting active involvement of the DNA repair processes in transgene integration.

TRANSGENE LOCI

Much can be deduced about the mechanism(s) of transgene locus formation from the characterization of transgene loci. In this section, the range of variation in transgene locus structure and the methods used for their analysis will be described to provide a framework to consider these mechanisms. Cases in which transgene locus structures directly point to specific mechanisms of transgene locus formation will be discussed in this section. Although the mechanisms of DNA delivery appear to be different between *Agrobacterium*-mediated transformation and microprojectile bombardment, there are more similarities than differences in the range of structural variation in transgene loci produced from both methods. Accordingly, the variation in complexity of transgene loci will be described for both methods such that differences in transgene loci produced by either method can be highlighted. It is important to emphasize that improvements in the ease and sensitivity of applying analytical methods to characterization of transgene loci are providing new insights into the structure of transgene loci. When appropriate, some historical documentation of the evolution of the current understanding of transgene locus structures and formation will be presented.

Transgene Locus Number

The conventional concept of transgene locus numbers is that one or a few loci are established in a transgenic genome. This view is primarily based on estimates derived from Southern (1975) DNA blot analysis of T_0 plants (Figure 2.1) and studies that employed segregation analysis of a transgene phenotype to determine locus numbers in

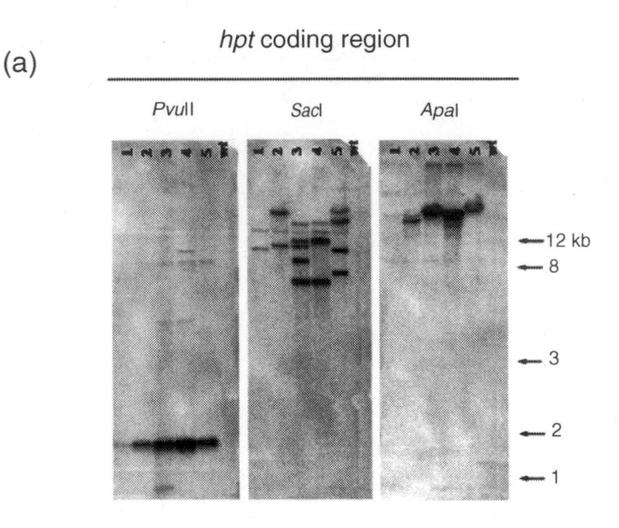

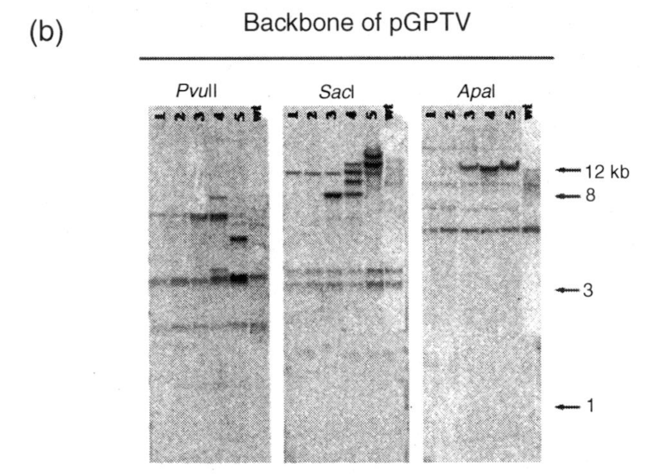

FIGURE 2.1. Southern blot analyses of five T_0 soybean plants and control nontransformed plant presented as examples of how different restriction digestion strategies and the choice of probes provide different types of information about transgene loci. The samples in panel A are probed with the selectable marker, *hpt,* whereas panel B was probed with the backbone of the binary plasmid used for *Agrobacterium*-mediated transformation. *Pvu*II digests the T-DNA to release the unit-length *hpt* gene, *Sac*I cuts the T-DNA roughly in half, and *Apa*I does not cleave the T-DNA.

transgenic plants produced by both direct DNA delivery (Potrykus et al., 1985, 1987; Spencer et al., 1992; Buising and Benbow, 1994; Cooley et al., 1995, Armstrong et al., 1995; Register et al., 1994) and *Agrobacterium*-based methods (Deroles and Gardner, 1988; Heberle-Bors et al., 1988; Clemente et al., 2000). Because phenotypic scoring reflects the segregation of functional transgenes only, more recent genotypic analyses using Southern analyses that follow the segregation of transgene-hybridizing fragments tend to suggest a somewhat higher frequency of individuals with multiple transgene loci (for example see Svitashev et al., 2000). The underestimation of transgene locus numbers based on phenotypic analysis is most likely attributable to problems with transgene expression such as production of nonexpressing loci or transgene silencing (Finnegan and McElroy, 1994; Matzke and Matzke, 1995).

Genotypic segregation analysis may also underestimate transgene locus number if segregation distortions due to chromosome aberrations are associated with the transgene. Fluorescent in situ hybridization analysis of transgenes (for review see Svitashev and Somers, 2001a) suggests that a high proportion of transformants produced via microprojectile bombardment may have more than one transgene locus. Combining phenotypic and genotypic segregation data with FISH analysis of 16 oat transformants produced via microprojectile bombardment indicated that half of the transgenic lines had two or more detectable loci (Svitashev et al., 2000; Figure 2.2a). Even this approach is subject to underestimation. If delivered DNA is fragmented either via DNA delivery or by plant cell processes, it seems likely that small fragments which do not comprise a functional transgene may be integrated throughout the genome. Integrated small fragments of delivered DNA (<500 bp) would be difficult to detect by Southern or other analyses, especially in large genomes. However, transgene cloning and sequencing have revealed the presence of small, integrated DNA fragments in the Roundup Ready locus in soybean (Palevitz, 2000) and shallot (Zheng et al., 2001). We also have observed the presence of small, unlinked delivered DNA fragments in a transgenic oat line (Makarevitch et al., 2003). Determination of the frequency at which these small integrations are detected in transgenic plants will be important because their presence suggests an alternative to the "all delivered DNA is integrated into one or a few loci" view of transgene locus formation. If transgene locus numbers actually represent a distribution of different sized fragments of the delivered DNA integrated throughout the genome, this would suggest that transgene locus number is controlled by the capacity of the recipient genome to withstand the mutation load imposed by the formation of multiple large and small transgene loci.

Cotransformations, in which different transgenes carried on separate plasmids or T-DNAs are codelivered into the same plant cell, provide further insight into transgene locus formation processes and the question of transgene locus numbers. In this section, factors that affect cotransformation will be discussed, while structural analysis of transgene loci originating from cotransformation will be covered in a later section.

Historically, cotransformation was primarily conducted using microprojectile bombardment and other direct DNA delivery methods (Figure 2.3; Schocher et al., 1986; Hadi et al., 1996; Chen et al., 1998; Jongsma et al., 1987). One of the plasmids usually carries a plant (positive) selectable marker gene, while the other plasmid is not selectable. Typically in such experiments, DNA from the different plasmids are found cointegrated into the same transgene locus. Chen and colleagues (1998) reported codelivery of 14 different plasmids in rice via microprojectile bombardment. The frequency of multiple transformants exhibited a skewed, normal distribution, suggesting that cointegration is a random process with each delivered DNA having an equal chance of integration. For example, 85 percent of the transgenic events cointegrated more than two different plasmids, and 17 percent had more than nine. None cointegrated all 14 plasmids. In two lines with nine and ten integrated transgenes, respectively, all delivered DNAs were cointegrated into the same locus, whereas in a third line an integrated transgene sequence was unlinked from the other ten delivered DNAs (Chen et al., 1998).

Cotransformation using *Agrobacterium*-mediated DNA delivery also produces plants that have both single and multiple transgene loci containing multiple different T-DNAs. Various factors including competence of both plant and *Agrobacterium* cells, the number of bacteria infecting a plant cell, and number of T-DNA copies produced per bacterium have been implicated in influencing T-DNA copy number, number of transgene loci, and their linkage in plant genomes. Ebinuma

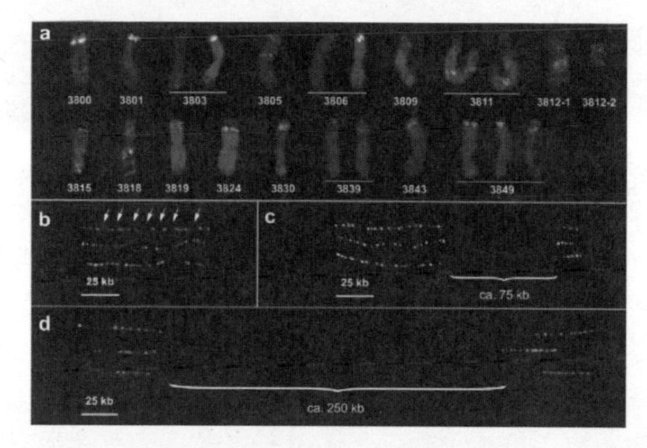

FIGURE 2.2. Examples of FISH to detect transgene loci in metaphase chromosomes and on extended DNA fibers. (a) Transgene loci in 16 independent transgenic oat lines. Red indicates the transgene loci. Blue indicates A or D genome chromosomes counterstained with DAPI. Green staining indicates C genome chromosomes. Fiber-FISH showing different transgene locus structures (stained green) ranging from (b) relatively compact loci with small genomic interspersions, and (c and d) two loci with much larger genomic interspersions.

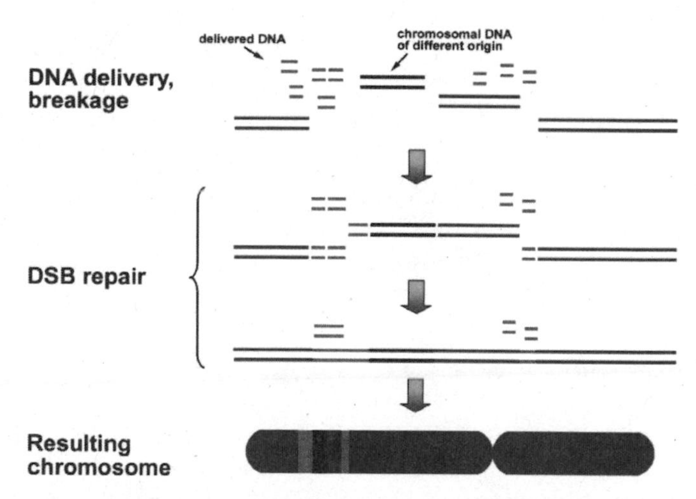

FIGURE 2.3. A model for transgene locus formation. The genomic target site and recipient chromosome are shown in green, the delivered DNA is represented by the pink lines, and the black lines are used to denote ectopic DNA to be integrated as a genomic interspersion.

and colleagues (2001) reviewed cotransformation of various plants using either mixed strain infections with each *Agrobacterium* carrying a single T-DNA binary vector, *Agrobacterium* containing two T-DNA binary plasmids, or with single bacterium carrying a single binary vector with two separated T-DNAs. They concluded that (1) several different *Agrobacterium* can deliver T-DNA into the same locus, (2) a single *Agrobacterium* can deliver and integrate several T-DNAs during a single infection, and (3) plants tended to integrate multiple T-DNAs from the same *Agrobacterium* rather than from multiple *Agrobacteria*. Interestingly, the choice of *Agrobacterium* strain may influence whether multiple transgenes are linked or unlinked in the plant genome (Ebinuma et al., 2001). *Agrobacterium*-mediated formation of multiple loci via cotransformation of different T-DNAs has been reported in a number of plants, indicating that it is a widespread phenomenon which offers a simple means of removing the selectable marker carrying T-DNA from the transgenic event by simple segregation in subsequent generations. In *Agrobacterium*-mediated cotransformation experiments, frequently more than 50 percent of the transformants may exhibit more than one transgene locus (Ebinuma et al., 2001).

Variation in the Internal Structure of Transgene Loci

Transgene loci vary from simple inserts to extensively rearranged structures. Internal structures of complex transgene loci may include (1) variation in copy numbers of full-length and truncated copies of the delivered DNA, (2) transgene rearrangements that generate hybridizing fragments differing in size from the full-length delivered DNA or T-DNA including direct and inverted repeats, and (3) genomic sequences interspersing tightly linked, clustered transgenes. As mentioned, these structures shed considerable light on the mechanisms of transgene locus formation.

The methods used to characterize the internal structures of transgene loci have had a significant impact on understanding these structures. For example, Southern analysis has been the primary method used to characterize transgene locus structures. The restriction enzyme digestion strategies and the choice of sequences for probes used to conduct a Southern analysis influence assessments of structural complexity of transgene loci. This is demonstrated in Figure 2.1. Genomic DNA of five transgenic soybean plants transformed using a modified cotyledonary-node method based on *Agrobacterium tumefaciens* (Olhoft et al., 2003) was digested with three restriction enzymes. *PvuII* cleaves the T-DNA to release the coding sequence of the *hpt* selectable marker gene. *SacI* restricts the T-DNA at one site to provide an estimate of the number of tandem repeats and rearrangements of the T-DNA. *ApaI* does not cut the T-DNA and thus provides an estimate of the number of independent integration events. Panel A is probed with the *hpt* coding sequence as is commonly performed to evaluate transgenic plants for the presence of a unit-length transgene. In the *PvuII* digest, all events exhibited the unit-length *hpt* and no other transgene hybridizing fragments were observed. However, the picture is entirely different when the *SacI* digests are probed with *hpt*. Most events have multiple transgene-hybridizing fragments, suggesting rearranged T-DNAs. Events three and four have similar-sized major bands, suggesting that these lines have tandem repeats of the T-DNA, which is supported by the intense hybridization of these events in the *PvuII* digests. The multiple bands observed in the *ApaI* digests in events two and four suggest either the presence of multiple loci in these lines or *ApaI* site-containing genomic DNA interspersing multiple T-DNAs. Probing these same blots with the plasmid backbone from the binary plasmid carrying the T-DNA shows that binary plasmid sequences may also frequently contribute to transgene locus structure (Figure 2.1b). Although the blots in Panel B are complicated by cross-hybridization between the plasmid backbone and plant DNA fragments, events three, four, and five exhibit plasmid backbone sequences integrated into their genomes. Incorporation of Ti or binary plasmid backbone sequences into transgene loci has been previously reported and may occur at frequencies as high as 40 percent in transgenic plants (Wolters et al., 1998; De Buck et al., 2000). FISH and fiber-FISH (examples shown in Figure 2.2) have also provided additional structural information on transgene loci (reviewed in Svitashev and Somers, 2001a). Obviously, sequencing transgene loci achieves the greatest resolution of structural features which provides information about transgene locus formation.

Rearrangements of Delivered DNA Appear to Be Mediated by IR and HR

The copy numbers of delivered DNA integrated into a transgene locus may range from a single truncated copy that is less than one unit length of the delivered plasmid or T-DNA to multiple copies of whole or truncated delivered DNA. Generally, reports indicate that the upper limit of copy numbers is around 20 copies (Pawlowski and Somers, 1996; Kumpatla and Hall, 1998); however, much higher numbers have been reported (Kohli et al., 1998; Svitashev and Somers, 2001a). The most common form of complex, multiple copy transgene locus, whether produced from single or codelivered DNAs, results from random integration of multiple whole and truncated delivered DNAs. The unordered integration of large and small noncontiguous and nonhomologous fragments of the delivered DNA has been shown in transgene loci produced from both direct DNA delivery and *Agrobacterium.* Rearrangement formation and genomic integration are both attributed to illegitimate recombination. Evidence for the involvement of IR in rearrangements of the delivered DNA is found in transgene locus sequence data from plants transformed using direct delivery methods including *Arabidopsis* (Sawasaki et al., 1998) and rice (Takano et al., 1997), specific PCR-amplified regions within integrated transgenes in rice (Kohli, Griffiths, et al., 1999; Kumpatla and Hall, 1999), a small, simple transgene locus from tobacco (Shimizu et al., 2001), ca. 160 kb of two complex transgene loci in oat (Svitashev et al., 2002), and the complete sequences of two simple transgene loci in oat (Makarevitch et al., 2003). The IR junctions between noncontiguous fragments of delivered DNA are characterized by short stretches of homology, referred to as microhomology, extending from one to several bp. In fact, apparent blunt-end ligations, which show no microhomology, are relatively rare. The presence of the microhomology indicates that some short end-sequence homology is necessary for IR junction formation. Sequence changes in the microhomologies are extremely rare, suggesting that the requirement for microhomology in IR junction formation is stringent.

The random integration of multiple truncated copies of delivered DNA via IR would account for both the multiple copies of a specific transgene hybridizing fragment and the high frequency of transgene-hybridizing fragments that deviate from the unit-length delivered DNA. Transgene fragment lengths incorporated into transgene loci via both *Agrobacterium* and direct DNA delivery methods vary from just a few bp long to greater than the unit length of the delivered DNA. For example, 50 out of 82 transgene fragments from more than 160 kb of sequence from two complex transgene loci in oat were less than 200 bp long, and six fragments were less than 15 bp (Svitashev et al., 2002). The presence of these small transgene fragments suggests that certain mechanisms extensively fragment directly delivered DNA. Interestingly, an *Agrobacterium*-derived transgene loci described by Jakowitsch and colleagues (1999) also exhibited scrambling of short T-DNA sequences, suggesting that similar mechanisms may function in both transformation systems.

Repeats of delivered DNA are frequently observed in transgene loci resulting from both direct DNA delivery (Czernilofsky, Hain, Baker, and Wirtz, 1986; Czernilofsky, Hain, Herrera-Estrella, et al., 1986; Riggs and Bates, 1986; Finer and McMullen, 1990; Kartzke et al., 1990) and *Agrobacterium* methods (De Neve et al., 1997; De Buck et al., 1999). For the most part these structures have been detected and characterized by Southern analyses as shown in Figure 2.1, which does not allow sufficient detail to determine the mechanism(s) of their formation. However, some repeats have been sequenced and will be discussed. Tandem T-DNA repeats are frequently detected in *Agrobacterium*-mediated transformants (De Neve et al., 1997). Since T-DNA is delivered to the plant nucleus as a linear and presumably ssDNA molecule and fragments of genomic DNA are frequently detected between both tandem and inverted T-DNA repeats, the formation of such end-to-end T-DNA repeats is likely also mediated by IR (De Neve et al., 1997; De Buck et al., 1999). Very long repeats of T-DNA molecules may also be produced by read through of the left T-DNA border, producing tandem repeats and incorporating the entire binary plasmid with the backbone separating the repeated T-DNAs (De Buck et al., 2000).

Perfect tandem repeats of directly delivered DNA, also called concatemers, are detected as perfect transgene locus sequences exceeding one complete copy of the delivered DNA. The presence of concatemers in transgene loci is generally implied by Southern analysis which shows multiple, apparently unit-length copies of the delivered DNA (Riggs and Bates, 1986). Sequences of concatemers have been reported in tobacco (Shimizu et al., 2001). We detected about 1.5 contiguous copies of the delivered DNA in a complex transgene locus in oat; however, Southern analysis indicates that this line likely had a number of copies arrayed as in tandem repeats (Svitashev et al., 2002). In a simple transgene locus from oat, we detected a plasmid copy just slightly longer than its unit length (Makarevitch et al., 2003). These results indicate that perfect tandem repeats are common features of plant transgene loci. Experiments with delivering linearized cassettes carrying only transgene sequences and not plasmid backbones have been shown to reduce concatemer formation (Fu et al., 2000; Breitler et al., 2002). These results suggest that delivery of either circular or linearized whole plasmids are required for concatemer formation. Perfect tandem repeats implicate HR (Folger et al., 1982) or possibly some form of synthesis-dependent mechanism such as rolling circle synthesis in transgene locus formation that can produce a perfect copy greater than the unit length of the delivered plasmid (Khan, 2000). Although HR has been demonstrated extrachromosomally, genomic HR is an extremely rare event in plant genomes, at least in T-DNA transformation (Offringa et al., 1990), suggesting that concatemer formation may proceed extrachromosomally via HR before genomic integration as proposed in mammalian systems (Folger et al., 1982). Although it is not understood at this time how much homologous sequence is necessary for extrachromosomal HR, no sequencing evidence from integrated transgene loci directly demonstrates HR involving homologous sequences carried on separate codelivered DNAs. Moreover, no sequencing evidence exists for homology-based deletion of integrated transgene sequence, but in contrast, evidence indicates that IR is the predominant mechanism (Muller et al., 1999). Although the absence of such evidence is by no means definitive, it does suggest that mechanism(s) other than HR may be responsible for formation of perfect tandem repeats of delivered DNAs.

Inverted T-DNA repeats also have been reported. These are by necessity formed either via IR of ds-T-DNA substrates as exemplified by left border inverted repeats containing filler DNA between the T-DNAs (De Buck et al., 1999) or some other synthesis-based mechanism from the ss-T-DNA. Right border inverted T-DNA repeats exhibit very precise end-joining without filler DNA (De Buck et al., 1999). Inverted repeats of unit-length delivered DNAs have also been detected in transgene loci resulting from direct DNA delivery methods using Southern analysis. Unfortunately, the lack of sequencing data from these structures precludes suggesting mechanisms for their formation because it is unclear whether these are perfect copies of the delivered DNA. It is difficult to understand how precise head-to-head repeats can be formed from a delivered circular DNA molecule. Either a specific breakpoint in the delivered plasmid is restricted by plant cell nucleases or some form of recombination may take place at a recombination hot spot to create interplasmid repeats as suggested from the results of Kohli, Griffiths, and colleagues (1999) and Kumpatla and Hall (1999) showing that a 19 bp palindrome in the CaMV (cauliflower mosaic virus) 35s promoter is prone to recombination. The sequence of a complex transgene locus in oat revealed a perfect inverted repeat composed of scrambled delivered and genomic DNA, indicating that repeat formation occurred after IR of delivered and genomic DNA fragments (Svitashev et al., 2000). This structure may have formed via the synthesis-dependent mechanism for indirect repeat formation in yeast proposed by Butler and colleagues (1996). This mechanism involves synthesis initiation around a small palindrome that undergoes intrastrand pairing and causes intramolecular recombination followed by bidirectional synthesis to form a large repeat. Shimizu and colleagues (2001) showed that fragments of the delivered DNA and genomic flanking DNA were arrayed as an inverted repeat flanking a core sequence of transgene locus in a transgenic tobacco plant. This structure also suggests that repeat formation occurred after IR of the delivered DNA with a genomic fragment either before or during transgene integration. Whether this mechanism of repeat formation might apply to the formation of inverted repeats of the unit-length delivered DNA is unknown at this time.

Imperfect repeats of small fragments of the delivered DNA in various orientations were observed within scrambled regions of transgene loci detected in oat transgene loci resulting from microprojectile bombardment (Svitashev et al., 2002). The probability that small fragments from the same region of two or more molecules of delivered DNA would be randomly integrated into the same region of a transgene locus seems to be very low, thus ruling out IR. More likely these short repeats are products of DNA synthesis involving template switching from the same region of a delivered DNA molecule.

Point Mutations in Transgene Loci

Although only a few transgene loci have been sequenced, the evidence indicates that sequence changes found in the DNAs integrated into the transgene locus from those of the delivered DNAs are relatively rare (Kohli, Griffiths, et al., 1999; Muller et al., 1999; Shimizu et al., 2001; Svitashev et al., 2002; Makarevitch et al., 2003). These results suggest that the mechanism(s) of transgene locus formation maintains the sequence integrity of the delivered DNA.

Breakpoint Locations in Delivered DNAs

Breakpoints by definition are the point at which the sequence of the DNA integrated into the transgene locus diverges from the delivered DNA sequence. In studies designed to determine transgene structure around the 35S promoter, the 19 bp palindrome has been identified as a recombination hot spot (Kohli, Griffiths, et al., 1999) and exhibits a propensity for fragmentation and illegitimate recombination (Kumpatla and Hall, 1999). To gain a more global view of breakpoint formation in directly delivered DNA, we analyzed locations of 155 breakpoints in the delivered plasmid DNAs integrated into two complex transgene loci in oat (Svitashev et al., 2002). The breakpoints were randomly distributed throughout the delivered DNA sequences. Only one breakpoint involved the 19 bp palindromic sequence in the 35S promoter (data not shown), creating a partial one-sided deletion of this palindrome through junction formation with a noncontiguous fragment of delivered DNA. The majority of transgene breakpoints occurred in short palindromes. Frequently, more than one breakpoint was clustered within a single palindrome. These observations are in close agreement with Muller and colleagues (1999) who showed that in a system designed to restore transgene function through illegitimate recombination all breakpoints were localized to palindromic elements. Muller and colleagues (1999) also showed that AT (adenine/thyamine)-rich regions were locations for breakpoints, but this was not observed in the oat transgene sequences.

Taken together these results suggest that palindromes in general are associated with breakpoint formation in directly delivered DNA. Certain structures in the delivered DNAs or in specific transgene loci may increase the probability of breakpoint formation or IR in their vicinity, as suggested by Kohli, Griffiths, and colleagues (1999). Topoisomerase sites have also been implicated in breakpoint formation (Kohli, Griffiths, et al., 1999; Muller et al., 1999). It is possible that palindromes are targets for topoisomerase II cleavage as suggested by Froelich-Ammon and colleagues (1994), but this has not been directly demonstrated in plants.

Interspersions of Genomic DNA in Transgene Loci

The presence of genomic interspersions among tightly linked, clustered fragments of delivered DNA in a single transgene locus was first reported in 1998 in both allohexaploid oat (Pawlowski and Somers, 1998) and diploid rice (Kohli et al., 1998) transformed using microprojectile bombardment. These studies relied on Southern analysis using restriction enzymes that did not cut the delivered DNA, followed with probing with the entire sequence of the delivered DNA. From 13 transgenic single transgene locus oat transformants, all lines exhibited interspersed transgene loci (Pawlowski and Somers, 1998), indicating that the frequency of genomic interspersions can be quite high. The number of interspersed transgene fragments ranged from two to 12 among the lines, and the sizes of the transgene fragments were shown by pulsed field gel electrophoresis to range from 3.6 to greater than 60 kb.

More recently, FISH and fiber-FISH have been used to show that genomic interspersions frequently occur in transgene loci and are highly variable in size, ranging from a few kb to more than 300 kb (Svitashev and Somers, 2001b; Figure 2.2). In wheat, three types of integration patterns were resolved using fiber-FISH, including (1) large tandem repeats of the transgene without apparent interspersions, (2) large tandem repeats interspersed with genomic DNA, and (3) small, possibly single-copy insertions (Jackson et al., 2001). Sequence data show that genomic interspersions in transgene loci may be very small, numbering only a few bases (Svitashev et al., 2002; Makarevitch et al., 2003). Transgene loci produced via *Agrobacterium* also show genomic interspersions in scrambled T-DNA copies (Jakowitsch et al., 1999), but this appears to be less common than in loci produced via direct DNA delivery. Genomic DNA is often detected between direct and inverted T-DNA repeats and is equated to filler DNA (Krizkova and Hrouda, 1998; De Buck et al., 1999).

Transgene Rearrangement Frequencies Vary Among Transformation Methods

Accurately assessing the differences between microprojectile bombardment and *Agrobacterium tumefaciens*-mediated transformation in regard to their resultant types and frequencies of transgene rearrangements is difficult because very few studies report direct comparisons. However, Zhao and colleagues (1998) provides a useful comparison of molecular analyses of maize plants transformed via microprojectile bombardment versus *Agrobacterium*. *Agrobacterium*-mediated transformation produced 58 percent events with low copy numbers of simple inserts compared to 8 percent via microprojectile bombardment. Furthermore, the *Agrobacterium* transformants exhibited better coexpression of a reporter gene carried on the same T-DNA as the selectable marker and a greater proportion of events with high reporter gene expression levels, both indicators of simple transgene loci, compared to the microprojectile bombardment transformants. The efficiency of production of events with low copy and simple transgene loci produced by *Agrobacterium*-mediated transformation was 62 desirable events from 100 treated embryos versus one desirable event per 100 embryos transformed using microprojectile bombardment (Zhao et al., 1998). Although this calculation is influenced by differences in the transformation frequency of the two systems, it clearly demonstrates why rapid adoption of *Agrobacterium* for transforming most plants has occurred.

Genomic Integration Target Sites

Transgene Locus Localization

Genetic mapping indicates that *Agrobacterium*-mediated T-DNA insertions in plant chromosomes are distributed randomly (Tinland, 1996). T-DNA integration occurs in transcribed, gene-rich regions as suggested by the high frequency of transformants that activate promoter and enhancer trap constructs (Koncz et al., 1989). However, integration appears to be equally distributed between transcribed and nontranscribed sequences in these regions (Tinland, 1996). This may be attributed to the previously mentioned plant nuclear trafficking proteins described by Tzfira and Citovsky (2002) which may target T-DNA to transcribing regions of the recipient genome. As larger T-DNA tagging populations in *Arabidopsis* and other plants with extensive genome sequence data are further analyzed for the location of T-DNA inserts, a comprehensive genomewide view of integration locations will emerge (Ortega et al., 2002).

In plants genetically engineered by microprojectile bombardment transgene locus formation in gene-rich regions also appears to be preferred. However, whether these regions are transcriptionally active during integration has not been addressed. For example, characterization of 26 transgene loci in 16 allohexaploid oat lines produced using microprojectile bombardment of a single plasmid using Southern analysis and FISH showed that the transgene loci were equally distributed among and within the three genomes, suggesting random integration (Svitashev et al., 2000; Figure 2.2a). In 18 out of 26 integration events transgene integration occurred in telomeric and subtelomeric regions. The tendency for transgene localization in telomeric regions has been reported by other authors in oat (Leggett et al., 2000), rice (Kharb et al., 2001), wheat, barley, triticale (Pedersen et al., 1997), and wheat (Chen et al., 1999; Jackson et al., 2001) and probably related to the fact that distal chromosomal areas are enriched with genes in most plants whereas heterochromatic regions are mainly located around the centromeres (Gill et al., 1993; 1996). Transgene integration into heterochromatic regions may be expected to result in poor or unstable transgene expression (Inglesias et al., 1997; Jakowitsch et al., 1999). As a result, the transgenic cells would not be selected or regenerated into plants. Thus, as previously discussed, it seems likely that the selection process may bias this assessment toward events exhibiting integration of delivered DNA into transcribed regions of the genome. To date no reports have described plant proteins that are involved in nuclear import and trafficking of directly delivered DNA. Furthermore, the host proteins involved in T-DNA targeting interact with the VirD2 and VirE2 proteins in the T-DNA complex (Tzfira and

Citovsky, 2002), suggesting the absence of a similar mechanism for directly delivered DNA.

In oat, extant translocations between the three homologous genomes in the allohexaploid can be readily visualized using FISH (Svitashev et al., 2000). Interestingly, transgene loci are frequently localized at the borders of these translocations. In combination, Leggett and colleagues (2000) and Svitashev and colleagues (2000) observed that 5 out of 28 transgene loci in 18 transgenic oat lines exhibited transgene loci apparently in the border of preexisting translocations. This higher than expected frequency of transgene integration in these regions suggests that they are hot spots for transgene integration.

Motifs Commonly Detected in Genomic Sequences Flanking Integrated DNA

Sequences of transgene loci resulting from direct DNA delivery suggest some structural similarities among genomic integration sites. In a study characterizing transgene-plant DNA junctions in two rice plants transformed using the calcium phosphate method, integration was found in AT-rich motifs which included scaffold/matrix attachment (S/MAR)-like regions, a retrotransposon, and telomere-repeats (Takano et al., 1997). There was also microhomology between the plasmid and the target DNA, implicating IR as well as massive rearrangements at the integration site. Recognition sites of topoisomerases were also detected in the rearranged sequences (Takano et al., 1997). In three *Arabidopsis* plants transformed using microprojectile bombardment, eight out of ten transgene-plant DNA, borders analyzed were AT-rich sequences that also exhibited characteristics of S/MAR regions (Sawasaki et al., 1998). Cleavage sites for topoisomerase I were also observed at or near the junction site. The authors speculated that the transgenes delivered on microprojectiles integrate in genomic regions near S/MAR which have a propensity for DNA curvature. A nucleotide near a short direct repeat on the delivered DNA is then joined near topoisomerase cleavage sites in the host DNA (Sawasaki et al., 1998). Both studies suggest that sequence or genomic region preference may occur in genomic integration of introduced DNA delivered by direct gene transfer. S/MAR-like regions and topoisomerase II sites were found in the target sites of simple transgene loci from tobacco (Shimizu et al., 2001) and oat (Makarevitch et al., 2003).

Target Site Modifications

Sequence analysis of transgene-plant genomic DNA border junctions from loci resulting from both *Agrobacterium*-produced transgenic plants (Gheysen et al., 1991; Mayerhofer et al., 1991; Hiei et al., 1994; Tinland, 1996; Matsumoto et al., 1990) and direct DNA delivery (Sawasaki et al., 1998; Takano et al., 1997; Shimizu et al., 2001) indicate that genomic integration, similar to rearrangement formation, is primarily mediated by IR. Some frequency of transgene loci produced via direct DNA delivery and *Agrobacterium*-mediated methods likely results from perfect integration into the recipient genome with minimal modification of the integration target site. The actual frequency of this class of events derived by the various methods has not been reported. However, analysis of the extensive genomic resources available in *Arabidopsis,* especially T-DNA-tagged lines, would provide some insight into the frequencies of target site modifications. The most frequently mentioned observations regarding modifications of the transgene integration target site are (1) small deletions of target site DNA (Gheysen et al., 1991; Tinland, 1996; Zheng et al., 2001; Kumar and Fladung, 2001), (2) apparent rearrangement and/or duplication of genomic DNA flanking the integrated DNA (Gheysen et al., 1991; Ohba et al., 1995), (3) the presence of short sequences of filler DNA between the transgene DNA and target site DNA that apparently originate from other genomic regions (Gheysen et al., 1991), and (4) larger target changes such as translocations (Nacry et al., 1998; Laufs et al., 1999; Tax and Vernon, 2001).

Less information is available on target site modifications in transformants produced by direct DNA delivery methods. Windels and colleagues (2001) demonstrated that genomic sequences adjacent to the Roundup Ready locus in soybean are apparently reorganized. In oat, analysis of genomic integration sites using PCR primers designed to amplify across the small transgene DNA inserts in cloned transgene fragments or along the genomic DNA flanking transgene inserts indicated extensive rearrangements of genomic DNA adjacent to the transgenes (Svitashev et al., 2002; Makarevitch et al., 2003).

Rearrangements in the genomic regions flanking the transgene would be expected to inhibit recombination and thus must be considered as part of the transgene locus.

Association of the Transgene Integration Sites with Chromosome Breakage

Evidence for the association of chromosome breakage with transgene loci is reported in both *Agrobacterium*-mediated transformants and transformants derived from direct DNA delivery methods. In *Arabidopsis,* two T-DNA tagged lines were found to have interchromosomal duplications/translocations caused by T-DNA integration (Tax and Vernon, 2001). In another study, a large paracentric inversion of 26 CentiMorgan (CM) was bordered by two T-DNAs in direct orientation, which argues that the inversion and integration were simultaneous events (Laufs et al., 1999). Castle and colleagues (1993) reported that 20 percent of the tagged embryonic mutants derived from *Agrobacterium*-treated seeds showed chromosomal translocations in addition to truncated, multiple, and tandem T-DNA inserts. Other examples of translocations have been reported by Fransz and colleagues (1996), Papp and colleagues (1996), and Nacry and colleagues (1998).

The only report of association of chromosome breakage with transgene loci produced via microprojectile bombardment comes from our work using FISH to characterize transgenic oat plants (Svitashev et al., 2000). In three lines, transgene loci were detected on two arms of the same A/D chromosomes (Figure 2.2a). In line 3800-1, the transgene locus on the short arm was detected on the border of a translocation with a large C genome chromosome fragment. All translocated A/D genome chromosomes in the parental genotype carry C genome translocations only on their long arms, indicating that this chromosome arm is rearranged. Two transgene integration sites were also detected on one A/D genome chromosome in 3812-1 and on a small, diminutive A/D chromosome in 3812-2 (Figure 2.2a).

Both lines were regenerated from the same transgenic tissue culture. From the similarities in Southern blot and FISH patterns it appears that the diminutive chromosome detected in 3812-2 was likely the product of breakage of the chromosome observed in 3812-1 after transgene integration. In line 3815-7, two transgene loci were localized on an A/D genome chromosome with two satellites, indicating that this chromosome was probably also the result of a translocation because there is no such chromosome in oat (Figure 2.2a). Line 3815-7 also was monosomic. Recombination between the transgene loci on the same chromosome was not observed in the progeny of these three lines using either phenotypic segregation, Southern analysis, or FISH, further suggesting that these were altered chromosomes (Svitashev et al., 2000). Although different chromosomal aberrations that result from chromosomal breakage are observed in plants regenerated from tissue culture (McCoy et al., 1982; Olhoft and Phillips, 1999), the probability that transgene integration sites were associated with rearranged chromosomes in these oat lines is much higher than expected for random transgene integration and random tissue culture-induced chromosome alterations.

We suggest that microprojectile bombardment and/or the process of transgene integration into the plant genome may lead to host DNA breakage and result in chromosomal aberrations. It also seems possible that the internal structure of tightly linked clusters of interspersed transgenes may lead to DNA breakage following transgene integration. Abranches and colleagues (2000) also observed a wheat line with two transgene loci on the same chromosome. However, no evidence was presented to suggest that this event was the result of chromosome breakage. Rather, it was suggested that the interphase orientation of chromosomes brought the two arms of the chromosome in close proximity, allowing DNA delivery into both arms. Alternatively, ectopic pairing mediated by the transgene loci may cause the interphase configuration reported (Abranches et al., 2000).

Transgene Locus Size

The size of a transgene locus is determined by (1) the number of complete and rearranged copies of delivered DNA, (2) the number and size of genomic interspersions, and (3) the extent of rearrangements in the genomic DNA flanking the transgene locus that may reduce meiotic recombination. A number of examples of transgenic plants produced using *Agrobacterium*-mediated transformation exhibit a single copy of a complete T-DNA integrated without any apparent alterations of the flanking genomic DNA. In some transformants produced using microprojectile bombardment-mediated transformation very simple loci are also reported (Shimizu et al., 2001). The upper range of transgene locus sizes is more difficult to determine. As mentioned, the number of copies of delivered DNA integrated into a transgene locus may vary from one to more than 20 transgene copies for transgenics produced either via *Agrobacterium* or microprojectile bombardment (Pawlowski and Somers, 1996); however, some loci are apparently far larger than this estimate (Svitashev and Somers, 2001b). For example, genomic interspersions in oat transgene loci range from very small fragments to hundreds of kilobases long (Svitashev and Somers, 2001b). Thus in some transgene loci in oat, clustered transgenes appear to be greater than 1 percent of the total prometaphase chromosome length, indicating that the size of these loci may exceed several megabases (Svitashev et al., 2000). Similar clustered transgene loci have been reported in wheat (Abranches et al., 2000) and triticale (Pedersen et al., 1997), suggesting that some transgene loci in these plants may be very large.

Transgene Locus Instability

Evidence of transgene locus instability is not abundant. This fact does not necessarily support the conclusion that transgene loci are stable, but more likely reflects the low sensitivity of analytical methods and lack of in-depth analysis applied to the question of transgene locus stability. The presence of multiple copies of the delivered DNA in a transgene locus would be expected to create instability during either mitosis or meiosis as shown for natural repeats in plant genomes. Moreover, the presence of translocations of the host genome flanking transgene loci may also be expected to create genome instability. As mentioned, different regenerants from the oat line 3812 described by Svitashev and colleagues (2000) exhibited loss of chromosome arms associated with a minor reduction in the transgene locus. More recently, Choffnes and colleagues (2001) described a transformed soybean line produced using microprojectile bombardment that exhibits loss of transgene-hybridizing fragments in its progeny. Locus instability is probably common in complex transgene loci and may occur immediately following transgene integration, during the multiple mitotic divisions during selection and plant regeneration, or during subsequent meioses. It thus seems likely that the locus structures observed in transgenic plants and their progeny are the results of the resolution of this instability and may only partially resemble the original locus created immediately following integration into the recipient genome.

Beyond instability related to the internal structure of the transgene locus, instability may also result from integration into mobile elements such as transposable elements. This phenomenon has not been reported.

MECHANISMS FOR TRANSGENE LOCUS FORMATION

Association of Transgene Locus Formation with Repair of Double-Strand Breaks in DNA

The various structural features of transgene loci and specifically the types of rearrangements in both the integrated delivered DNA and genomic sequences flanking the inserts strongly implicate IR, HR, and possibly other mechanism(s) in transgene locus formation. Evidence of chromosome breakage associated with transgene loci produced from both T-DNA and directly delivered DNA indicates that some transgene loci are created at DSB. The major pathways for DSB repair in eukaryotes also involve IR and HR. Because of this apparent overlap and since the genes and enzymes involved in DSB repair have been better defined compared to those involved in transgene integration, it is worthwhile to briefly review the state of understanding of eukaryotic DSB repair to provide a framework for assessing the role of this process in transgene integration.

In the somatic cells of plants and mammals, DSBs are primarily repaired by IR with minor contributions from HR (Gorbunova and Levy, 1999; Haber, 2000; Karran, 2000). This is likely the reason that HR-mediated integration of transgenes occurs at very low frequencies in plant genomes (Vergunst and Hooykaas, 1999). IR may be mediated by homology-dependent single-strand annealing of noncontiguous DNA fragments (SSA-like), synthesis-dependent strand annealing (SDSA) (Gorbunova and Levy, 1999), and a double-stranded end-joining mechanism, referred to as nonhomologous end-joining (NHEJ) (Haber, 2000; Karran, 2000). SSA-like repair proceeds via degradation of DNA ends to form protruding 3' ends, which anneal at microhomologies shared on two noncontiguous strands. Nonannealed 3' nucleotides are degraded and ss gaps are filled in to repair the break (Gorbunova and Levy, 1999). SDSA is initiated by end-degradation to produce protruding 3' ends that ectopically invade ds-DNA at a microhomology.

New DNA synthesized from the ectopic template would be displaced from the template, producing complementary 3' tails which, when annealed, would be filled in to complete the repair. *Cis* and *trans* template switching during SDSA produces complex filler DNA structures inserted into the DSB. NHEJ is primarily described in the mammalian and yeast DSB repair literature (Haber, 2000). Because IR is sometimes equated with NHEJ (Gorbunova and Levy, 1999), NHEJ will be used to exclusively refer to the Ku pathway described in the following discussion. DSB repair via NHEJ involves a DNA-dependent protein kinase complex consisting of a DNA-binding Ku70/80 heterodimer and a catalytic center (for review see Tuteja and Tuteja, 2000). The Ku protein complex binds to the termini of ds-DNA fragments, protects them from end-degradation, concentrates the broken DNA ends at a DSB site, and positions the DNA ends for ligation by ligase IV creating NHEJ junctions. As described, SSA-like, SDSA, and NHEJ all may result in IR junctions with short stretches of microhomology between the noncontiguous fragments, a feature frequently observed in transgene IR junctions.

In vivo investigations of DSB repair indicate that DSBs may be directly religated but more frequently undergo end-degradation, creating deletions at the repaired site followed by patching through the addition of filler DNA (Gorbunova and Levy, 1997). Filler DNA detected in DSB repair sites include nuclear and organellar DNA, retroelement cDNAs, or DNA from the plasmids used to create the DSB repair system (Moore and Haber, 1996; Gorbunova and Levy, 1997; Ricchetti et al., 1999; Yu and Gabriel, 1999) and T-DNA (Salomon and Puchta, 1998). Investigations of DSB repair in plants suggest that the products of IR in repaired DSB sites are more complex than those from other organisms (Gorbunova and Levy, 1999). Observations of extensively scrambled regions in transgene loci and the deletions, rearrangements, and incorporation of filler DNA in the genomic sequences flanking the locus closely resemble both the products of extrachromasomal (Gorbunova and Levy, 1997) and genomic (Salomon and Puchta, 1998) DSB repair sites in plants, providing structural evidence that DSB repair processes are the primary mechanism(s) of transgene locus formation.

Plant orthologues of the genes involved in eukaryotic DSB repair exist in plant gene databases (Britt, 1999; van Attikum et al., 2001). Functional characterization of their gene products will represent important contributions to the understanding of both DSB repair and the mechanisms of transgene integration and rearrangement in plants. Recent progress along these lines has been achieved in characterization of plant genes involved in T-DNA integration. Thus, the literature on this subject provides direct demonstration of the mechanism(s) of transgene integration and will be reviewed before direct delivery transformation, which is not as clearly understood. Because DSB repair may proceed via multiple processes, it is very likely that transgene locus formation also involves the same range of processes. For example, HR-mediated transgene integration has been demonstrated in plants (Lee et al., 1990) but, as in DSB repair, occurs at frequencies several orders of magnitude lower than IR-mediated integration (Vergunst and Hooykaas, 1999). Furthermore, NHEJ- and SDSA-mediated transgene integration and rearrangement probably both occur. Thus, evidence for one particular mechanism does not necessarily preclude any other mechanism.

Formation of Transgene Loci Produced via Agrobacterium

The status of understanding the integration of T-DNA into the plant genome has been recently reviewed (Gelvin, 2000). Current evidence suggests a model for T-DNA integration that does not involve integration at a DSB but rather template invasion followed by repair synthesis (Tinland, 1996). This model is similar to the SDSA model described by Gorbunova and Levy (1999) for repair of DSBs. Briefly, the ss-T-DNA covalently attached at its 5' end to the VirD2 protein and protected by VirE2 is transported to the nucleus. Apparently, while in the ss state, the 3' end of the T-DNA invades the upper strand of the plant DNA, annealing at a region of microhomology displacing the bottom strand (Tinland, 1996) which, with the 3' overhang of the T-DNA, is degraded by nucleases. The 5' nucleotides of the T-DNA, including the one covalently attached to VirD2, anneal to microhomology in the upper strand of genomic DNA and are ligated to the 3' end of the bottom strand. Repair synthesis produces the upper T-DNA strand to complete locus formation (Ziemienowicz et al., 2000). This model accounts for the frequently observed conservation of the 5' end of the T-DNA and truncations observed in the 3' end as well as small deletions in the genomic integration site. Moreover, it is supported by observations that the majority of T-DNA integration sites analyzed have little to no homology between the plant sequence and inserted T-DNA at the integration junction other than microhomology, indicating that IR, most likely via SDSA, is the main mechanism of T-DNA integration (Gheyson et al., 1991; Mayerhofer et al., 1991; Zambryski, 1992; Vergunst and Hooykass, 1999). Although it was previously postulated that VirD2 covalently attached to the 5' end of the T-DNA may be involved in creation of DSBs and ligation of the T-DNA into the recipient genome, this does not seem to be the case (Ziemienowicz et al., 2000). Rather, it appears that plant proteins are responsible for IR of the T-DNA into the target site. This model would primarily account for simple transgene loci, but it is possible that multiple template switches during SDSA may lead to rearrangements in the integrated T-DNA which resemble filler DNA and repeated initiation of SDSA at the integration site may result in formation of T-DNA repeats which often exhibit genomic filler DNA between the adjacent T-DNA borders. However, these filler DNAs would be expected to originate from the genomic target site. It is also possible that some aspect of the T-DNA integration complex, perhaps the breaks required to resolve the structure or the possibility that the T-DNA integration structure resembles a DSB repair intermediate, induces repair enzymes to recruit ectopic DNA sequences to the vicinity of the forming locus, allowing template switching to these sequences and their incorporation into the locus. Clearly, further examination of the origins of the filler DNA sequences should resolve the remaining questions concerning this model.

Other structures found in T-DNA loci and their flanking genomic sequences argue for alternative or additional mechanisms for locus formation. One of the main questions is whether the T-DNA becomes double stranded at the genomic integration site and, if so, whether ds-T-DNA is the substrate for integration (Gelvin, 2000). The possibility that some portion of the T-DNA is presented to the integration machinery as ds-T-DNA does not necessarily rule out the model proposed by Tinland (1996) because 3' strand invasion may still occur from a double-stranded T-DNA. However, the observed heterogeneity at the right T-DNA border is much less than would be expected from ds-T-DNA because strand invasion from the 3' end of the complement T-DNA strand would generate heterogeneity similar to levels observed at left border junctions. Although this reasoning favors ss-T-DNA as the integration substrate, the presence of inverted repeats argues for end-joining of ds-T-DNAs (De Buck et al., 1999). Inverted repeats of T-DNA differ in their structure. Right border repeats are precise whereas left border repeats have filler DNA between them (De Buck et al., 1999). This is presumably because the right border is protected by covalently attached VirD2 protein. Left border T-DNA repeats could be products of IR via NHEJ of ds-T-DNAs, whereas right border precise repeats could be formed either by IR of ds-T-DNA or a synthesis-dependent mechanism from ss-T-DNA, as proposed by Butler and colleagues (1996). Interestingly, only the right border inverted repeat would be created by SDSA from ss-T-DNA, whereas precise repeats from both borders would be formed from ds-T-DNA because the polarity of the left border repeat from ss-T-DNA would not allow DNA synthesis in a direction appropriate for inverted repeat formation. Taken together the evidence suggests that both ss- and ds-T-DNA may serve as substrates for genomic integration and rearrangements and that multiple mechanisms are involved in locus formation.

The observations of rearranged T-DNAs, rearrangements and filler DNA in genomic DNA flanking the transgene, chromosomal translocations in the genomic DNA flanking some T-DNA loci, and the demonstration that T-DNA can be captured in genomic DSBs more directly indicate that DSB repair is also a pathway for T-DNA integration. The involvement of NHEJ in T-DNA integration via IR was directly demonstrated in yeast (van Attikum et al., 2001). These authors showed that yeast mutants deficient in NHEJ proteins, and specifically Yku70, were resistant to *Agrobacterium*-mediated transformation. Yku70 mutants were completely incapable of T-DNA integration, while a second pathway for T-DNA integration in telomeric regions that was accompanied with gross chromosomal rearrangements operated in the absence of Rad50, Mre11, and Xrs2.

Because transgene loci produced in plants via *Agrobacterium* are structurally similar to those in yeast, and plant orthologues of Ku and its associated proteins are found in plant gene databases, it is extremely tempting based on these results to propose that Ku-mediated NHEJ is involved in T-DNA integration in plants. However, the role of Ku in DSB repair or its relation to SDSA in plants has not been elucidated. Therefore, further investigation will be required to determine the role of plant NHEJ proteins in transgene locus formation from T-DNA (van Attikum et al., 2001). Accordingly, several strategies are being used to determine what plant proteins are active during the *Agrobacterium*-plant interaction as well as for T-DNA integration. Ditt and colleagues (2001) isolated plant genes that are both up- and down-regulated at various times after *Agrobacterium* infection.

In this study, sequences were identified with similarities to genes involved in signal perception, signal transduction, and plant defense. To identify plant genes involved in the integration of the T-DNA, several groups are screening populations of *Arabidopsis* for sensitivity to DNA damaging agents, since plants with inefficient DNA repair or recombination machinery may also be recalcitrant to T-DNA integration (Gelvin, 2000). T-DNA insertion mutant populations in *Arabidopsis* have been screened for resistance to *Agrobacterium* transformation (*rat* mutants) and tagged genes characterized. Such studies resulted in the identification of RAT5, which is a member of the histone H2A multigene family (Mysore et al., 2000). RAT5 interacts with VirD2 in a yeast two-hybrid assay, suggesting an interaction between the integration of the T-DNA strands and histones presumably directing the T-DNA complex to specific chromatin conformations (Gelvin, 2000). Recently, H2AX from mice, an isoform of the H2A gene family that differs from other H2A family members by a C-terminal phosphorylation sequence, was shown to direct repair enzymes to genomic DSBs (Celeste et al., 2002). Although H2A-1 from *Arabidopsis* does not have this C-terminal sequence, this observation for H2AX raises the possibility that H2A-1 and other H2A family members may also be involved in some aspect of DSB repair.

Formation of Transgene Loci Produced via Direct DNA Delivery

Genomic integration and rearrangements of transgenes delivered by microprojectile bombardment and other direct delivery methods appear to be mediated by the interplay of processes involved in DSB repair, of which IR is the main mechanism (Gorbunova and Levy, 1999; Vergunst and Hooykass, 1999). A general model for transgene locus formation based on DSB repair that encompasses all reported features of the transgene loci produced via direct DNA delivery is presented to provide a framework for the following discussion (Figure 2.3). Transgene loci produced via direct DNA delivery methods differ from T-DNA-derived loci in two aspects. First, they more often consist of multiple copies of full-length, truncated, and rearranged transgene DNA (Pawlowski and Somers, 1996; Klein and Jones, 1999). Second, a substantial proportion of transgene loci produced using microprojectile bombardment exhibit interspersion with genomic DNA (Kohli et al., 1998; Pawlowski et al., 1998; Sawasaki et al., 1998), and scrambled and ectopic genomic DNA frequently is found flanking the transgene locus and the integration target site. Moreover, investigations of DSB repair processes in plants show that transgene DNA, delivered by electroporation as part of the experimental system, is frequently incorporated as filler DNA in repaired DSB sites (Gorbunova and Levy, 1997). Thus, it seems likely that both transgene and genomic DNA are incorporated as filler DNA into a DSB in the recipient genome. It is also possible that large interspersions, including chromosome fragments, may be incorporated as filler DNA.

Genomic integration of directly delivered DNA likely proceeds via some of the steps outlined in the T-DNA integration model proposed by Tinland (1996). SDSA-mediated transgene locus formation is supported by the majority of observations from sequencing transgene loci. Of particular significance is the preponderance of very small fragments of delivered and genomic DNA revealed by sequencing transgene loci from oat (Svitashev et al., 2000). For efficient DNA binding by Ku, the DNA fragment must be longer than 15 to 20 bp (Walker et al., 2001). We detected 14 transgene and genomic fragments less than 20 bp that probably are not products of NHEJ (Svitashev et al., 2002). However, it is also possible that NHEJ for delivered DNA fragments may be followed by end-degradation to create a small fragment before the next DNA fragment is aligned and ligated by the NHEJ complex. Our observations of imperfect duplications involving short transgene fragments (Svitashev et al., 2000) also suggest that mechanisms such as SDSA may be responsible for junction formation with very short DNA fragments (Gorbunova and Levy, 1999) as shown in Ku-deficient cell lines (Feldmann et al., 2000; Ramsden and Gellert, 1998). Clearly further research on this issue specifically involving Ku-deficient plant mutants will be required to determine the role of NHEJ in transgene locus formation. The main divergence from the T-DNA model is that transgene loci produced via direct delivery methods exhibit ectopic genomic (filler) DNA flanking the integrated delivered DNA and interspersing the locus (Figure 2.2) at higher frequencies than T-DNA derived loci. Unlike the T-DNA complex, directly delivered DNA is not protected by proteins such as VirE2 or VirD2, thus SDSA may proceed with frequent template switches between both the fragmented delivered DNA molecules and genomic filler DNA concentrated at the integration site. As mentioned, the breakpoints in IR junctions between integrated delivered DNA fragments frequently occur in sequence palindromes. This suggests that either these sequences are targets for cellular nucleases or they assume a secondary structure that either is more susceptible to breakage, cleavage by enzymes, or strand invasion, or may be endpoints for degradation by cellular exonucleases. The formation of secondary structures such as stem loops would suggest that, similar to T-DNA, some portion of the directly delivered DNA assumes single-stranded form capable of undergoing intrastrand interactions.

The origin of genomic DSBs that create transgene integration sites is unclear. DSBs occur in genomes during replication and as a result of DNA damage. Certainly, penetration of the nucleus with a DNA-coated microprojectile would be expected to break genomic DNA and to concentrate delivered DNA in a specific location, resulting in integration of scrambled delivered and genomic DNA into a single locus (Hunold et al., 1994). However, complex transgene locus structures localized at a single site have been reported in plants transformed using Ca-phosphate to deliver DNA into plant protoplasts (Takano et al., 1997), which would not necessarily cause DSBs or target the delivered DNA to a specific region in the recipient genome. Microprojectile bombardment would also be expected to shear and fragment delivered DNA. However, the other direct DNA delivery methods would not fragment DNA. The similarity of transgene locus structures resulting from the different direct DNA delivery methods suggests that plant cell processes are more likely responsible for creating genomic DSBs and fragmenting the delivered DNA into the short sequences described in scrambled transgene loci. It seems likely that the presence of fragmented delivered DNA would induce DSB-repair pathways and concentrate fragments of extrachromosomal delivered and genomic DNA at the genomic DSB site, creating an integration hot spot as suggested in Figure 2.3. It is possible that the presence of delivered DNA free in the nucleus could catalyze DSB formation. The association of palindromes, S/MARs, topoisomerase II sites, and other features with breakpoints in both the delivered DNA integrated into and genomic DNA flanking transgene loci supports the concept that some portion of genomic integration sites are created by enzyme-mediated processes. Specific genomic structures also may create DSBs for transgene integration as suggested by the observation in oat that a high frequency of transgene loci were located at translocation borders (Leggett et al., 2000; Svitashev et al., 2000). It is possible that these regions of oat chromosomes may act as fragile sites described in mammalian systems which are target sites for DNA integration (Rassool et al., 1992, 1996; Mishmar et al., 1998). Finally, formation of DSBs is also likely caused by DNA damage by the DNA delivery processes per se but more likely due to the elevated frequency of chromosome breakage observed in plant cells in tissue culture (for review see Olhoft and Phillips, 1999).

It is interesting to speculate whether formation of complex transgene locus structures occurs before or during genomic integration. Models proposed for cointegration of codelivered plasmids address this question (Chen et al., 1998; Gelvin, 1998). It was proposed that codelivered DNAs undergo HR among homologous plasmid backbone sequences extrachromosomally as described by Offringa and colleagues (1990) and for concatemer formation in mammalian systems (Folger et al., 1982) and that these transgene arrays were later integrated into the genome. Incidentally, an option not mentioned by Gelvin (1998) is that the plasmids could also be recombined extrachromosomally by IR (Bates et al., 1990; Gorbunova and Levy, 1997). Alternatively, simultaneous transgene rearrangement and integration could be mediated by an initially integrated plasmid serving as a HR-mediated integration hot spot or by integration of all plasmids into a DSB by repair enzyme complexes presumably via IR

(Gelvin, 1998). As previously mentioned, no published evidence exists for HR among cointegrated delivered DNAs, although concatemers of the same plasmid have been described. Thus, the bulk of the evidence points to IR as the means of cointegration. The observations of inverted repeats that involve genomic and transgenic sequences (Shimizu et al., 2001) argue that some portion of transgene locus rearrangements are created before repeat formation, but these repeats could also be produced via SDSA or other synthesis-dependent mechanisms during genomic integration.

Two models have been proposed to account for genomic interspersions in clustered transgene copies. Kohli and colleagues (1998) hypothesized that the structure of transgene loci results from a "two-phase" process similar to those proposed for mammalian systems. The "preintegration" phase occurs extrachromosomally and involves ligation of transgene fragments into complex multicopy arrays of transgenes, presumably via simple ligation or NHEJ. These transgene arrays are then integrated into the genome at a single site, which initiates a hot spot that facilitates integration of other arrays of delivered and genomic DNA in a genomic region, resulting in plant DNA separating the transgenes in a tightly linked cluster. Although not stated, it is also possible that extrachromosomal array formation could incorporate genomic DNA. Presumably, the recombination hot spot identified in the 35S promoter (Kohli, Griffiths, et al., 1999) would provide an integration target site as described in the "two phase model." Pawlowski and colleagues (1998) proposed a "synthesis model" in which integration of transgene DNA at a cluster of active replication forks may result in the interspersion of the integrated transgenes with host DNA (Pawlowski et al., 1998). Replication of eukaryotic DNA proceeds at multiple, closely localized replication forks (Mills et al., 1989).

Template switching to or ligation of the delivered DNA to plant DNA at the replication fork could result in interspersed transgene integration patterns. Both models predict that some of the genomic interspersions would originate from the integration target site unless ectopic genomic DNA was available for incorporation either into the multicopy arrays of the two-phase model or through template switching evoked in the synthesis model. It is likely that some features of both models are correct; however, further investigations are needed to determine the origin of genomic interspersions to verify them. Both models could take place in concert with DSB repair processes, most likely SDSA with multiple template switches between delivered DNA and ecotopic genomic DNA in which the delivered DNA and genomic DNAs are integrated as filler DNA (Figure 2.3). Either random preintegration IR of DNA fragments in an extrachromosomal pool would produce interspersed structures containing transgene and host DNA or direct integration of DNA from this pool into the DSB via SDSA would result in formation of all features described for transgene loci. While we believe that the bulk of the evidence supports the latter model that transgene integration and rearrangements occur simultaneously, further research is required to resolve this issue. Furthermore, it should be emphasized that, as in DSB repair, multiple mechanisms, some of which probably have yet to be elucidated, likely interact to orchestrate transgene locus formation resulting in the diverse array of transgene locus structures.

REFERENCES

Abranches, R., A.P. Santos, E. Wegel, S. Williams, A. Castillo, P. Christou, P. Shaw, and E. Stoger (2000). Widely separated multiple transgene integration sites in wheat chromosomes are brought together at interphase. *The Plant Journal* 24:713-723.

Ambros, P.F., M.A. Matzke, and A.J.M. Matzke (1986). Detection of a 17 kb unique sequence (T-DNA) in plant chromosomes by in situ hybridization. *Chromosoma* 94:11-18.

Armstrong, C.L., G.B. Parker, J.C. Pershing, S.M. Brown, P.R. Sanders, D.R. Duncan, T. Stone, D.A. Dean, D.L. DeBoer, J. Hart, et al. (1995). Cell biology and molecular genetics: Field evaluation of European corn borer control in progeny of 173 transgenic corn events expressing an insecticidal protein from *Bacillus thuringiensis*. *Crop Science* 35:550-557.

Barcelo, P., C. Hagel, D. Becker, A. Martin, and H. Lorz (1994). Transgenic cereal (tritordeum) plants obtained at high frequency by microprojectile bombardment of inflorescence tissue. *The Plant Journal* 5:583-592.

Bates, G.W. (1994). Genetic transformation of plants by protoplast electroporation. *Molecular Biotechnology* 2:135-145.

Bates, G.W., S.A. Carle, and W.C. Piastuch (1990). Linear DNA introduced into carrot protoplasts by electroporation undergoes ligation and recircularization. *Plant Molecular Biology* 14:899-908.

Bechtold, N., J. Ellis, and G. Pelletier (1993). In planta *Agrobacterium*-mediated gene transfer by infiltration of adult *Arabidopsis thaliana* plants. Comptes Rendus de l'Academe des Sciences, Paris. *Life Sciences* 316:1194-1199.

Becker, D., R. Brettschneider, and H. Lorz (1994). Fertile transgenic wheat from microprojectile bombardment of scutellar tissue. *The Plant Journal* 5:299-307.

Birch, R.G. (1997). Plant transformation: Problems and strategies for practical application. *Annual Review of Plant Physiology and Plant Molecular Biology* 48:297-326.

Brar, G.S., B.A. Cohen, C.L. Vick, and G.W. Johnson (1994). Recovery of transgenic peanut (*Arachis hypogaea* L.) plants from elite cultivars utilizing ACCELL technology. *The Plant Journal* 5:745-753.

Breitler, J.-C., A. Labeyrie, D. Meynard, and T. Legavre (2002). Efficient microprojectile bombardment-mediated transformation of rice using gene cassettes. *Theoretical and Applied Genetics* 104:709-719.

Britt, A.B. (1999). Molecular genetics of DNA repair in higher plants. *Trends in Plant Science* 4(1):20-25.

Buising, C.M. and R.M. Benbow (1994). Molecular analysis of transgenic plants generated by microprojectile bombardment: Effect of petunia transformation booster sequence. *Molecular and General Genetics* 243:71-81.

Butler, D.K., L.E. Yasuda, and M.C. Yao (1996). Induction of large DNA palindrome formation in yeast: Implications for gene amplification and genome stability in eukaryotes. *Cell* 87:1115-1122.

Castle, L.A., D. Errampalli, T. Atherton, L. Franzmann, E. Yoon, and D.W. Meinke (1993). Genetic and molecular characterization of embryonic mutants identified following seed transformation in *Arabidopsis*. *Molecular and General Genetics* 241:504-514.

Celeste, A., S. Petersen, P.J. Romanienko, O. Fernandez-Capteillo, H.T. Chen, O.A. Sedelnikova, B. Reina-San-Martin, V. Coppola, E. Meffre, M.J. Difilippantonio, et al. (2002). Genomic instability in mice lacking histone H2AX. *Science* 296(5569):922-927.

Chang, S.S., S.K. Park, B.C. Kim, B.J. Kang, D.U. Kim, and H.G. Nam (1994). Stable genetic transformation of *Arabidopsis thaliana* by *Agrobacterium* inoculation in planta. *The Plant Journal* 5:551-558.

Chen, L., P. Marmey, N.J. Taylor, J. Brizard, C. Espinoza, P. D'Cruz, H. Huet, S. Zhang, A. de Kochko, R.N. Beachy, and C.M. Fauquet (1998). Expression and inheritance of multiple transgenes in rice plants. *Nature Biotechnology* 16:1060-1064.

Chen, W.P., P.D. Chen, D.J. Liu, R. Kynast, B. Friebe, R. Velazhahan, S. Muthukrishnan, and B.S. Gill (1999). Development of wheat scab symptoms is delayed in transgenic wheat plants that constitutively express a rice thaumatin-like protein gene. *Theoretical and Applied Genetics* 99:755-760.

Cheng, M., J.E. Fry, S. Peng, H. Zhou, C.M. Hironaka, D.R. Duncan, T.W. Conner, and Y. Wan (1997). Genetic transformation of wheat mediated by *Agrobacterium tumefaciens*. *Plant Physiology* 115:971-980.

Choffnes, D.S., R. Philip, and L.O. Vodkin (2001). A transgene locus in soybean exhibits a high level of recombination. *In Vitro Cellular and Developmental Biology—Plant* 37:756-762.

Choi, H.W., P.G. Lemaux, and M.-J. Cho (2000). Increased chromosomal variation in transgenic versus nontransgenic barley (*Hordeum vulgare* L.) plants. *Crop Science* 40:524-533.

Christou, P. (1992). Genetic transformation of crop plants using microprojectile bombardment. *The Plant Journal* 2:275-281.

Christou, P., T.L. Ford, and M. Kofron (1991). Production of transgenic rice (*Oryza sativa* L.) plants from agronomically important indica and japonica varieties via electric discharge particle acceleration of exogenous DNA into mature zygotic embryos. *Bio/Technology* 9:957-962.

Christou, P., W.F. Swain, N.S. Yang, and D.E. McCabe (1989). Inheritance and expression of foreign genes in transgenic soybean plants. *Proceedings of the National Academy of Sciences of the United States of America* 86:7500-7504.

Clemente, T.E., B.J. LaVallee, A.R. Howe, D. Conner-Ward, R.J. Rozman, P.E. Hunter, D.L. Broyles, D.S. Kasten, and M.A. Hinchee (2000). Progeny analysis of glyphosate selected transgenic soybeans derived from *Agrobacterium*-mediated transformation. *Crop Science* 40:797-803.

Clough, S.J. and A.F. Bent (1998). Floral dip: A simplified method for *Agrobacterium*-mediated transformation of *Arabidopsis thaliana*. *The Plant Journal* 16:735-743.

Cooley, J., T. Ford, and P. Christou (1995). Molecular and genetic characterization of elite transgenic rice plants produced by electric-discharge particle acceleration. *Theoretical and Applied Genetics* 90:97-104.

Czernilofsky, A.P., R. Hain, B. Baker, and U. Wirtz (1986). Studies of structure and functional organization of foreign DNA integrated into the genome of *Nicotiana tabacum*. *DNA* 5:473-482.

Czernilofsky, A.P., R. Hain, L. Herrera-Estrella, H. Lorz, E. Goyvaerts, B.J. Baker, and J. Schell (1986). Fate of selectable marker DNA integrated into the genome of *Nicotiana tabacum*. *DNA* 5:101-113.

De Block, M., L. Herrera-Estrella, M. Van Montagu, J. Schell, and P. Zambryski (1984). Expression of foreign genes in regenerated plants and in their progeny. *The EMBO Journal* 3:1681-1689.

De Buck, S., C. De Wilde, M. Van Montagu, and A. Depicker (2000). T-DNA vector backbone sequences are frequently integrated into the genome of transgenic plants obtained by *Agrobacterium*-mediated transformation. *Molecular Breeding* 6:459-468.

De Buck, S., A. Jacobs, M. Van Montagu, and A. Depicker (1999). The DNA sequences of T-DNA junctions suggest that complex T-DNA loci are formed by a recombination process resembling T-DNA integration. *The Plant Journal* 20:295-304.

De Neve, N., S. De Buck, A. Jacobs, M. Van Montagu, and A. Depicker (1997). T-DNA integration patterns in cotransformed plant cells suggest that T-DNA repeats originate from cointegration of separate T-DNAs. *The Plant Journal* 11:15-29.

Deroles, S.C. and R.C. Gardner (1988). Analysis of the T-DNA structure in a large number of transgenic petunias generated by *Agrobacterium*-mediated transformation. *Plant Molecular Biology* 11:365-377.

D'Halluin, K., E. Bonne, M. Bossut, M. De Beuckeleer, and J. Leemans (1992). Transgenic maize plants by tissue electroporation. *The Plant Cell* 4:1495-1505.

Ditt, R.F., E.W. Nester, and L. Comai (2001). Plant gene expression response to *Agrobacterium tumefaciens*. *Proceedings of the National Academy of Sciences of the United States of America* 98:10954-10959.

Ebinuma, H., K. Sugita, E. Matsunaga, S. Endo, K. Yamada, and A. Komamine (2001). Systems for the removal of a selection marker and their combination with a positive marker. *Plant Cell Reports* 20:383-392.

Feldman, K.A. and M.D. Marks (1987). *Agrobacterium* mediated transformation of germinating seeds of *Arabidopsis thaliana:* A non-tissue culture approach. *Molecular and General Genetics* 208:1-9.

Feldmann, E., V. Schmiemann, W. Goedecke, S. Reichenberger, and P. Pfeiffer (2000). DNA double-strand break repair in cell-free extracts from Ku80-deficient cells: Implications for Ku serving as an alignment factor in non-homologous DNA end joining. *Nucleic Acids Research* 28:2585-2596.

Finer, J.J. and M.D. McMullen (1990). Transformation of cotton (*Gossypium hirsutum* L.) via particle bombardment. *Plant Cell Reports* 8:586-589.

Finnegan, J. and D. McElroy (1994). Transgene inactivation: Plants fight back! *Bio/Technology* 12:883-888.

Folger, K.R., E.A. Wong, G. Wahl, and M.R. Capecchi (1982). Patterns of integration of DNA microinjected into cultured mammalian cells: Evidence for homologous recombination between injected plasmid DNA molecules. *Molecular and Cellular Biology* 2:1372-1387.

Fransz, P.F., M. Stam, B. Montijn, R. ten Hoopen, J. Wiegant, J.M. Kooter, O. Oud, and N. Nanninga (1996). Detection of single-copy genes and chromosome rearrangements in *Petunia hybrida* by fluorescence in situ hybridization. *The Plant Journal* 9:767-774.

Froelich-Ammon, S.J., K.C. Gale, and N. Osheroff (1994). Site-specific cleavage of a DNA hairpin by topoisomerase II: DNA secondary structure as a determinant of enzyme recognition/cleavage. *Journal of Biological Chemistry* 269:7719-7725.

Fromm, M.J., F. Morrish, C. Armstrong, R. Williams, J. Thomas, and T.M. Klein (1990). Inheritance and expression of chimeric genes in the progeny of transgenic maize plants. *Bio/Technology* 8:833-839.

Fu, X., L. Tan Duc, S. Fontana, B. Ba Bong, P. Tinjuangjun, D. Sudhakar, R.M. Twyman, P. Christou, and A. Kohli (2000). Linear transgene constructs lacking vector backbone sequences generate low-copy-number transgenic plants with simple integration patterns. *Transgenic Research* 9:11-19.

Gelvin, S.B. (1998). Multigene plant transformation: More is better. *Nature Biotechnology* 16:1009-1010.

Gelvin, S.B. (2000). *Agrobacterium* and plant genes involved in T-DNA transfer and integration. *Annual Review of Plant Physiology and Plant Molecular Biology* 51:223-256.

Gharti-Chhetri, G.B., W. Cherdshewasart, J. Dewulf, J. Paszkowski, M. Jacobs, and I. Negrutiu (1990). Hybrid genes in the analysis of transformation conditions: 3. Temporal/spatial fate of NPTII gene integration, its inheritance and factors affecting these processes in *Nicotiana plumbaginifolia. Plant Molecular Biology* 14:687-696.

Gheysen, G., R. Villarroel, and M. Van Montagu (1991). Illegitimate recombination in plants: A model for T-DNA integration. *Genes and Development* 5:287-297.

Gill, K.S., B.S. Gill, and T.R. Endo (1993). A chromosome region-specific mapping strategy reveals gene-rich telomeric ends in wheat. *Chromosoma* 102:374-381.

Gill, K.S., B.S. Gill, T.R. Endo, and E.V. Boyko (1996). Identification and high-density mapping of gene-rich regions in chromosome group 5 of wheat. *Genetics* 143:1001-1012.

Gorbunova, V. and A.A. Levy (1997). Non-homologous DNA end joining in plant cells is associated with deletions and filler DNA insertions. *Nucleic Acids Research* 25:4650-4657.

Gorbunova, V. and A.A. Levy (1999). How plants make ends meet: DNA double-strand break repair. *Trends in Plant Science* 4:263-269.

Gordon-Kamm, W.J., T.M. Spencer, M.L. Mangano, T.R. Adams, R.J. Daines, W.G. Start, J.V. O'Brien, S.A. Chambers, W.R. Adams Jr., N.G. Willetts, et al. (1990). Transformation of maize cells and regeneration of fertile transgenic plants. *The Plant Cell* 2:603-618.

Grevelding, C., V. Rantes, E. Kemper, J. Schell, and R. Masterson (1993). Single-copy T-DNA insertions in *Arabidopsis* are the predominant form of integration in root-derived transgenics, whereas multiple insertions are found in leaf discs. *Plant Molecular Biology* 23:847-860.

Haber, J.E. (2000). Partners and pathways repairing a double-strand break. *Trends in Genetics* 16:259-264.

Hadi, M.Z., M.D. McMullen, and J.J. Finer (1996). Transformation of 12 different plasmids into soybean via particle bombardment. *Plant Cell Reports* 15:500-505.

Hansen, G., R.D. Shillito, and M.D. Chilton (1997). T-strand integration in maize protoplasts after codelivery of a T-DNA substrate and virulence genes. *Proceedings of the National Academy of Sciences of the United States of America* 94:11726-11730.

Heberle-Bors, E., B. Charvat, D. Thompson, J.P. Schernthaner, A. Barta, A.J.M. Matzke, and M.A. Matzke (1988). Genetic analysis of T-DNA insertions into the tobacco genome. *Plant Cell Reports* 7:571-574.

Hiei, Y., S. Ohta, T. Komari, and T. Kumashiro (1994). Efficient transformation of rice (*Oryza sativa* L.) mediated by *Agrobacterium* and sequence analysis of the boundaries of the T-DNA. *The Plant Journal* 6:271-282.

Hooykaas, P.J. and R.A. Schilperoort (1992). *Agrobacterium* and plant genetic engineering. *Plant Molecular Biology* 19:15-38.

Horsch, R., R. Fraley, S. Rogers, J. Fry, H. Klee, D. Shah, S. McCormick, J. Niedermeyer, and H. Hoffman (1987). *Agrobacterium*-mediated transformation of plants. In *Plant Tissue and Cell Culture* (pp. 317-329), Eds. C.E. Green, D.A. Somers, W.P. Hackett, and D.D. Biesboer. New York: Alan R. Liss.

Horsch, R.B., R.T. Fraley, S.G. Rogers, P.R. Sanders, A. Lloyd, and N. Hoffmann (1984). Inheritance of functional foreign genes in plants. *Science* 223:496-498.

Hunold, R., R. Bronner, and G. Hahne (1994). Early events in microprojectile bombardment: Cell viability and particle location. *The Plant Journal* 5:593-604.

Iglesias, V.A., E.A. Moscone, I. Papp, F. Neuhuber, S. Michalowski, T. Phelan, S. Spiker, M. Matzke, and A.J.M. Matzke (1997). Molecular and cytogenetic analyses of stably and unstably expressed transgene loci in tobacco. *The Plant Cell* 9:1251-1264.

Ishida, Y., H. Saito, S. Ohta, Y. Hiei, T. Komari, and T. Kumashiro (1996). High efficiency transformation of maize (*Zea mays* L.) mediated by *Agrobacterium tumefaciens*. *Nature Biotechnology* 14:745-750.

Jackson, S.A., P. Zhang, W.P. Chen, R.L. Phillips, B. Friebe, S. Muthukrishnan, and B.S. Gill (2001). High-resolution structural analysis of biolistic transgene integration into the genome of wheat. *Theoretical and Applied Genetics* 103:56-62.

Jakowitsch, J., I. Papp, E.A. Moscone, J. van der Winden, M. Matzke, and A.J.M. Matzke (1999). Molecular and cytogenetic characterization of a transgene locus that induces silencing and methylation of homologous promoters in *trans*. *The Plant Journal* 17:131-140.

Jones, J.D., S.C. Weller, and P.B. Goldsbrough (1994). Selection for kanamycin resistance in transformed petunia cells leads to the coamplification of a linked gene. *Plant Molecular Biology* 24:505-514.

Jongsma, M., M. Koornneef, P. Zabel, and J. Hille (1987). Tomato protoplast DNA transformation: Physical linkage and recombination of exogenous DNA sequences. *Plant Molecular Biology* 8:383-394.

Kaeppler, H.F., G.K. Menon, and R.W. Skadsen (2000). Transgenic oat plants via visual selection of cells expressing green fluorescent protein. *Plant Cell Reports* 19:661-666.

Kapila, J., R. De Rycke, M. Van Montagu, and G. Angenon (1997). An *Agrobacterium*-mediated transient gene expression system for intact leaves. *Plant Science* 122:101-108.

Karran, P. (2000). DNA double strand break repair in mammalian cells. *Current Opinion in Genetics and Development* 10:144-150.

Kartzke, S., H. Saedler, and P. Meyer (1990). Molecular analysis of transgenic plants derived from transformations of protoplasts at various stages of the cell cycle. *Plant Science* 67:63-72.

Khan, S.A. (2000). Plasmid rolling-circle replication: Recent developments. *Molecular Microbiology* 37:477-484.

Kharb, P., J. Dong, M.N. Islam-Faridi, D.M. Stelly, and T.C. Hall (2001). Fluorescence in situ hybridization of single copy transgenes in rice chromosomes. *In Vitro Cellular and Developmental Biology—Plant* 37:1-5.

Klein, T.M. and T.J. Jones (1999). Methods of genetic transformation: The gene gun. *Molecular Improvement of Cereal Crops, Advances in Cellular and Molecular Biology of Plants,* Volume 5(1), pp. 21-42, Ed. I. K. Vasil. Dordrecht, the Netherlands: Kluwer Academic Publishers.

Klein, T.M., L. Kornstein, J.C. Sanford, and M.E. Fromm (1989). Genetic transformation of maize cells by particle bombardment. *Plant Physiology* 91:440-444.

Klein, T.M., E.D. Wolf, R. Wu, and J.C. Sanford (1987). High-velocity microprojectiles for delivering nucleic acids into living cells. *Nature* 327:70-73.

Kohler, F., G. Cardon, M. Pohlman, and O. Schieder (1989). Enhancement of transformation rates in higher plants by low-dose irradiation: Are DNA repair systems involved in the incorporation of exogenous DNA into plant genome? *Plant Molecular Biology* 12:189-199.

Kohli, A., D. Gahakwa, P. Vain, D.A. Laurie, and P. Christou (1999). Transgene expression in rice engineered through particle bombardment: Molecular factors controlling stable expression and transgene silencing. *Planta* 208:88-97.

Kohli, A., S. Griffiths, N. Palacios, R.M. Twyman, P. Vain, D.A. Laurie, and P. Christou (1999). Molecular characterization of transforming plasmid rearrangements in transgenic rice reveals a recombination hotspot in the CaMV 35S promoter and confirms the predominance of microhomology mediated recombination. *The Plant Journal* 17:591-601.

Kohli, A., M. Leech, F. Vain, D. Laurie, and P. Christou (1998). Transgene organization in rice engineered through direct DNA transfer supports a two-phase integration mechanism mediated by the establishment of integration hot spots. *Proceedings of the National Academy of Sciences of the United States of America* 95:7203-7208.

Koncz, Z., N. Martini, R. Mayerhofer, Z. Koncz-Kalman, H. Korber, G.P. Redei, and J. Schell (1989). High-frequency T-DNA-mediated gene tagging in plants. *Proceedings of the National Academy of Sciences of the United States of America* 86:8467-8471.

Koprek, T., R. Hansch, A. Nerlich, R.R. Mendel, and J. Schulze (1996). Fertile transgenic barley of different cultivars obtained by adjustment of bombardment conditions to tissue response. *Plant Science* 119:79-91.

Krizkova, L. and M. Hrouda (1998). Direct repeats of T-DNA integrated in tobacco chromosome: Characterization of junction regions. *The Plant Journal* 16:673-680.

Kumar, S. and M. Fladung (2001). Controlling transgene integration in plants. *Trends in Plant Science* 6(4):155-159.

Kumpatla, S.P. and T.C. Hall (1998). Recurrent onset of epigenetic silencing in rice harboring a multi-copy transgene. *The Plant Journal* 14:129-135.

Kumpatla, S.P. and T.C. Hall (1999). Organizational complexity of a rice transgene locus susceptible to methylation-based silencing. *Life* 48:459-467.

Laufs, P., D. Autran, and J. Traas (1999). A chromosomal paracentric inversion associated with T-DNA integration in *Arabidopsis*. *The Plant Journal* 18:131-139.

Lee, K.Y., P. Lund, K. Lowe, and P. Dunsmuir (1990). Homologous recombination in plant cells after *Agrobacterium*-mediated transformation. *The Plant Cell* 2:415-425.

Leggett, J.M., S.J. Perret, J. Harper, and P. Morris (2000). Chromosomal localization of cotransformed transgenes in the hexaploid cultivated oat *Avena sativa* L. using fluorescence in situ hybridization. *Heredity* 84:46-53.

Loc, N.T., P. Tinjuangjun, A.M.R. Gatehouse, P. Christou, and J.A. Gatehouse (2002). Linear transgene constructs lacking vector backbone sequences generate transgenic rice plants which accumulate higher levels of proteins conferring insect resistance. *Molecular Breeding* 9:231-244.

Makarevitch, I., S.K. Svitashev, and D.A. Somers (2003). Complete sequence analysis of transgene loci from plants transformed via microprojectile bombardment. *Plant Molecular Biology* 52:421-432.

Matsumoto, S., Y. Ito, T. Hosoi, Y. Takahashi, and Y. Machida (1990). Integration of *Agrobacterium* T-DNA into a tobacco chromosome: Possible involvement of DNA homology between T-DNA and plant DNA. *Molecular and General Genetics* 224:309-316.

Matzke, M.A. and A.J.M. Matzke (1995). How and why do plants inactivate homologous *(trans)* genes? *Plant Physiology* 107:679-685.

Mayerhofer, R., Z. Koncz-Kalman, C. Nawrath, G. Bakkeren, A. Crameri, K. Angelis, G. Redei, J. Schell, B. Hohn, and C. Koncz (1991). T-DNA integration: A mode of illegitimate recombination in plants. *The EMBO Journal* 10:697-704.

McCabe, D.E., W.F. Swain, B.J. Martinell, and P. Christou (1988). Stable transformation of soybean *(Glycine max)* by particle acceleration. *Bio/Technology* 6:923-926.

McCoy, T.J., R.L. Phillips, and H.W. Rines (1982). Cytogenetic analysis of plants regenerated from oat *(Avena sativa)* tissue culture; high frequency of partial chromosome loss. *Canadian Journal of Genetics and Cytology* 24:37-50.

Mills, A.D., J.J. Blow, J.G. White, W.B. Amos, D. Wilcock, and R.A. Laskey (1989). Replication occurs at discrete loci spaced throughout nuclei replicating in vitro. *Journal of Cell Science* 94:471-474.

Mishmar, D., A. Rahat, S.W. Scherer, G. Nyakatura, B. Hinzmann, Y. Kohwi, Y. Mandel-Gutfroind, J.R. Lee, B. Drescher, D.E. Sas, et al. (1998). Molecular characterization of a common fragile site (*FRA7H*) on human chromosome 7 by the cloning of a simian virus 40 integration site. *Proceedings of the National Academy of Sciences of the United States of America* 95:8141-8146.

Moore, J.K. and J.E. Haber (1996). Capture of retrotransposon DNA at the sites of chromosomal double-strand breaks. *Nature* 383:644-646.

Moscone, E.A., M.A. Matzke, and A.J.M. Matzke (1996). The use of combined FISH/GISH in conjunction with DAPI counterstaining to identify chromosomes containing transgene inserts in amphidiploid tobacco. *Chromosoma* 105:231-236.

Mouras, A. and I. Negrutiu (1989). Localization of the T-DNA on marker chromosomes in transformed tobacco cells by in situ hybridization. *Theoretical and Applied Genetics* 78:715-720.

Mouras, A., M.W. Saul, S. Essad, and I. Potrykus (1987). Localization by in situ hybridization of a low copy chimaeric resistance gene introduced into plants by direct gene transfer. *Molecular and General Genetics* 207:204-209.

Muller, A.E., Y. Kamisugi, R. Gruneberg, I. Niedenhof, R.J. Horold, and P. Meyer, (1999). Palindromic sequences and A+T-rich DNA elements promote illegitimate recombination in *Nicotiana tabacum*. *Journal of Molecular Biology* 291:29-46.

Mysore, K.S., J. Nam, and S.B. Gelvin (2000). An *Arabidopsis* histone H2A mutant is deficient in *Agrobacterium* T-DNA integration. *Proceedings of the National Academy of Sciences of the United States of America* 97(2):948-953.

Nacry, P., C. Camilleri, B. Courtial, M. Caboche, and D. Bouchez (1998). Major chromosomal rearrangements induced by T-DNA transformation in *Arabidopsis. Genetics* 149:641-650.

Nehra, N.S., R.N. Chibbar, N. Leung, K. Caswell, C. Mallard, L. Steinhauer, M. Baga, and K.K. Kartha (1994). Self-fertile transgenic wheat plants regenerated from isolated scutellar tissues following microprojectile bombardment with two distinct gene constructs. *The Plant Journal* 5:285-297.

Offringa, R., M.J.A. de Groot, H.J. Haagsman, M.P. Does, P. van den Elzen, and P.J.J. Hooykaas (1990). Extrachromosomal homologous recombination and gene targeting in plant cells after *Agrobacterium* mediated transformation. *The EMBO Journal* 9:3077-3084.

Ohba, T., Y. Yoshioka, C. Machida, and Y. Machida (1995). DNA rearrangement associated with the integration of T-DNA in tobacco: An example for multiple duplications of DNA around the integration target. *The Plant Journal* 7(1):157-164.

Olhoft, P.M., L.E. Flagel, C.M. Donovan, and D.A. Somers (2003). Efficient soybean transformation using using hygromicin B selection in the cotyledonary-node method. *Planta* 216:723-735.

Olhoft, P.M. and R.L. Phillips (1999). Genetic and epigenetic instability in tissue culture and regenerated progenies. In *Plant Response to Environmental Stresses: From Phytohormones to Genomic Reorganization* (pp. 111-148), Ed. H.R. Lerner. New York: M. Dekker Inc., pp. 111-148.

Ortega, D., R. Monique, M. Laudié, C. Llauro, R. Cooke, M. Devic, S. Genestier, G. Picard, P. Abad, P. Contard, et al. (2002). Flanking sequence tags in *Arabidopsis thaliana* T-DNA insertion lines: A pilot study. *Comptes Rendus Biologies* 325:773-780.

Palevitz, B.A. (2000). DNA surprise: Monsanto discovers extra sequences in its Roundup Ready soybeans. *The Scientist* 14(15):20.

Papp, I., V.A. Iglesias, E.A. Moscone, S. Michalowski, S. Spiker, Y.-D. Park, M.A. Matzke, and A.J.M. Matzke (1996). Structural instability of a transgene locus in tobacco is associated with aneuploidy. *The Plant Journal* 10:469-478.

Paszkowski, J., R.D. Shillito, M. Saul, V. Mandak, T. Hohn, B. Hohn, and I. Potrykus (1984). Direct gene transfer to plants. *The EMBO Journal* 3:2717-2722.

Pawlowski, W.P. and D.A. Somers (1996). Transgenic inheritance in plants genetically engineered using microprojectile bombardment. *Molecular Biotechnology* 6:17-30.

Pawlowski, W.P. and D.A. Somers (1998). Transgenic DNA integrated into the oat genome is frequently interspersed by host DNA. *Proceedings of the National Academy of Sciences of the United States of America* 95:12106-12110.

Pawlowski, W.P., K.A. Torbert, H.W. Rines, and D.A. Somers (1998). Irregular patterns of transgene silencing in allohexaploid oat. *Plant Molecular Biology* 38:597-607.

Pedersen, C., J. Zimmy, and D. Becker (1997). Localization of introduced genes on the chromosomes of transgenic barley, wheat and triticale by fluorescence in situ hybridization. *Theoretical and Applied Genetics* 94:749-757.

Phillips, R.L., D.A. Somers, and K.A. Hibberd (1988). Cell tissue culture and in-vitro manipulation. In *Corn and Corn Improvement* (pp. 345-387), Eds. G.F. Sprague and J.W. Dudle. Madison, WI: American Society of Agronomy.

Potrykus, I. (1990). Gene transfer to cereals: An assessment. *Bio/Technology* 8:535-542.

Potrykus, I., J. Paszkowski, M.W. Saul, I. Negrutiu, and R.D. Shillito (1987). Direct gene transfer to plants: Facts and future. In *Plant Tissue and Cell Culture* (pp. 289-302), Eds. C.E. Green, D.A. Somers, W.P. Hackett, and D.D. Biesboer. New York: Alan R. Liss.

Potrykus, I., J. Paszkowski, M.W. Saul, J. Petruska, and R.D. Shillito (1985). Molecular and general genetics of a hybrid foreign gene introduced into tobacco by direct gene transfer. *Molecular and General Genetics* 199:169-177.

Ramsden, D.A. and M. Gellert (1998). Ku protein stimulates DNA end joining by mammalian DNA ligase: A direct role of Ku in repair of DNA double-strand breaks. *The EMBO Journal* 17:609-614.

Rassool, F.V., M.M. Le Beau, M.E. Neilly, E. van Melle, R. Espinoza III, and T.W. McKeithan (1992). Increased genetic instability of the genome fragile site at 3p14 after integration of exogenous DNA. *American Journal of Human Genetics* 50:1243-1251.

Rassool, F.V., M.M. Le Beau, M.-L. Shen, M.E. Neilly, R. Espinoza III, S.T. Ong, F. Boldog, H. Drabkin, R. McCarroll, and T.W. McKeithan (1996). Direct cloning of DNA sequences from the common fragile site region at chromosome band 3p14.2. *Genomics* 35:109-117.

Register, J.C., III, D.J. Peterson, P.J. Bell, W.P. Bullock, I.J. Evans, B. Frame, A.J. Greenland, N.S. Higgs, I. Jepson, S. Jiao, et al. (1994). Structure and function of selectable and non-selectable transgenes in maize after introduction by particle bombardment. *Plant Molecular Biology* 25:951-961.

Ricchetti, M., C. Fairhead, and B. Dujon (1999). Mitochondrial DNA repairs double-strand breaks in yeast chromosomes. *Nature* 402:96-100.

Riggs, C.D. and G.W. Bates (1986). Stable transformation of tobacco by electroporation: Evidence for plasmid concatenation. *Proceedings of the National Academy of Sciences, USA* 83:5602-5606.

Ritala, A., K. Aspegren, U. Kurten, M. Salmenkallio-Marttila, L. Mannonen, R. Hannus, V. Kauppinen, T.H. Teeri, and T.-M. Enari (1994). Fertile transgenic barley by particle bombardment of immature embryos. *Plant Molecular Biology* 24:317-325.

Russell, D.R., K.M. Wallace, J.H. Bathe, B.J. Martinell, and D.E. McCabe (1993). Stable transformation of *Phaseolus vulgaris* via electric-discharge mediated particle acceleration. *Plant Cell Reports* 12:165-169.

Salomon, S. and H. Puchta (1998). Capture of genomic and T-DNA sequences during double-strand break repair in somatic plant cells. *The EMBO Journal* 17:6086-6095.

Sanford, J.C. (1988). The biolistic process. *Trends in Biotechnology* 6:299-302.

Sanford, J.C. (1990). Biolistic plant transformation. *Physiologia Plantarum* 79:206-209.

Sanford, J.C., M.J. DeVit, J.A. Russell, F.D. Smith, P.R. Harpending, M.K. Roy, and S.A. Johnston (1991). An improved, helium-driven biolistic device. *Techniques* 3:3-16.

Sanford, J.C., T.M. Klein, E.D. Wolf, and N. Allen (1987). Delivery of substances into cells and tissues using a particle bombardment process. *Particle Science and Technology* 5:27-37.

Sanford, J.C., F.D. Smith, and J.A. Russell (1993). Optimizing the biolistic process for different biological applications. *Methods in Enzymology* 217:483-509.

Sawasaki, T., M. Takahashi, N. Goshima, and H. Morikawa (1998). Structures of transgene loci in transgenic *Arabidopsis* plants obtained by particle bombardment: Junction regions can bind to nuclear matrices. *Gene* 218:27-35.

Schocher, R.J., R.D. Shillito, M.W. Saul, J. Paszkowski, and I. Potrykus (1986). Cotransformation of unlinked foreign genes into plants by direct gene transfer. *Bio/Technology* 4:1093-1096.

Sheng, J. and V. Citovsky (1996). *Agrobacterium*-plant cell DNA transport: Have virulence proteins, will travel. *The Plant Cell* 8:1699-1710.

Shillito, R.D., M.W. Saul, J. Paszkowski, M. Müller, and I. Potrykus (1985). High efficiency direct gene transfer to plants. *Bio/Technology* 3:1099-1103.

Shimizu, K., M. Takahashi, N. Goshima, S. Kawakami, K. Irifune, and H. Morikawa (2001). Presence of an SAR-like sequence in junction regions between an introduced transgene and genomic DNA of cultured tobacco cells: Its effect on transformation frequency. *The Plant Journal* 26:375-384.

Somers, D.A., H.W. Rines, W. Gu, H.F. Kaeppler, and W.R. Bushnell (1992). Fertile, transgenic oat plants. *Bio/Technology* 10:1589-1594.

Songstad, D.D., D.A. Somers, and R.J. Griesbach (1995). Advances in alternative DNA delivery techniques. *Plant Cell, Tissue and Organ Culture* 40:1-15.

Southern, E.M. (1975). Detection of specific sequences among DNA fragments separated by gel electrophoresis. *Journal of Molecular Biology* 98:503-517.

Spencer, T.M., J.V. O'Brien, W.G. Start, T R. Adams, W.J. Gordon-Kamm, and P.G. Lemaux (1992). Segregation of transgenes in maize. *Plant Molecular Biology* 18:201-210.

Stewart Jr., C.N. (2001). The utility of green fluorescent protein in transgenic plants. *Plant Cell Reports* 20:376-382.

Svitashev, S., E. Ananiev, W.P. Pawlowski, and D.A. Somers (2000). Association of transgene integration sites with chromosome rearrangements in hexaploid oat. *Theoretical and Applied Genetics* 100:872-880.

Svitashev, S.K. , W.P. Pawlowski, I. Makarevitch, D.W. Plank, and D.A. Somers (2002). Complex transgene locus structures implicate multiple mechanisms for plant transgene rearrangement. *The Plant Journal* 32:433-445.

Svitashev, S.K. and D.A. Somers (2001a). Characterization of transgene loci in plants using FISH: A picture is worth a thousand words. *Plant Cell, Tissue and Organic Culture* 65:205-214.

Svitashev, S.K. and D.A. Somers (2001b). Genomic interspersions determine the size and complexity of transgene loci in transgenic plants produced by microprojectile bombardment. *Genome* 44:691-697.

Takano, M., H. Egawa, J.-E. Ikeda, and K. Wakasa (1997). The structures of integration sites in transgenic rice. *The Plant Journal* 11:353-361.

Tax, F.E. and D.M. Vernon (2001). T-DNA-associated duplication/translocations in *Arabidopsis:* Implications for mutant analysis and functional genomics. *Plant Physiology* 126:1527-1538.

Tingay, S., D. McElroy, R. Kalla, S. Fieg, M. Wang, S. Thornton, and R. Brettell (1997). *Agrobacterium tumefaciens*-mediated barley transformation. *The Plant Journal* 11:1369-1376.

Tinland, B. (1996). The integration of T-DNA into plant genomes. *Trends in Plant Science* 1:178-184.

Torbert, K.A., H.W. Rines, and D.A. Somers (1998). Transformation of oat using mature embryo-derived tissue cultures. *Crop Science* 38:226-231.

Tuteja, R. and N. Tuteja (2000). Ku autoantigen: A multifunctional DNA-binding protein. *Critical Reviews in Biochemistry and Molecular Biology* 35:1-33.

Tzfira, T. and V. Citovsky (2002) Partners-in-infection: Host proteins involved in the transformation of plant cells by *Agrobacterium. Trends in Cell Biology* 12:121-129.

Uze, M., I. Potrykus, and C. Sautter (1999). Single-stranded DNA in the genetic transformation of wheat (*Triticum aestivum* L.): Transformation frequency and integration pattern. *Theoretical and Applied Genetics* 99:487-495.

Vain, P., J. De Buyser, V. Bui Trang, R. Haicour, and Y. Henry (1995). Foreign gene delivery into monocotyledonous species. *Biotechnology Advances* 13:653-671.

Vain, P., V.A. James, B. Worland, and J.W. Snape (2002). Transgene behaviour across two generations in a large random population of transgenic rice plants produced by particle bombardment. *Theoretical and Applied Genetics* 105:878-889.

van Attikum, H., P. Bundock, and P.J. Hooykaas (2001). Non-homologous end-joining proteins are required for *Agrobacterium* T-DNA integration. *The EMBO Journal* 20:6550-6558.

Vasil, I.K. (1999). Molecular improvement of cereal crop—An introduction. In *Molecular Improvement of Cereal Crops* (pp. 1-8), Ed. I.K. Vasil. Dordrecht, the Netherlands: Kluwer Academic Publishers.

Vasil, V., A.M. Castillo, M.E. Fromm, and I.K. Vasil (1992). Herbicide resistant fertile transgenic wheat plants obtained by microprojectile bombardment of regenerable embryogenic callus. *Bio/Technology* 10:667-674.

Vasil, V., V. Srivastava, A.M. Castillo, M.E. Fromm, and I.K. Vasil (1993). Rapid production of transgenic wheat plants by direct bombardment of cultured immature embryos. *Bio/Technology* 11:1553-1558.

Vergunst, A.C. and P.J.J. Hooykaas (1999). Recombination in the plant genome and its application in biotechnology. *Critical Reviews in Plant Science* 18:1-31.

Walker, J.R., R.A. Corpina, and J. Goldberg (2001). Structure of the Ku heterodimer bound to DNA and its implications for double-strand break repair. *Nature* 412:607-614.

Walters, D.A., C.S. Vetsch, D.E. Potts, and R.C. Lundquist (1992). Transformation and inheritance of a hygromycin phosphotransferase gene in maize plants. *Plant Molecular Biology* 18:189-200.

Wan, Y. and P.G. Lemaux (1994). Generation of large numbers of independently transformed fertile barley plants. *Plant Physiology* 104:37-48.

Ward, D.V., J.R. Zupan, and P.C. Zambryski (2002). *Agrobacterium* VirE2 gets the VIP1 treatment in plant nuclear import. *Trends in Plant Science* 7:1-3.

Weeks, J.T., O.D. Anderson, and A.E. Blechl (1993). Rapid production of multiple independent lines of fertile transgenic wheat *(Triticum aestivum). Plant Physiology* 102:1077-1084.

Weising, K., J. Schell, and G. Kahl (1988). Foreign genes in plants: Transfer, structure, expression, and applications. *Annual Review of Genetics* 22:421-477.

Wilmink, A. and J.J.M. Dons (1993). Selective agents and marker genes for use in transformation of monocotyledonous plants. *Plant Molecular Biology Reporter* 11:165-185.

Windels, P., I. Taverniers, A. Depicker, E. Van Bockstaele, and M. De Loose (2001). Characterization of the Roundup Ready soybean insert. *European Food Research and Technology* 213:107-112.

Wolters, A.A., L.M. Trindade, E. Jacobsen, and G.F. Visser (1998). Fluorescence in situ hybridization on extended DNA fibres as a tool to analyse complex T-DNA loci in potato. *The Plant Journal* 13(6):837-847.

Yu, X. and A. Gabriel (1999). Patching broken chromosomes with extranuclear cellular DNA. *Molecular Cell* 4:873-881.

Zambryski, P. (1992). Chronicles from the *Agrobacterium*-plant cell DNA transfer story. *Annual Review of Plant Physiology and Plant Molecular Biology* 43:465-490.

Zambryski, P., J. Tempe, and J. Schell (1989). Transfer and function of T-DNA genes from *Agrobacterium* Ti and Ri plasmids in plants. *Cell* 56:193-201.

Zhao, Z.Y., W. Gu, T. Cai, L.A. Tagliani, D. Hondred, D. Bond, S. Krell, M.L. Rudert, W.B. Bruce, and D.A. Pierce (1998). Molecular analysis of T_0 plants transformed by *Agrobacterium* and comparison of *Agrobacterium*-mediated transformation with bombardment transformation in maize. *Maize Genetics Cooperation Newsletter* (University of Missouri) 72:34-38.

Zheng, S-J., B. Henken, E. Sofiari, E. Jacobsen, F.A. Krens, and C. Kik (2001). Molecular characterization of transgenic shallots (*Allium cepa* L.) by adaptor ligation PCR (AL-PCR) and sequencing of genomic DNA flanking T-DNA borders. *Transgenic Research* 10:237-245.

Ziemienowicz, A., B. Tinland, J. Bryant, V. Gloeckler, and B. Hohn (2000). Plant enzymes but not *Agrobacterium* VirD2 mediate T-DNA ligation in vitro. *Molecular and Cellular Biology* 20:6317-6322.

Zupan, J., T.R. Muth, O. Draper, and P. Zambryski (2000). The transfer of DNA from *Agrobacterium tumefaciens* into plants: A feast of fundamental insights. *The Plant Journal* 23:11-28.

Chapter 3

Gene Stacking Through Site-Specific Integration

David W. Ow

INTRODUCTION

The directing of exogenous DNA to a chosen location in a host genome requires either homology-dependent recombination or recombinase-mediated recombination. In plants, the site-specific integration of DNA by homologous recombination has had limited successes (for review, see Puchta, 2002). However, a recent article reported ~1 percent of the transformed rice lines harbored the introduced DNA at a chosen site (Terada et al., 2002). Despite this encouraging report, this still means generating hundreds of transgenic lines to recover a few site-directed insertions. More important, the precise placement of a transgene in itself does not guarantee suitable transgene expression, as current knowledge cannot predict how a chromosome location affects newly introduced DNA. Hence, even if homologous recombination were practical, the screening of a collection of random integration events may still be a preferred option. Given that a "favorable" integration site is found empirically and through considerable labor, this begs the question of whether subsequent DNA deliveries can also be directed to a previously characterized location.

Recombinase-mediated site-specific integration of DNA has been possible for quite a few years. This approach requires two steps. In the first step, a recombination site is introduced into the plant genome by conventional transformation. Hence, the integrated recombination site, or the target site, resides at random locations. In a second step, a second DNA molecule with a second recombination site is introduced into the plant genome. In the presence of a recombinase, the first and second recombination sites recombine to integrate the new DNA. Depending on the site-specific recombination system, the first and second recombination sites can be identical, nearly identical, or quite dissimilar in sequence. In theory, a wide variety of site-specific recombination systems can perform the site-specific integration steps, but only a few have been tested in plants.

A similar review on recombinase-mediated plant transformation has recently appeared (Ow, 2002), and inevitably, there are some overlaps with the materials presented in this chapter. Some of the materials presented previously are not covered here. On the other hand, a most important component of site-specific integration, gene stacking, is presented in this chapter but is absent in that review. Before gene-stacking strategies can be introduced, the relevant site-specific recombination systems and the reported cases of site-specific integration in plants are reviewed.

HETEROLOGOUS SITE-SPECIFIC RECOMBINATION SYSTEMS

The refereed literature reports that four of the freely reversible types of site-specific recombination systems have been tested in plants. They are the Cre-*lox* system from bacteriophage P1 (Dale and Ow, 1990, 1991; Odell et al., 1990; Bayley et al., 1992; Russell et al., 1992), the FLP-*FRT* system of *Saccharomyces cerevisiae* (Lyznik et al., 1993; Lloyd and Davis, 1994; Sonti et al., 1995; Kilby et al., 1995; Bar et al., 1996), the R-*RS* system of *Zygosaccharomyces rouxii* (Onouchi et al., 1991, 1995), and a modified Gin-*gix* system from bacteriophage Mu (Maeser and Kahmann, 1991). Cre, FLP, R, and Gin are the recombinases, and *lox, FRT, RS,* and *gix* the respective recombination sites (Figure 3.1a). Each of these sites consists of a short asymmetric spacer sequence of 2 to 8 bp, where strand exchange takes place, and a set of inverted repeats of 12 or 13 bp, where the recombinases bind. As the two sites participating in a recombination reaction, or the substrate sites, have the same or nearly the same sequence, the two product sites generated by the recombination reaction are identical to the original substrate sites, and are recognized by the recombinase for further recombination. No other protein factor

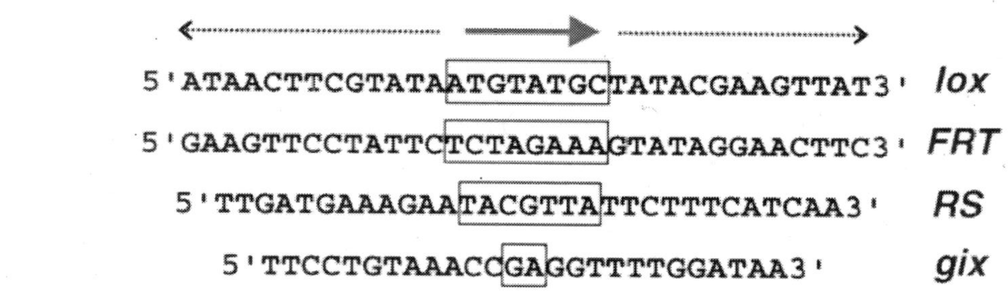

FIGURE 3.1. Recombination sites that function in plants. (a) The *lox*, *FRT*, *RS*, and *gix* sequences are structurally similar in that an asymmetric spacer sequence (boxed) is flanked by inverted repeats. (b) The ϕC31 *attP* and *attB* sequences share only 16 bp of identity in a 50 bp stretch centered at TTG (boxed), the recombination crossover point.

aside from the single-polypeptide recombinase is required for the recombination reaction.

Recently, a member of the irreversible type of site-specific recombination systems has shown function in plants (Ow et al., 2001). Derived from *Streptomyces* phage ϕC31 (Thorpe and Smith, 1998), this system consists of a 68 kD_a single-polypeptide integrase, encoded by the *int* gene, and a set of bacterial and phage attachment sites, known respectively as *attB* and *attP,* that share only 3 bp of homology at the point of crossover, or a total of 16 bp of identity within 50 bp (Figure 3.1b). The products of the recombination reaction, *attL* and *attR,* differ in sequence from *attB* and *attP* and are not recognized by the integrase, at least not without an accompanying excisionase.

TARGETING VIA CRE-LOX

Tobacco

A circular DNA can fuse into the chromosome through a single recombination reaction. The chromosomal product of this cointegration reaction would have two directly oriented recombination sites flanking the newly introduced DNA. The subsequent recombination between these two sites can excise the newly introduced DNA. To minimize the excision of integrated molecules, Albert and colleagues (1995) used mutant *lox* sequences with mutations in each of the flanking 13 bp inverted repeats of the *lox* sequence, the region where a Cre monomer binds. These mutant sites can be referred to as binding-element mutant sites (in contrast to spacer mutant sites described in a later section). Because the two monomers bind cooperatively, a mutation in only one binding element has less of an effect on dimer occupation of the site. Hence, recombination can occur between a left element (LE) mutant and a right element (RE) mutant (Figure 3.2a). The reaction regenerates a wild type site, and a new left and right element (LE+RE) mutant site. Mutations on both elements can substantially reduce recognition by Cre. Therefore, as one of the product sites is no longer efficiently recognized by the recombinase, the reaction is much less reversible.

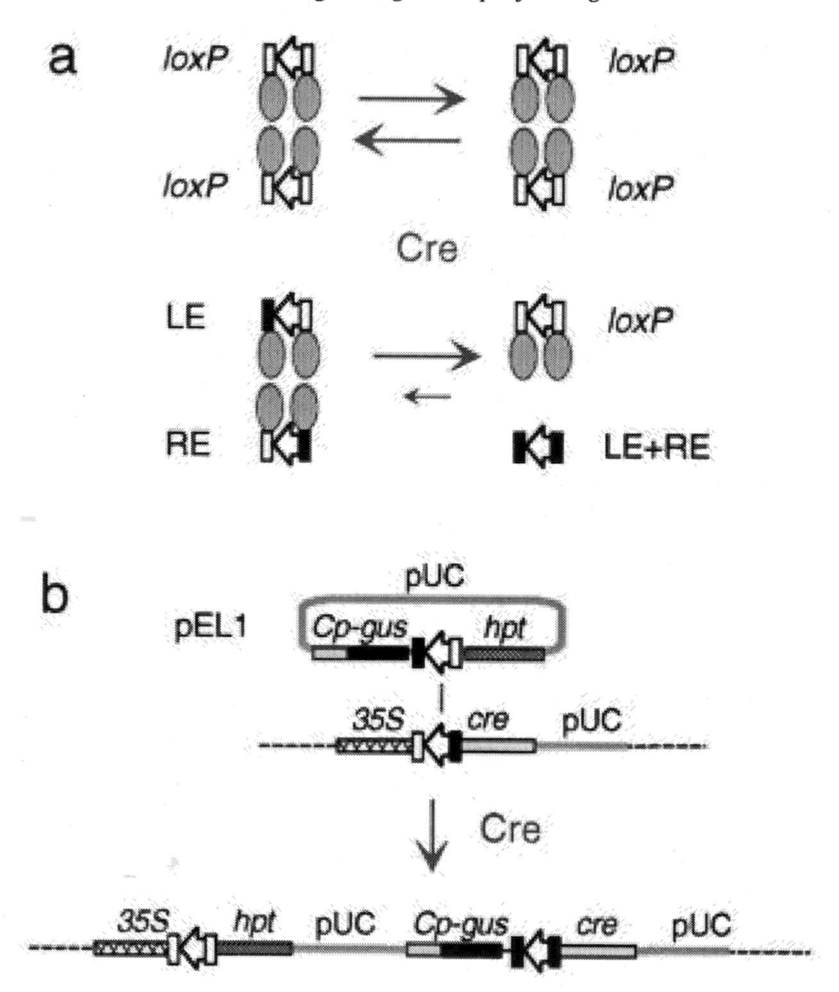

FIGURE 3.2. Site-specific integration using *lox* sites mutated in recombinase binding elements. (a) The *lox* site is depicted by rectangles (13 bp inverted repeats) flanking an open arrowhead (8 bp asymmetric spacer). Filled rectangles mutant sequences with reduced recognition by Cre. Unlike *loxP* × *loxP* reactions, an LE × RE reaction yields an LE + RE double mutant site that is poorly recognized by the recombinase (shaded ovals). (b) DNA integration in tobacco using mutant *lox* sites in conjunction with disruption of *cre* expression. Dotted lines point to recombination between sites. Abbreviations: *35S,* CaMV 35S RNA promoter; *hpt,* hygromycin resistance coding region; *Cp-gus,* Commelina yellow mottle virus promoter fused to β-glucuronidase coding region. In this and other figures, gene terminators are not shown as separate DNA fragments.

In addition to using binding element mutant sites, two separate strategies were tested to limit the availability of Cre recombinase after site-specific integration. In a first set of experiments, Cre was produced from a *35S-lox-cre* construct placed into the plant genome, where *35S* is the CaMV 35S RNA promoter. Subsequent integration by a circular *lox-hpt* plasmid, where *hpt* is a promoterless hygromycin phosphotransferase coding region, displaces the *cre* coding region from its promoter to terminate *cre* transcription (Figure 3.2b). In a second set of experiments, plant lines were generated with a *35S-lox-luc* target, where *luc* is the firefly luciferase gene. The Cre protein was provided by a *35S-cre* plasmid cotransformed with a *lox-hpt* plasmid. When a cell depletes its supply of transiently synthesized Cre proteins, a *lox-hpt* insertion should fail to excise.

The integrating construct was delivered into plant protoplasts through treatment with polyethylene glycol (PEG). Because random integration of the *lox-hpt* plasmid would not be detected unless it integrated behind a genomic promoter, the frequency of random integration was deduced from a parallel transformation with a *35S-hpt* construct. To confirm that the *lox-hpt* plasmid integrated at the *35S-lox* target, molecular analysis was conducted on representative calluses and regenerated plants. Eight different target sites in the tobacco genome were tested. In conjunction with the two strategies to eliminate postintegration recombinase activity, a comparison was made between wild type and two mutant *lox* pairs, *lox66* × *lox71* and *lox76* × *lox75*. Considerable variation in transformation efficiencies was found. In particular, there was a tendency for a lack of site-specific integration with protoplast preparations that showed low transformation competency. For example, among six attempts on two *35S-lox66-cre* lines, two of them failed to produce site-specific events. In these two experiments, the transformation efficiency of the *35S-hpt* plasmid averaged only 3.5×10^{-5}, an order of magnitude below expectation. With the other four experiments, the *35S-hpt* plasmid had an average transformation efficiency of 3.6×10^{-4}. In these same experiments, site-specific recombination by the *lox71-hpt* plasmid averaged 2.5×10^{-4}. Likewise, for four out of six experiments from two different *35S-lox76-cre* lines, the transformation efficiencies of the *35S-hpt* plasmid and the *lox75-hpt* plasmid averaged 5.1×10^{-4} and 3.8×10^{-4}, respectively. In contrast, from a total of eight attempts, the transformation of two *35S-loxP-cre* lines by a *loxP-hpt* plasmid failed to produce a single integrant, whereas the control *35S-hpt* plasmid averaged 1.3×10^{-4}. Thus, for this integration strategy, there was a clear advantage in mutant over wild-type sites.

In the transient expression approach, however, the mutant sites were only slightly better. Counting only experiments in which both the *lox75-hpt* plasmid and the *35S-hpt* yielded transformants (three of seven experiments), the average transformation efficiency obtained from a single *35S-lox76-luc* line by either the *lox75-hpt* plasmid or the *35S-hpt* plasmid was 1.6×10^{-4} or 1.4×10^{-4}, respectively. The average transformation efficiency, from three out of seven experiments, on a single *35S-loxP-luc* line by the *loxP-hpt* plasmid or the *35S-hpt* plasmid was 7.9×10^{-5} or 2.9×10^{-4}, respectively. The difference in *loxP* × *loxP* integration with the two experimental approaches can be attributed to the relative abundance of Cre. In the promoter displacement strategy, Cre is readily available within each cell. As a *loxP* × *loxP* excision occurs more readily than a *loxP* × LE+RE excision, this could account for a less stable *loxP* × *loxP* integration. With the transient expression approach, each cell within a transformed population experiences a different level of Cre concentration depending on the amount of *cre* DNA taken up. Those with a rapid depletion of the recombinase will more likely trap an integrated molecule.

In a separate set of integration experiments in tobacco, Choi and colleagues (2000) tested the site-specific integration of large BAC clones. A *lox-hpt* fragment was placed into a BAC vector for cointegration at a genomic *35S-lox-cre* target. Three BAC clones, containing 50, 100, or 150 kb of cotton DNA, were biolistically transformed into a *35S-lox-cre* tobacco line. Between two to 14 independently derived hygromycin resistant plants were obtained from each BAC clone. Polymerase chain reaction (PCR) and Southern analyses showed that these plants have the correct recombination junctions. Unfortunately, it proved difficult to determine whether the large inserts were intact. The cotton DNA cross-hybridized extensively to tobacco DNA, making such determination impractical.

Rice

Using the most effective strategy described for tobacco (Albert et al., 1995), namely combining both the disruption of *cre* transcription along with a pair of *lox75* and *lox76* alleles, Cre-*lox* site-specific integration was tested in rice (Srivastava and Ow, 2002). Two target cell lines were generated harboring a *P2-lox76-cre* construct, where *P2* is the maize ubiquitin promoter. Calluses were biolistically transformed with pVS55, a plasmid with a promoterless neomycin phosphotransferase coding region *(npt)* and a β-glucuronidase coding region *(gus)* under the control of *P2*. This *npt-P2-gus* fragment is flanked by a *loxP* upstream of *npt* and a mutant *lox75* allele downstream of *gus*. Cre-mediated intramolecular recombination between *loxP* and *lox75* separates pVS55 into two molecules: a circular plasmid backbone with a *loxP* site, and a circular fragment consisting of *lox75-npt-P2-gus*. Insertion of the latter molecule into the genomic *lox76* target forms a *P2-loxP-npt* linkage to confer geneticin resistance. As before, *cre* transcription terminates through displacement of *cre* from its promoter, resulting in a *gus-lox75/76-cre* junction.

From the PCR analysis of 36 geneticin resistant calluses collected from the two target cell lines, the selected *P2-npt* junction was found in every line. The unselected *gus-cre* junction was also found in all but three clones. These three failed to express *gus,* possibly due to the DNA acquiring a mutation, or was silenced. However, a PCR product corresponding to the *P2-cre* target junction was detected in every callus line, indicating that the geneticin resistant clones were chimeric, with target cells growing among integrant cells. Due to the chimeric calluses, further analysis was confined to regenerated plants that express the *gus* gene. Only three GUS$^+$ integrant plants regenerated, and hybridization analysis indicated site-specific integration in each case. However, only one of the three plants showed the site-specifically integrated molecule as the sole copy in the genome. The other two plants harbored additional molecules that are most likely the result of random integration elsewhere in the genome. Moreover, one of the two plants also retained the plasmid backbone that was resolved from the intramolecular recombination of pVS55.

Arabidopsis

Although direct DNA delivery methods are effective in a wide range of crop species, many plants are most easily transformed via *Agrobacterium.* Recombinase-catalyzed reactions require a double-stranded (ds) DNA substrate. Therefore, the T-DNA must first acquire second-strand synthesis before it can be a substrate for recombination. It is not clear whether the T-DNA must first integrate into the plant genome before a double-stranded form emerges. After all, the initial burst of high-level expression by T-DNA encoded transgenes is thought to arise from nonintegrated, presumably duplex T-DNA templates.

Agrobacterium-mediated delivery was explored with the Cre-*lox* system in *Arabidopsis*. In one attempt, the *Arabidopsis* genome was transgenic for a promoterless *lox-npt* construct (Vergunst and Hooykaas, 1998). Two insertion vectors were used, one with a *35S-lox* fragment, and a second with a *lox-35S-lox* fragment. The Cre recombinase was provided through *cre* transient expression by a cotransforming vector. Site-specific insertion into the chromosomal target would confer kanamycin resistance from formation of a *35S-lox-npt* linkage. One precise integration event was obtained by the *lox-35S-lox* construct, none by the *35S-lox* construct. This suggests that a circular *35S-lox* molecule, excised from the *lox-35S-lox* construct, might have been the integration substrate. In another attempt, a *35S-lox-cre* construct was placed into the *Arabidopsis* genome and was targeted by a *lox-npt-lox* construct (Vergunst et al., 1998). With this strategy, kanamycin-resistant clones appeared at a frequency of 1 to 2 percent of the frequency of random integration. The simplest interpretation would be a *lox-npt* circular molecule produced by site-specific recombination prior to insertion into the genomic target, although a double chromosome translocation event can also yield the same product.

TARGETING VIA FLP-FRT

Tobacco

FLP-*FRT*-mediated gene targeting has been reported in tobacco and maize. Interestingly, both reports describe using a DNA exchange strategy. Nucleotide changes within the 8 bp spacer of the *FRT* can create new recognition sites for FLP (Schlake and Bode, 1994). The same holds true for the Cre-*lox* system (Hoess et al., 1986). These spacer mutant sites are commonly known as heterospecific sites. Tuttle and colleagues (1999) described target lines harboring constructs flanked by a set of heterospecific *FRT* sites. An example is the recombination between the genome target *FRT-Pu-bar-Pa-FLP-FRTx[int]-npt3'*, and the integrating construct *FRT-Pn-FRTx[int]-5'npt,* where *FRTx[int]* is an *FRT*-heterospecific site situated within an intron, *Pu, Pa,* and *Pn* are promoters, and *5'npt* and *npt3'* encode, respectively, the NH_3 and COOH halves of neomycin phosphotransferase (Figure 3.3a). The integrating construct was delivered into the target cell by microparticle bombardment or by *Agrobacterium.* Recombination between the two sets of heterospecific sites restored a functional *npt* gene and replaced the target site DNA. Calluses resistant to kanamycin, sensitive to Basta, and with PCR-detectable junction fragments were reported.

Maize

Baszczynski and colleagues (Baszczynski, Bowen, Peterson, and Tagliani, 2001) described very similar events obtained in maize. Maize cell lines were transformed with a construct with *P2* upstream of two heterospecific *FRT* sites. Upon retransformation with a promoterless *gus* construct flanked by the same set of heterospecific *FRT* sites (Figure 3.3b), GUS^+ calluses were recovered for 18 target lines, at very high frequencies that range from 2 to 67 percent of the total transformation frequency, as scored by a *bar* marker also present on the construct. The high efficiency might be attributed to an improved FLP recombinase, as a separate patent described codon changes in the *FLP* gene that enhanced its recombination activity in maize (Baszczynski, Bowen, Drummond, et al., 2001). As molecular data on the downstream junction were not mentioned, what relative proportion of the events is replacement and not cointegration is not known.

EXPRESSION OF TARGETED TRANSGENES

Day and colleagues (2000) sought to address whether expression of a transgene at a target site would be similar among independently transformed lines and whether expression would differ among different chromosome locations. The targeting strategy was analogous to

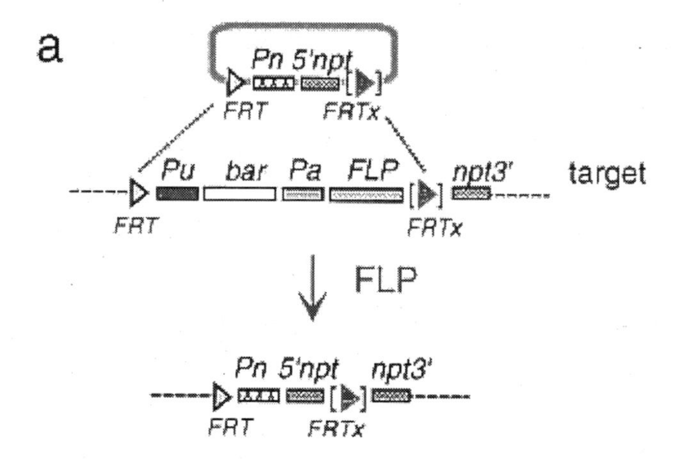

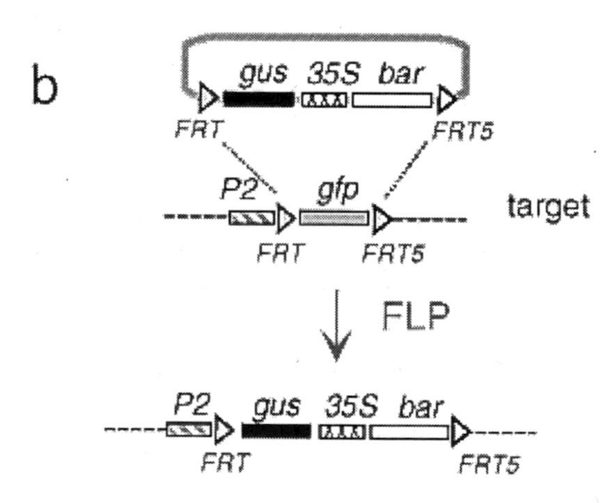

FIGURE 3.3. DNA replacement reactions. (a) Example in tobacco cells shows an *FRT*-DNA1-*FRTx* fragment, replaced by a double recombination event with an *FRT*-DNA2-*FRTx* fragment. As *FRTx* is embedded within an intron (indicated by brackets), recombination at this site restores the two halves of the neomycin phosphotransferase gene, *npt.* (b) Example in maize cells shows an *FRT-gfp-FRT5* fragment, replaced by an *FRT-gus-35S-bar-FRT5* fragment from a plasmid. Fusion between *gus* and *P2* produces a scorable phenotype, while *35S-bar* confers resistance to bialaphos. Site-specific or random DNA integrations were estimated from expression of *gus* or *bar,* respectively. *Pu, Pa, Pn* are promoters; *gfp* is green fluorescent protein gene.

that described earlier (Albert et al., 1995) except that the revised *lox-hpt* plasmid, pEL1, carried a *Cp-gus* fragment, where *Cp* is the Commelina yellow mottle virus promoter with activity in vascular and reproductive tissue (Figure 3.2b). Selection for active expression was imposed on *35S-lox-hpt* but not on the linked *Cp-gus* reporter. Several independent integrant plants from each of the five sites, a total of 22 lines, were shown to contain the correct single-copy integration structure.

Gene Imprinting from Site-Specific Integration

When examined for *Cp-gus* expression, approximately half of the integrant lines in each of four target sites showed the expected pattern of expression, namely expression throughout vascular tissues. The remaining integrant lines, however, showed various degrees of gene silencing. The GUS staining pattern of the F1 plants showed sharp boundaries that define blue sectors, suggesting a clonal origin. The expression patterns were maintained through plant development and were transmitted to subsequent generations. Thus, this implicated a stable and heritable imprinting phenomenon. Both PCR and the Southern analyses concluded that the integrated DNA remained in place. Hence, inactivity of the *Cp-gus* transgene was not due to excision of pEL1 from the target site.

Methylation sensitive and insensitive enzymes were used to probe the extent of DNA methylation in active and silenced *Cp-gus* transgenes at the different loci. From this analysis, a difference in the degree of DNA methylation was found. For the lines that showed full expression in vascular tissues, there were fewer methylated sites. For the silenced lines, however, hypermethylation was observed. Most surprising was the finding that the DNA outside of the newly integrated transgene did not show a change in the methylation pattern. It was as though the "old" DNA was fixed into a certain state of methylation, whereas different degrees of imprinting were imposed on each independently integrated "new" DNA.

The mechanism for this imprinting is not known. However, amidst the current models of gene silencing, the RNA-directed DNA methylation model (Wassenegger et al., 2000) seems most plausible for this phenomenon. During DNA uptake by competent cells, aberrant or duplex RNAs may be produced by transient expression of the transformed DNA, especially from rearranged and/or concatameric extrachromosomal molecules. Although in each case a single molecule integrated site specifically, the amount of initial DNA uptake can differ considerably among individual recipient cells. Those with the highest initial copies may therefore be more prone to an RNA-directed DNA methylation effect. In this model, the imprinting may occur prior to or soon after DNA integration.

The biolistic targeting in rice followed a similar pattern (Srivastava and Ow, 2002). Among the 36 geniticin-resistant calli collected from the two target cell lines, 17 of them tested negative for *gus* expression. Only two of these lacked a correct downstream junction to suggest imprecise integration. The rest appeared to be correct. In light of the results of Day and colleagues (2000), hypermethylation-associated gene silencing seems to be a probable cause.

Position Effect on Gene Expression

The second finding from the Day and colleagues (2000) study is derived from a comparison of the full-expression integrant lines. These plants all showed the expected vascular-specific expression pattern. The intensity of expression was similar among independent lines of the same site, but dissimilar among different integration sites. For example, the level of GUS activity was about tenfold higher at one integration site than at another, although both sites could give rise to the same expression pattern. This supports the hypothesis that the chromosome location can exert an effect on the expression of the integrated transgene.

PROSPECTS FOR GENE STACKING

The available data show that recombinase-direct site integration can place a single-copy nonrearranged DNA fragment into the target site at a practical frequency. Moreover, a high percentage of those insertions express the transgene at a predictable and reproducible level. The problem that lies ahead is whether a target site, once it is found favorable for transgene expression, can accept additional transgene molecules. As mentioned earlier, site-specific recombination is a two-step process: the first step is to create a site, and the second step is to deliver the desired transgene. In the absence of homologus recombination, single-copy target lines must be generated by conventional random integration. These target lines must be physically characterized and functionally tested for transgene integration and expression. If the target site can accommodate only a single delivery event, it is difficult to justify investing the time and labor into screening for a suitable target. However, if additional DNA can be appended onto the existing target site, not only would it make the initial screening for target sites worth the effort, but it would also permit the construction of transgene clusters where collections of desirable traits reside. Clustering transgenes, as opposed to scattering them at various places in the genome, would facilitate the introgression of large gene sets to field cultivars.

The idea of gene stacking rests on the concept that the integrating DNA brings along a different recombination site, such that after insertion of the new recombination site into the genome, the new recombination then becomes the new target for the next round of integration (Ow and Medberry, 1995). The following section describes two strategies for stacking through the introduction of fresh target sites.

Stacking via a Reversible Recombination System

Baszczynski and colleagues (Baszczynski, Bowen, Drummond, et al., 2001) described a strategy that uses heterospecific recombination sites. These sites, with mutations in the spacer region, recombine with sites of identical spacer sequence, but not with sites of different spacer sequence. Figure 3.4 shows six heterospecific *FRT* sites, *FRT, FRT5, FRT*, FRT2, FRT2*, FRTz,* a collection of promoters, *Px, Py, Pz,* a collection of trait genes, *G1, G2, G3,* and two selectable markers, *M1* and *M2*. Here, *G1, G2,* and *G3* are promoter-containing genes, but *M1* and *M2* represent promoterless coding regions.

The process begins with a genomic target consisting of *Px-FRT-M1-FRT5-G1,* using *Px-M1* for selection (Figure 3.4a). The stacking of *G2* to the *G1* locus is accomplished with a first cassette: *FRT-M2-FRT*-G2-Py-FRT2-FRT5* (Figure 3.4b). In an exchange reaction involving *FRT* × *FRT* and *FRT5* × *FRT5,* the resulting genomic structure would be *Px-FRT-M2-FRT*-G2-Py-FRT2-FRT5-G1* (Figure 3.4c).

The FLP recombinase can be provided, for example, by transient expression from a *FLP*-expressing plasmid. To stack *G3* onto the *G2-G1* locus, a second cassette would be used, bearing *FRT2-M1-FRT2*-G3-Pz-FRTz-FRT5* (Figure 3.4d). Upon *FRT2* × *FRT2* and *FRT5* × *FRT5* recombination, the genomic structure would be *Px-FRT-M2-FRT*-G2-Py-FRT2-M1-FRT2*-G3-Pz-FRTz-FRT5-G1* (Figure 3.4e).

At this point, the locus is not only stacked with *G1, G2,* and *G3,* but also with *Px, Py, Pz, M1,* and *M2*. The authors proposed using chimeraplasty to convert the sequences of *FRT** and *FRT2** to *FRT* and *FRT2,* respectively. Chimeraplasty is a technique in which RNA-DNA chimeric oligonucleotides are introduced into the cell to mutate homologous target DNA (for review, see Rice et al., 2001). Subsequent FLP-mediated *FRT* × *FRT* and *FRT2* × *FRT2* recombinations would delete, respectively, *M1* and *M2,* to generate the structure *Px-FRT-G2-Py-FRT2-G3-Pz-FRTz-FRT5-G1* (Figure 3.4f). To deliver another trait gene to the locus, a third cassette would have to make use of *FRTz* and *FRT5,* and so on.

There are several concerns about this strategy. One is the limited availability of heterospecific sites. Even if *lox* and *RS* sites were incorporated into this strategy to expand the repertoire of recombination targets, it would not be long before new sites become unavailable. Second, chimeraplasty occurs at a frequency of 10^{-4} in tobacco and maize. It was successful for converting herbicide-sensitive genes to herbicide-resistant alleles (Beetham et al., 1999; Zhu et al., 1999). However, unlike herbicide resistance, *FRT* spacer mutations are not selectable phenotypes. Therefore, this marker removal step should be viewed with some skepticism. Third, Figure 3.4 shows the clustering of quite a bit of extra DNA fragments in addition to the chosen trait genes. The *FRT* sequences are rather short and may be considered negligible, but the promoters are of considerable size.

Stacking via an Irreversible Recombination System

Although some recombination systems catalyze freely reversible reactions, many do not. Instead, the substrate sites, often known as *attB* and *attP,* are nonidentical. This necessitates that the product sites generated from an *attB* × *attP* reaction, *attL* and *attR,* are dissimilar in sequence to *attB* and *attP.* The recombination enzyme that promotes

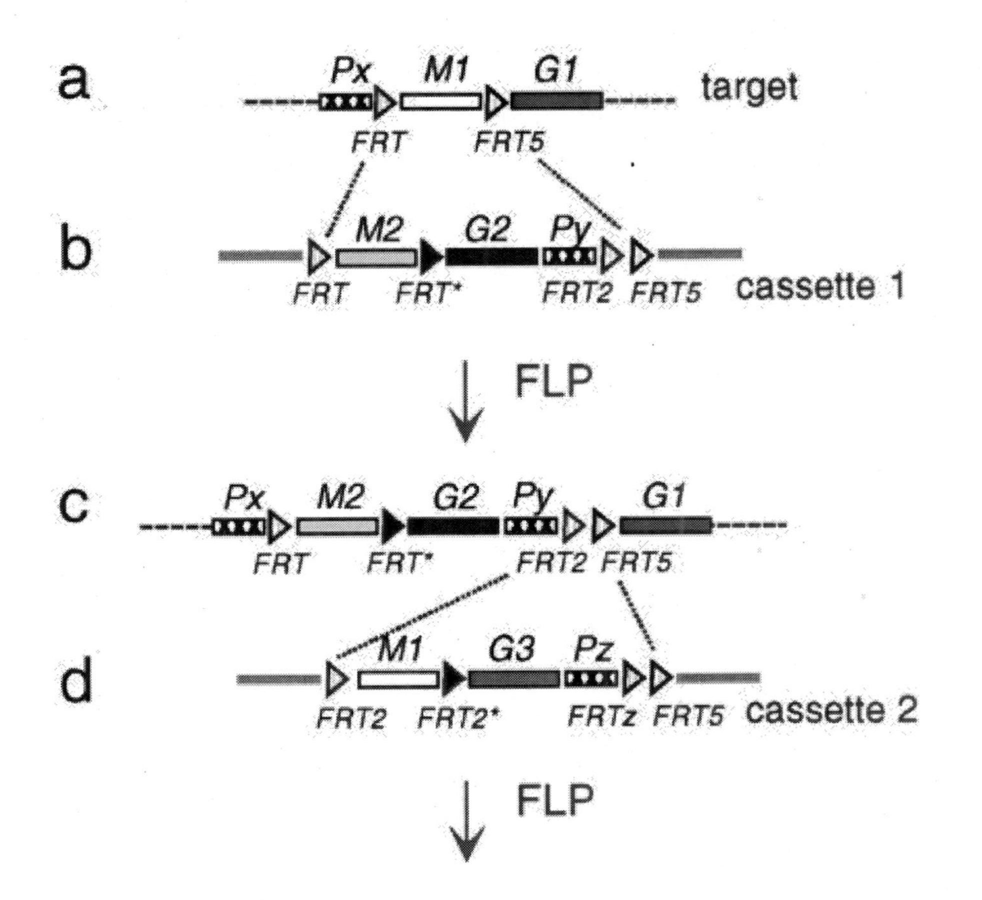

a
Px
M1
G1
target
FRT
FRT5
b
M2
G2
Py
FRT
FRT*
FRT2
FRT5
cassette 1
FLP
c
Px
M2
G2
Py
G1
FRT
FRT*
FRT2
FRT5
d
M1
G3
Pz
FRT2
FRT2*
FRTz
FRT5
cassette 2
FLP

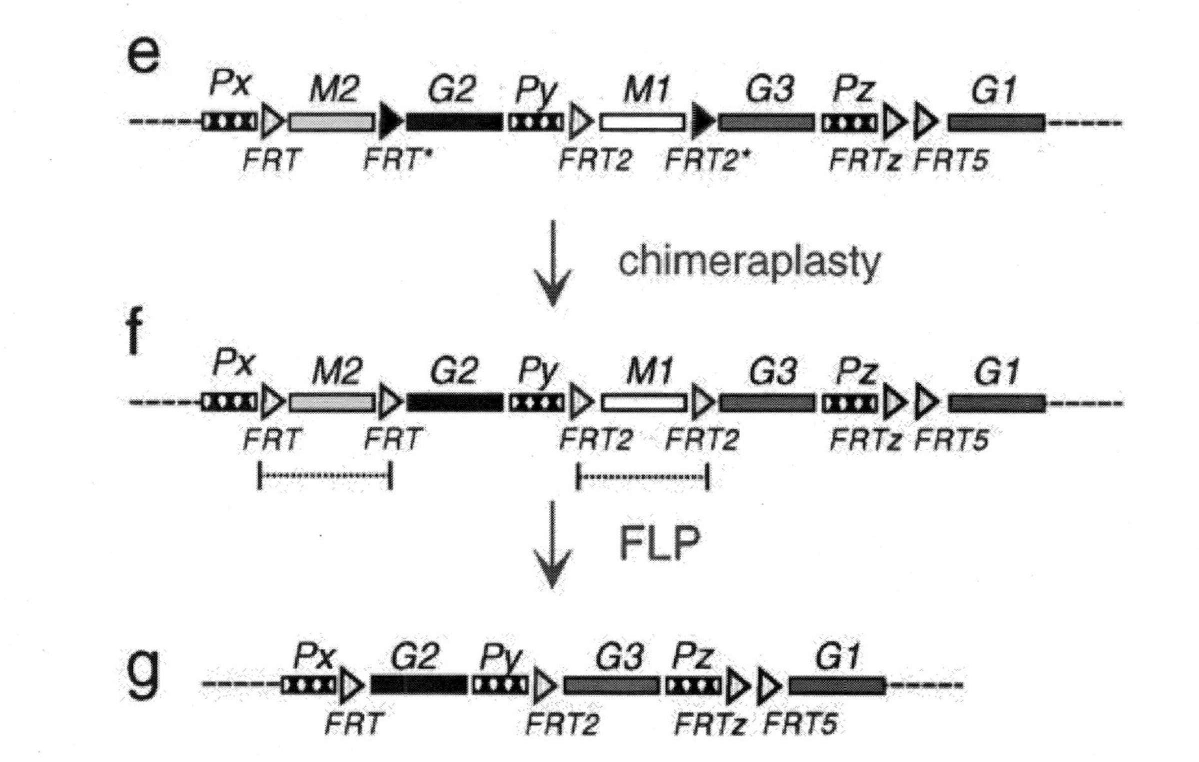

FIGURE 3.4. Gene stacking strategy with a reversible recombination system. *FRT, FRT5, FRT*, FRT2, FRT2*, FRTz* are heterospecific recombination sites. *Px, Py, Pz* are promoters, *G1, G2, G3* are trait genes, and *M1, M2* are selectable markers. Baszczynski and colleagues (Baszczynski, Bowen, Drummond, et al., 2001) described *M1, M2* as lacking the ATG start codon, which is provided by fusion to *Px, Py,* or *Pz*. For simplicity, neither transcriptional nor translational fusions are specified here. Dotted lines indicate recombination between pairs of sites.

the *attB* × *attP* reaction, often referred to as the integrase, does not recombine *attL* × *attR.* The lack of a readily reversible reaction gives a distinct advantage for employing such a system in DNA integration since integrated molecules are stable. Most important, however, an irreversible system permits a novel gene stacking strategy that is not achievable using only freely reversible systems. In fact, this is the underlying reason for this laboratory's interest in the ϕC31 recombination system.

Figure 3.5 shows BB', PP', BP', and PB' as *attB, attP, attL,* and *attB*, respectively, *G1, G2, G3, G4,* and *G5,* as trait genes, and *M1* and *M2* as markers. However, unlike in Figure 3.4, *M1* and *M2* include functional promoters. The process begins with a single-copy trait gene linked to a marker: *lox-M1-lox-G1*-BB'-(inverted *lox*). The single-copy locus may be obtained by molecular screening. Alternatively, a complex multicopy integration pattern may be resolved by Cre-*lox* site-specific recombination into a single-copy state (Srivastava et al., 1999; Srivastava and Ow, 2001). If a resolution-based strategy were used, the marker *M1* would have been deleted, leaving a configuration consisting of *lox-G1*-BB'- (inverted *lox*). To append *G2* to the *G1* locus, the integrating plasmid with the structure PP'-*G2*-PP'-*lox-M2* recombines with the genomic BB' target. The recombination enzyme for the reaction, or the integrase, can be provided, for example, by transient expression from a cotransformed plasmid. Since either PP' can recombine with the single BB', two different integration structures would arise that are distinguishable by molecular analysis. Figure 3.5b shows only the structure useful for further stacking, consisting of *lox-M1-lox-G1*-BP'-*G2*-PP'-*lox-M2*-plasmid backbone-PB'-(inverted *lox*). The Cre recombinase is introduced into the system to remove the unneeded DNA. The resulting structure becomes *lox-G1*-BP'-*G2*-PP'-(inverted *lox*). To stack *G3,* the construct BB'-*G3*-BB'-*lox-M2* is introduced (Figure 3.5c). Analogous to the previous steps, the genome has only a single PP' site to recombine with either of the BB' sites on the plasmid. Recombination with the *G3* upstream site produces the structure shown in Figure 3.5d. After removing the unneeded DNA, the locus containing *G1, G2,* and *G3* is ready for the stacking of *G4* (Figure 3.5e). In another variation, sets of inverted *attB* and *attP* sites, rather than sets of directly oriented sites, can also

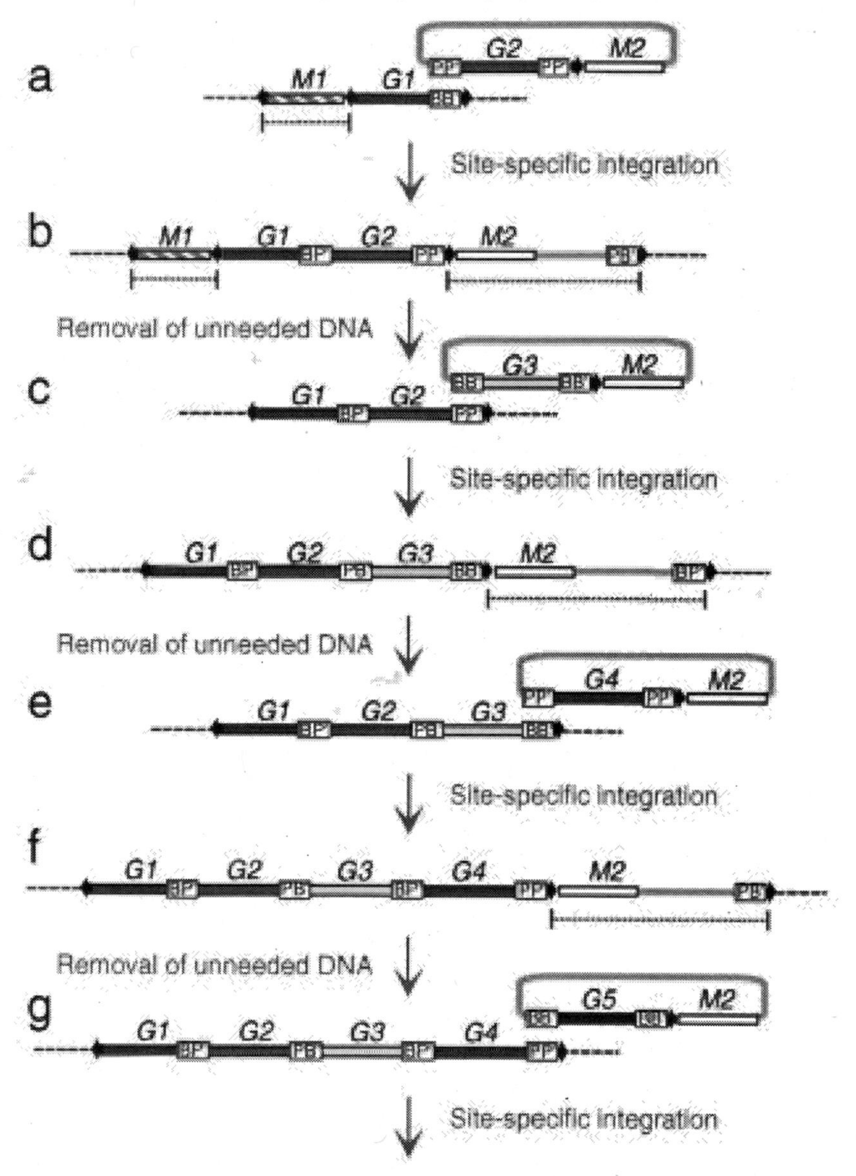

FIGURE 3.5. Gene stacking strategy with irreversible recombination system. Recombination sites *attB, attP, attL,* and *attB* are shown as BB', PP', BP' and PB', respectively. *G1, G2, G3, G4* and *G5* are trait genes, and *M1* and *M2* are markers. However, unlike in Figure 3.4, *M1* and *M2* include functional promoters. Dotted lines indicate recombination between pairs of sites.

be used. The sequence of events is analogous to those described for Figure 3.5.

Several features are worth noting. First, the vector for delivery of *G4* is the same as the vector for delivery of *G2*. Likewise, the vector for delivery of *G5* (Figure 3.5g) is the same as the vector for delivery of *G3*. In principle, the stacking process can be repeated indefinitely, alternating between the uses of two simple vectors. Second, the stacking from *G2* onward requires only a single marker gene, and this is critical as it bypasses the need to continually develop new selectable markers. Third, the trait genes, such as *G1, G2,* and so on, should not be narrowly interpreted as a single promoter-coding region-terminator fragment. Each DNA fragment could not only be composed of multiple transgenes, but could also include border DNA that insulate its (their) expression from surrounding regulatory elements. This may be useful when clustering transgenes that bring with them dominant *cis*-regulatory elements.

ϕC31 SITE-SPECIFIC RECOMBINATION SYSTEM

As mentioned earlier, the stacking strategy requires the deployment of an irreversible type of recombination system. The quest for such a system led us to examine the ϕFC31 system in fission yeast, mammalian cells, and transgenic plants. In the fission yeast *Schizosaccharomyces pombe,* Thomason and colleagues (2001) placed a ϕC31 *attP* (84 bp) site into the *leu1* locus of chromosome II. This strain was transformed with a nonreplicating plasmid pLT45, which harbors *ura4*$^{+}$ and an *attB* (280 bp) sequence. When pLT45 was introduced by itself, Ura^{+} colonies appeared from random integration of the DNA at a frequency of 6.3×10^{-6}. When introduced along with pLT43 that produced ϕFC31 integrase, the number of Ura^{+} transformants increased up to 22-fold at optimal integrase concentration, and this increase was due mainly (94 percent) to site-specific recombination. Southern analysis showed that 88 percent of the transformants were perfect integration events, 6 percent were correct integrants, but with additional random integrations, and only 6 percent showed an integration pattern that could not be attributed to site-specific integration. For two site-specific integration events, sequence analysis showed precise *attP* × *attB* recombination. Relative to the transformation of a replicating plasmid, site-specific insertion into the genome was ~20 percent as efficient at optimal integrase concentration.

An exchange reaction was also examined in a strain that harbors two *attP* sites in direct orientation. The replacement substrate was either pLT50, or a linear *attB-ura4+-attB* fragment purified or synthesized by PCR from pLT50. The recombination between *attP* and *attB* sites replaced the *attP* flanked genomic DNA with the *attB* flanked *ura4+* marker. This exchange reaction yielded a much greater number of Ura+ transformants than did the insertion of circular substrates. At optimal integrase concentration, the enhancement over background integration was up to 150-fold. This is close to the frequency obtained by an autonomously replicating plasmid. Since the ϕFC31 reaction is not reversible, the difference between an exchange reaction and an insertion reaction cannot be attributed to the rate of reaction reversal. Rather, this difference may be due to substrate availability. Direct DNA delivery methods are known to yield the concatamerization of extrachromosomal molecules, which will reduce the number of the single-copy circular substrates for the cointegration reaction. For an exchange reaction, concatamers may still be as effective, as all that is required in a substrate is two directly oriented recombination sites.

Following the experiments in *S. pombe,* cointegration and replacement types of integration were examined in mammalian cells (Ow et al., 2001). Chinese hamster ovary cell lines were generated with stably integrated *attB* target constructs pFY12, pFY14, or pFY15 (Figure 3.6a). These plasmids harbor an *attB* site of various lengths located between *Pc,* the human cytomegalovirus promoter, and the *lacZ* coding region (Figure 3.6a). Plasmids pFY12, pFY14, and pFY15 contain, respectively, 90, 50, and 30 bp of the *attB* sequence. Four lines of each construct were used for integration experiments to average out the possibility of chromosome position effects. Each of the 16 lines was transfected with pFY6, a ϕC31 integrase expression plasmid, along with an integration vector, pFY17, pFY19, or pFY20. The plasmids pFY17, pFY19, and pFY20 harbor an *attP* sequence of lengths 90, 50, and 32 bp, respectively. The *attP* sequence is situated upstream of the *hpt* open reading frame. Recombination between *attP* and *attB* should place the target site promoter upstream of *hpt* to

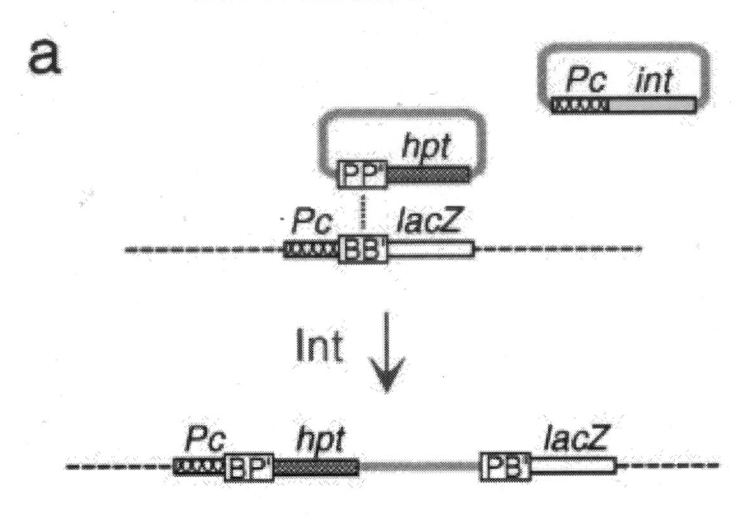

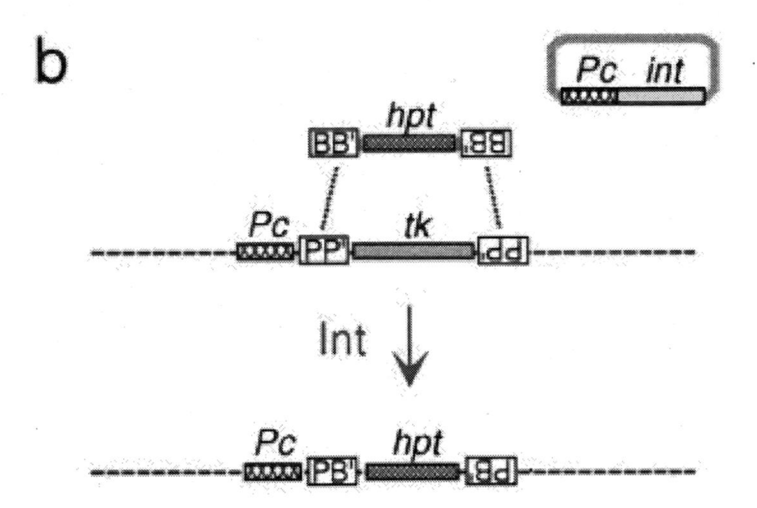

FIGURE 3.6. ϕC31-mediated targeting in mammalian cells. (a) The PP'-*hpt* plasmid inserts into chromosomal *Pc*-BB'-*lacZ* construct to form a *Pc-hpt* linkage and a hygromycin-resistant phenotype. *Pc:* human cytomegalovirus promoter. Integrase is provided by a cotransfected plasmid. (b) A linear BB'-*hpt*-B'B fragment replaces the genomic *tk* marker to confer resistance to hygromycin and ganiciclovir. *tk:* human thymidine kinase gene.

confer hygromycin resistance. For control, pBSK-hpt, isogenic to the test plasmids but lacking an *attP* sequence, was used to monitor the frequency of random promoter fusion to *hpt*.

The control plasmid pBSK-hpt yielded at most two hygromycin-resistant colonies per million cells ($< 2 \times 10^{-6}$). In contrast, pFY17, pFY19, and pFY20 yielded up to a thousandfold higher number of colonies, depending on the particular integration plasmid and the particular cell line. Although different target lines could exhibit intrinsic differences resulting from the site of integration, an overall trend was seen. Most efficient combinations appear to be 90 bp *attB* × 50 bp *attP* (9×10^{-4}) and 90 bp *attB* × 90 bp *attP* (6×10^{-4}). A number of other combinations, 50 bp *attB* × 50 bp *attP*, 90 bp *attB* × 32 bp *attP*, and 30 bp *attB* × 50 bp *attP* were slightly less efficient ($\sim 2 \times 10^{-4}$), and inefficient combinations were 50 bp *attB* × 90 bp *attP* and 50 bp *attB* × 32 bp *attP* ($<6 \times 10^{-5}$). The shorter sites were still functional, but only in combinations with certain longer sites. PCR detected the expected recombination junction from representative clones.

To test a replacement reaction, human kidney cell lines were generated harboring a single *Pc*-PP'-*tk*-P'P target, where *tk* is the human thymidine kinase gene, and P'P is the *attP* site in an inverse orientation. Transfection with a plasmid containing BB'-*hpt*-B'B, or with a PCR-derived linear BB'-*hpt*-B'B fragment led to the replacement of *tk* with *hpt*. As expression of *tk* confers sensitivity to the nucleoside analog ganiciclovir, the replacement event conferred resistance to both hygromycin and glaniciclovir.

The ϕC31 system has also been tested by another laboratory. Groth and colleagues (2000) conducted intra- and intermolecular recombination on episomic vectors in a human cell line. For the intramolecular reaction, a plasmid containing a marker gene flanked by *attP* and *attB* sites was transfected along with a ϕC31 integrase-expressing plasmid. For the intermolecular reaction, stable cell lines were used that harbor an Epstein-Barr viral vector with an *attB* target. A second *attP*-containing plasmid was subsequently cotransfected into these cells, along with a ϕC31 integrase-expressing plasmid. To assay for recombination, plasmid DNA was recovered from the transfected cells for introduction into *Escherichia coli*. The expected deletion, or cointegration of two plasmids, was detected and confirmed by molecular analyses. Episomal excision and cointegration efficiencies, using full-length *att* sites, were highest at 54 percent and 7.5 percent, respectively. Minimal sites of 34 bp for *attB* and 39 bp for *attP* were still functional but less efficient.

Human and mouse chromosome integrations were also reported (Thyagarajan et al., 2001). An *attP* target was engineered into the genome. Upon transfecting the target line with an *attB* plasmid along with an integrase-expressing construct, the number of stably transfected colonies increased 5- to 17-fold over background random integration (without cointroduced integrase construct). Surprisingly, however, only 15 percent of the transfected colonies were integrations at the *attP* target, which means that site-specific integration was only 0.75- to 2.6-fold above background. A majority of the background events were insertions into pseudo-*attP* sequences that share only limited primary sequence identity with ϕC31 *attP*. The authors estimated ~100 different pseudo *attP* sites in these mammalian genomes. This finding suggests that similar pseudo-*attP* sites may also be present in plant DNA.

In plants, intra- and intermolecular recombination have been observed respectively in tobacco and *Arabidopsis*. The CHO results suggested that 50 bp sites would suffice. As synthetic sites of this length are more convenient and faithful to produce, 50 bp sites were chosen for the plant experiments. In tobacco, intramolecular recombination was tested with two types of transgenic lines (Ow et al., 2001). The first type is transgenic for *35S-int,* and the second type for a *35S-attB-npt-attP-gus* construct (Figure 3.7a). Integrase-promoted site-specific recombination between *attB* and *attP* should delete the *npt* marker and fuse *35S* to *gus,* separated by a hybrid *attB/attP* sequence. This linkage should produce GUS active plants. Indeed, when the hemizygous parents were crossed, GUS active progeny were obtained in a quarter of the progeny, but in some combination of crosses, the progeny were clearly chimeric, indicating late somatic recombination (Figure 3.7b). In other crosses, progeny plants showed uniform staining to suggest an early recombination event (Figure 3.7c). Surprisingly, however, germinal transmission of the recombination event was not found. This may be a reflection of the particular *35S-int* line used, and some fine-tuning may be needed to get better expression in germline cells. Nonetheless, PCR analysis detected the

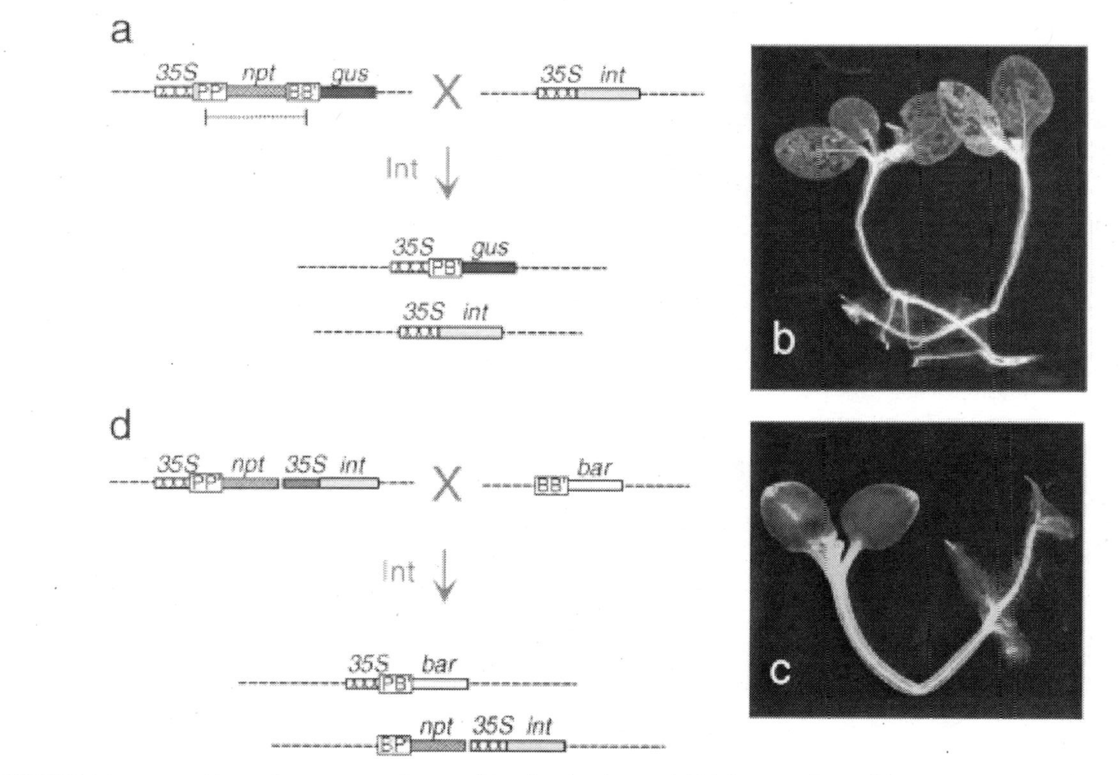

FIGURE 3.7. Intra- and interchromosomal recombination in plants. (a) Tobacco plant with a test construct is crossed with an integrase-expressing plant. Excision of *npt* and fusion between promoter and *gus* results from *attB* × *attP* intramolecular recombination. (b,c) Examples of progeny plants derived from crosses described in (a). (d) *Arabidopsis* plant with an *attP* site and an integrase-expressing gene is crossed with a plant harboring an *attB-bar* construct. Intermolecular *attB* × *attP* recombination links *35S* to *bar* to yield Basta resistance.

recombination junction, and sequencing of the PCR product confirmed a precise recombination event.

In *Arabidopsis,* site-specific recombination between nonhomologous chromosomes was detected in the progeny of crosses between plants containing a *35S-attP* construct and plants with an *attB-bar* construct (Figure 3.7d). The *35S-attP* plants also harbor a *35S-int* fragment. Some of the progeny from these crosses exhibited resistance to the herbicide Basta. Southern analysis confirmed the presence of the recombination junction. As Cre-*lox*-mediated recombination between nonhomologous chromosomes is a rare event (Qin et al., 1994; Koshinsky et al., 2000; Vergunst et al., 2000), the ϕC31 *attB* × *attP* reaction may offer improvement for chromosome engineering within and between plant species (Ow, 1996).

CONCLUDING REMARKS

The stacking protocol described in Figure 3.5 addresses a key issue in the commercial utilization of genetically modified plants. The presence of selectable markers in commercial products has been controversial in recent years (Hohn et al., 2001; Ow, 2001). Avoiding the use of antibiotic resistance genes should alleviate public concerns, but alternative markers may not necessarily be free of public scrutiny. For genes that are not relevant to the intended traits to be introduced, a prudent approach in dealing with the controversy is to just get rid of them. A decade ago, we described removing marker genes through site-specific recombination (Dale and Ow, 1991). Over the years, recombinase-mediated marker removal has been achieved by numerous laboratories and for many plant species, including the major crops maize, wheat, rice, cotton, and soybean (Russell et al., 1992; Lyznik et al., 1996; Gleave et al., 1999; Srivastava et al., 1999; Sugita et al., 2000; Corneille et al., 2001; Gilbertson et al., 2001; Hajdukiewicz et al., 2001; Zuo et al., 2001). The marker removal feature is an integral part of the stacking strategy, which can eliminate as much as possible the DNA not needed for an engineered trait. Only short recombination sequences are necessarily cointroduced along with the trait genes, but most become nonrecombinogenic BP' or PB' sites.

The immediate task ahead is to test the efficacy of the stacking strategy. Providing that it is successful, suitable target lines in crop plants would need to be generated. This could be a major undertaking given the large number of different crop plants in which this technology may be applicable. A concerted effort by interested parties would be much more preferable to independent efforts. How a target site is built dictates future stacking options. If engineered with common elements, they can be shared among research and commercial communities.

REFERENCES

Albert, H., E.C. Dale, E. Lee, and D.W. Ow (1995). Site-specific integration of DNA into wild-type and mutant *lox* sites placed in the plant genome. *Plant Journal* 7:649-659.

Bar, M., B. Leshem, N. Gilboa, and D. Gidoni (1996). Visual characterization of recombination at FRT-gusA loci in transgenic tobacco mediated by constitutive expression of the native FLP recombinase. *Theoretical and Applied Genetics* 93:407-413.

Baszczynski, C.L., B.A. Bowen, B. Drummond, W.J. Gordon-Kamm, D.J. Peterson, G.A. Sandahl, L.A. Tagliani, and Z.-Y. Zhao (2001). Novel nucleic acid sequence encoding FLP recombinase. US patent No. 6,175,058 B1.

Baszczynski, C.L., B.A. Bowen, D.J. Peterson, and L.A. Tagliani (2001). Composition and methods for genetic modification of plants. US patent No. 6,187,994 B1.

Bayley, C.C., M. Morgan, E.C. Dale, and D.W. Ow (1992). Exchange of gene activity in transgenic plants catalyzed by the Cre-*lox* site-specific recombination system. *Plant Molecular Biology* 18:353-361.

Beetham, P.R., P.R. Kipp, Z.L. Sawycky, C.J. Arntzen, and G.D. May (1999). A tool for functional plant genomics: Chimeric RNA/DNA oligonucleotides cause in vivo gene-specific mutations. *Proceedings of the National Academy of Sciences of the United States of America* 96:8774-8778.

Choi, S., D. Begum, H. Koshinsky, D.W. Ow, and R.A. Wing (2000). A new approach for the identification and cloning of genes: The pBACwich system using Cre/*lox* site-specific recombination. *Nucleic Acids Research* 28:e19, i-vii.

Corneille, S., K. Lutz, Z. Svab, and P. Maliga (2001). Efficient elimination of selectable marker genes from the plastid genome by the Cre-*lox* site-specific recombination system. *Plant Journal* 27:171-178.

Dale, E.C. and D.W. Ow (1990). Intra- and intermolecular site-specific recombination in plant cells mediated by bacteriophage P1 recombinase. *Gene* 91:79-85.

Dale, E.C. and D.W. Ow (1991). Gene transfer with the subsequent removal of the selection gene from the host genome. *Proceedings of the National Academy of Sciences of the United States of America* 88:10558-10562.

Day, C.D., E. Lee, J. Kobayashi, L.D. Holappa, H. Albert, and D.W. Ow (2000). Transgene integration into the same chromosomal location can produce alleles that express at a predictable level, or alleles that are differentially silenced. *Genes and Development* 14:2869-2880.

Gilbertson, L., P. Addae, C. Armstrong, N. Bernabe, J. Ekena, G. Keithly, M. Neuman, V. Peschke, M. Petersen, S. Subbarao, et al. (2001). Cre/*lox* mediated marker gene excision in transgenic crop plants. *In Vitro Cellular and Developmental Biology—Animal* 37(3 Part 2):26.A.

Gleave, A.P., D.S. Mitra, S.R. Mudge, and B.A.M. Morris (1999). Selectable marker-free transgenic plants without sexual crossing: Transient expession of *cre* recombinase and use of a conditional lethal dominant gene. *Plant Molecular Biology* 40:223-235.

Groth, A.C., E.C. Olivares, B. Thyagarajan, and M.P. Calos (2000) A phage integrase directs efficient site-specific integration in human cells. *Proceedings of the National Academy of Sciences of the United States of America* 97:5995-6000.

Hajdukiewicz, P.T., L.A. Gilbertson, and J.M. Staub (2001). Multiple pathways for Cre/*lox*-mediated recombination in plastids. *Plant Journal* 27:161-170.

Hoess, R.H., A. Wierzbicki, and K. Abremski (1986). The role of the *loxP* spacer region in P1 site-specific recombination. *Nucleic Acids Research* 14:2287-2300.

Hohn, B., A.A. Levy, and H. Puchta (2001). Elimination of selection markers from transgenic plants. *Current Opinion in Biotechnology* 12:139-143.

Kilby, N.J., G.J. Davies, M.R. Snaith, and J.A.H. Murray (1995). FLP recombinase in transgenic plants: Constitutive activity in stably transformed tobacco and generation of marked cell clones in *Arabidopsis*. *Plant Journal* 8:637-652.

Koshinsky, H.A., E. Lee, and D.W. Ow (2000). Cre-*lox* site-specific recombination between *Arabidopsis* and tobacco chromosomes. *Plant Journal* 23:715-722.

Lloyd, A.M. and R.W. Davis (1994). Functional expression of the yeast FLP/*FRT* site-specific recombination system in *Nicotiana tabacum*. *Molecular and General Genetics* 242:653-657.

Lyznik, L.A., J.C. Mitchell, L. Hirayama, and T.K. Hodges (1993). Activity of yeast FLP recombinase in maize and rice protoplasts. *Nucleic Acids Research* 21:969-975.

Lyznik, L.A., K.V. Rao, and T.K. Hodges (1996). FLP-mediated recombination of *FRT* sites in the maize genome. *Nucleic Acids Research* 24:3784-3789.

Maeser, S. and R. Kahmann (1991). The Gin recombinase of phage Mu can catalyze site-specific recombination in plant protoplasts. *Molecular and General Genetics* 230:170-176.

Odell, J., P. Caimi, B. Sauer, and S. Russell (1990). Site-directed recombination in the genome of transgenic tobacco. *Molecular and General Genetics* 223:369-378.

Onouchi, H., R. Nishihama, M. Kudo, Y. Machida, and C. Machida (1995). Visualization of site-specific recombination catalyzed by a recombinase from *Zygosaccharomyces rouxii* in *Arabidopsis thaliana*. *Molecular and General Genetics* 247:653-660.

Onouchi, H., K. Yokoi, C. Machida, H. Matzuzaki, Y. Oshima, K. Matsuoka, K. Nakamura, and Y. Machida (1991). Operation of an efficient site-specific recombination system of *Zygosaccharomyces rouxii* in tobacco cells. *Nucleic Acids Research* 19:6373-6378.

Ow, D.W. (1996). Recombinase-directed chromosome engineering in plants. *Current Opinion in Biotechnology* 7:181-186.

Ow, D.W. (2001). The right chemistry for marker gene removal? *Nature Biotechnology* 19:115-116.

Ow, D.W. (2002). Recombinase-directed plant transformation for the post genomic era. *Plant Molecular Biology* 48:183-200.

Ow, D.W., R. Calendar, and L. Thomason (2001). DNA recombination in eukaryotic cells by the bacteriophage phiC31 recombination system. International patent filing, WO 01/07572.

Ow, D.W. and S.L. Medberry (1995). Genome manipulation through site-specific recombination. *Critical Reviews in Plant Sciences* 14:239-261.

Puchta, H. (2002). Gene replacement by homologous replacement in plants. *Plant Molecular Biology* 48:173-182.

Qin, M., C. Bayley, T. Stockton, and D.W. Ow (1994). Cre recombinase mediated site-specific recombination between plant chromosomes. *Proceedings of the National Academy of Sciences of the United States of America* 91:1706-1710.

Rice, M.C., K. Czymmek, and E.B. Kmiec (2001). The potential of nucleic acid repair in functional genomics. *Nature Biotechnology* 19:321-326.

Russell, S.H., J.L. Hoopes, and J.T. Odell (1992). Directed excision of a transgene from the plant genome. *Molecular and General Genetics* 234:49-59.

Schlake, T. and J. Bode (1994). Use of mutated FLP recognition target (*FRT*) sites for the exchange of expression cassettes at defined chromosomal loci. *Biochemistry* 33:12746-12751.

Sonti, R.V., A.F. Tissier, D. Wong, J.-F. Viret, and E.R. Signer (1995). Activity of the yeast FLP recombinase in *Arabidopsis*. *Plant Molecular Biology* 28:1127-1132.

Srivastava, V., O.A. Anderson, and D.W. Ow (1999). Single-copy transgenic wheat generated through the resolution of complex integration patterns. *Proceedings of the National Academy of Sciences of the United States of America* 96:11117-11121.

Srivastava, V. and D.W. Ow (2001). Single copy primary transformants of maize obtained through the cointroduction of a recombinase-expressing construct. *Plant Molecular Biology* 46:561-566.

Srivastava, V. and D.W. Ow (2002). Biolistic mediated site-specific integration in rice. *Molecular Breeding* 8:345-350.

Sugita, K., T. Kasahara, E. Matsunaga, and H. Ebinuma (2000). A transformation vector for the production of marker-free transgenic plants containing single copy transgene at high frequency. *Plant Journal* 22:461-469.

Terada, R., H. Urawa, Y. Inagaki, K. Tsugane, and S. Iida (2002). Efficient gene targeting by homologous recombination in rice. *Nature Biotechnology* 20:1031-1034.

Thomason, L.C., R. Calendar, and D.W. Ow (2001). Gene insertion and replacement in *Schizosacchromyces pombe* mediated by the *Streptomyces* bacteriophage ϕC31 site-specific recombination system. *Molecular Genetics and Genomics* 265:1031-1038.

Thorpe, H.M. and M.C. Smith (1998). In vitro site-specific integration of bacteriophage DNA catalyzed by a recombinase of the resolvase/invertase family. *Proceedings of the National Academy of Sciences of the United States of America* 95:5505-5510.

Thyagarajan, B., E.C. Olivares, R.P. Hollis, D.S. Ginsburg, and M.P. Calos (2001). Site-specific genomic integration in mammalian cells mediated by phage ϕC31 integrase. *Molecular and Cellular Biology* 21:3926-3934.

Tuttle, A.B., E.J. Pascal, J.L. Suttie, and M.-D. Chilton (1999). Site-directed transformation of plants. International patent filing WO 99/55851.

Vergunst, A.C. and P.J.J. Hooykaas (1998). Cre/*lox*-mediated site-specific integration of *Agrobacterium* T-DNA in *Arabidopsis thaliana* by transient expression of *cre*. *Plant Molecular Biology* 38:393-406.

Vergunst, A.C., L.E.T. Jansen, P.F. Fransz, J.H. de Jong, and P.J.J. Hooykaas (2000). Cre/*lox*-mediated recombination in *Arabidopsis:* Evidence for a transmission of a translocation and a deletion event. *Chromosoma* 109:287-297.

Vergunst A.C., L.E.T. Jansen, and P.J.J. Hooykaas (1998). Site-specific integration of *Agrobacterium* T-DNA in *Arabidopsis thaliana* mediated by Cre recombinase. *Nucleic Acids Research* 26:2729-2734.

Wassenegger, M. (2000). RNA-directed DNA methylation. *Plant Molecular Biology* 43:203-220.

Zhu, T., D.J. Peterson, L. Tagliani, G. St. Clair, C.L. Baszczynski, and B. Bowen (1999). Targeted manipulation of a maize genes in vivo using chimeric RNA/DNA oligonucleotides. *Proceedings of the National Academy of Sciences of the United States of America* 96:8768-8773.

Zuo, J., Q.-W. Niu, S.G. Moller, and N.-H. Chua (2001). Chemical-regulated, site-specific DNA excision in transgenic plants. *Nature Biotechnology* 19:157-161.

Chapter 4

Transgenics of Plant Hormones and Their Potential Application in Horticultural Crops

Yi Li
Hui Duan
Yan H. Wu
Richard J. McAvoy
Yan Pei
Degang Zhao
John Wurst
Qi Li
Keming Luo

INTRODUCTION

The plant hormones auxin, cytokinin, gibberellin (GA), abscisic acid (ABA), and ethylene control many important processes during plant growth and development. Synthetic hormones or growth regulators have been widely used in horticultural crops to produce desirable characteristics and to improve their performance. For instance, to restrict stem elongation and to increase branching in ornamentals, more than 40,000 pounds (active ingredient) of growth regulators are applied to greenhouse and nursery crops in the United States each year (Norcini et al., 1996). However, exogenous applications of growth regulators are expensive, carry the risk of environmental contamination and worker exposure, and sometimes produce undesirable

This is scientific contribution number 2104 of the Storrs Agricultural Experiment Station.

side effects. Recent advances in the areas of molecular cloning and plant transformation have made it possible to manipulate hormone concentration and tissue sensitivity to hormones in plants. In this chapter, we will focus our discussion on the application and potential application of genetic manipulations of plant hormone concentration and plant sensitivity in horticultural crops.

AUXINS

Auxins play a critical role in cell elongation, phototropism, geotropism, apical dominance, organ abscission, flower initiation and development, root initiation, fruit growth and development, tuber and bulb formation, and seed germination. In horticultural practice, synthetic auxins have been widely used in promoting root initiation, apical dominance, and fruit set (Weaver, 1972). With genes that can modulate auxin concentration, transgenic techniques have been used to manipulate endogenous levels of auxin in plants (Klee and Lanahan, 1995). It has been shown that expression of the *Agrobacterium* tryptophan monooxygenase gene *(iaaM)* alone or in combination with the *Agrobacterium* indoleacetamide hydrolase gene *(iaaH)* leads to conversion of the amino acid tryptophan to indoleacetic acid (IAA) in plants (Klee et al., 1987; Romano et al., 1995; Guilfoyle et al., 1993). Tobacco, petunia, and *Arabidopsis* plants constitutively expressing the *iaaM* gene displayed increased apical dominance and reduced stem growth (Klee et al., 1987; Romano et al., 1995; Guilfoyle et al., 1993). Adventitious roots were formed from unwounded leaves of the *iaaM* transgenic tobacco plants, a phenomenon not normally observed in wild-type plants treated with exogenous auxins (Klee et al., 1987; Guilfoyle et al., 1993). These findings demonstrate that overproduction of auxin in the plant is more effective than exogenous auxin application in promoting root initiation. Rooting is a key and often difficult step in asexual propagation of many horticultural crops. Expression of the *iaaM* gene, if under the control of appropriate gene promoters, could enhance formation of adventitious roots of cuttings of hard-to-root horticultural crops.

The most commercially valuable trait observed in auxin-overproducing transgenic plants is most likely parthenocarpy (Varoquaux et al., 2002). Parthenocarpic generation of seedless fruits from vegetable and fruit tree crops is projected to be one of the most important technologies in the agricultural industry over the next 20 years (Ortiz, 1998). Traditionally, seedless fruits are produced from mutants, triploid plants, or flowers treated with exogenous growth regulators. Triploid plants and seedless fruit mutants are difficult to breed. The use of synthetic growth regulators to produce seedless fruits often causes environmental and health concerns and is also ineffective in some fruit species because of insufficient uptake and translocation of the applied hormones to the target tissues. Parthenocarpy was initially observed in auxin-overproducing petunia (Klee et al., 1987) and tobacco plants (Guilfoyle et al., 1993). More recently, with ovary-specific gene promoters to direct expression of the *iaaM* gene, seedless fruits have been produced in eggplant (Rotino et al., 1997; Spena, 1998), tomato (Barg and Salts, 1996; Li, 1997), and watermelon (Y. Li Lab, unpublished data). In all these cases, because overproduction of auxin was restricted to the ovary and developing fruit, no obvious deleterious pleiotropic effects were evident. Poor pollination is a major cause of inadequate fruit set and undersized fruits in many vegetables and fruit trees, including greenhouse- and field-grown tomatoes. Diseases and environmental stresses such as low and high temperatures, low light intensity, and drought cause poor pollination and reduction of fruit yield. Production of fruits independent of pollination may therefore reduce or eliminate yield reduction problems associated with poor pollination.

In the case of the transgenic seedless tomato, the size, weight, acid, and sugar contents of fruits were significantly higher than in seeded fruits from wild-type plants (Li, 1997). Also, a significant increase in total solids content was observed in the transgenic seedless tomato fruits, which could benefit the tomato processing industry. Tomato fruits contain 95 percent water, and the remaining 5 percent consists of pulp, seeds, and soluble products (mostly sugars, organic acids, and flavor compounds). Most of the water is removed during processing of tomato products such as ketchup, paste, and sauce. To a large degree, the price of these tomato products is determined by the cost of removing water. It is estimated that a 25 percent increase in total solids would save the U.S. tomato industry at least $75 million a year (Chrispeels and Sadava, 1994).

Reducing the concentration of endogenous auxin could be used to improve horticultural plants. A bushy phenotype, for instance, may be achieved by reducing auxin concentration in the plant. The *iaaL* gene cloned from *Pseudomonas syringae,* encoding an indoleacetic acid-lysine synthase, is capable of reducing free, active auxin concentrations in plants (Romano et al., 1991). The iaaL enzyme converts IAA to IAA-lysine, a biologically inactive auxin in plants. IAA-lysine conjugates are not synthesized in plants, and it appears that the conjugation is not reversible. With a strong constitutive and globally active *CaMV 35S* promoter to control the expression of the *iaaL* gene, IAA concentration in transgenic tobacco plants was 19-fold lower than in wild-type plants (Romano et al., 1991). Phenotypically, tobacco plants expressing the *iaaL* gene exhibited reduced apical dominance or increased branching, but they also had underdeveloped root and vascular systems. Recently, we produced transgenic tobacco, petunia, and chrysanthemum plants that expressed the *iaaL* gene under the control of a gene promoter which is active only in aboveground organs. Some of the resulting transgenic plants were compact, with reduced stem height growth and/or more branches, and the leaves were dark green in color (Y. Li Lab, unpublished data). However, reduction in number of lateral roots, an undesirable side effect, has been observed in a number of the *iaaL* plants. Expression of the *iaaL* gene in a more organ/tissue- or stage-specific manner may eliminate the poor root symptom.

CYTOKININS

Cytokinins stimulate cell division, reduce apical dominance and delay senescence, including postharvest leaf deterioration and low temperature-induced leaf yellowing, and maintain vigor in many vegetable and floriculture plants (Ludford, 1995; Weaver, 1972). Several synthetic cytokinins have been registered for use on carnations and roses to increase lateral bud formation and branching, but large-scale commercial applications of cytokinins to horticultural crops have not been possible because of inefficient absorption of the hormone by plants and high costs associated with exogenous application (Gianfagna, 1995).

Overproduction of cytokinins in transgenic plants has been achieved by expression of the *ipt* gene from *Agrobacterium tumefaciens*. The *Agrobacterium* isopentenyl transferase catalyzes the rate-limiting step for de novo cytokinin biosynthesis (McGaw and Burch, 1995), i.e., addition of isopentenyl pyrophosphate to the N6 of 5'-adenosine monophosphate (AMP) to form isopentenyl AMP (Chen, 1997). Isopentenyl AMP is the precursor of all other cytokinins, of which the three most commonly detected and physiologically active forms in plants are isopentenyl adenine, zeatin, and dihydrozeatin. With a wide range of constitutive, tissue-specific, or inducible gene promoters to control the expression of the *ipt* gene, potential utilities of the *ipt* gene have been demonstrated.

Phenotypes observed in cytokinin-overproducing transgenic plants include increases in the number of lateral shoots, shorter internodes, formation of adventitious shoots directly from unwounded leaves and roots, altered source sink relations, and delayed senescence. For example, Estruch and colleagues (1991) reported that tobacco plants that were somatic mosaics for expression of a *CaMV 35S* promoter-*ipt* gene frequently produced adventitious shoots on veins at the unwounded leaf tip. With a *SAUR* promoter-*ipt* fusion gene expressed in tobacco plants, Li and colleagues (1992) observed adventitious shoots developed on the unwounded leaf petioles and veins, where the *SAUR* promoter is highly active. We also observed initiation of shoots from unwounded roots of *CaMV 35S* promoter-*ipt* tobacco plants (Y. Li Lab, unpublished data). Because direct shoot formation from unwounded tobacco leaves and roots has not previously been reported with exogenously applied cytokinins, these results suggest that increases in endogenous cytokinins are more effective than exogenous application of the hormones. The *ipt* gene may therefore be used for inducing shoot organogenesis in horticultural species that are difficult to shoot. In addition, it has been reported that transgenic plants that overproduced cytokinin were resistant to biotic and abiotic stresses. Smigocki and colleagues (1993) reported that expression of a wound-inducible proteinase inhibitor II promoter-*ipt* gene in tobacco plants led to resistance to the tobacco hornworm and inhibited normal development of green peach aphid nymphs. Zhang and colleagues (2000) showed that endogenously produced cytokinin can regulate senescence caused by flooding stress, thereby increasing plant tolerance to flooding.

The most beneficial trait observed in cytokinin-overproducing plants so far is probably the delay of senescence of detached organs and whole plants (Smart et al., 1991; Li et al., 1992; Gan and Amasino, 1995). For example, excised leaves of cytokinin-overproducing transgenic tobacco plants showed a significantly prolonged retention of chlorophyll and delayed senescence when incubated for a prolonged period in a dark, moist chamber at room temperature (Li et al., 1992). There were no signs of senescence in these detached leaves after 14 weeks of incubation. Even after six months of incubation, the leaves still remained green, healthy, and turgid. Gan and Amasino (1995) produced transgenic tobacco plants by using a senescence-specific promoter *(PSAG12)* to control expression of the *ipt* gene. Transgenic tobacco plants expressing *ipt* under the control of the *PSAG12* promoter did not exhibit the developmental abnormalities usually associated with constitutive *ipt* expression. The leaves of the transgenic tobacco plants exhibited a prolonged photosynthetically active life span. Because the *PSAG12-ipt* gene was activated only at the onset of senescence in the lower mature leaves of tobacco, the use of the *PSAG12-ipt* gene resulted in cytokinin biosynthesis in the leaves, which inhibited leaf senescence and, consequently, attenuated activity of the *PSAG12-ipt* gene, preventing cytokinin overproduction.

When the *PSAG12-ipt* gene was introduced into lettuce plants, developmental and postharvest leaf senescence in mature heads were significantly delayed (McCabe et al., 2001). There were no significant differences in head diameter or fresh weight of leaves and roots compared to wild-type plants. At the stages of bolting and preflowering, relative to the wild-type controls, there was little decrease in chlorophyll, total protein, and Rubisco content in transgenic leaves. However, the transgenic *PSAG12-ipt* lettuce plants showed a four- to eight-week delay in flowering and premature senescence of their upper leaves, a somewhat surprising phenotype. In another study, Schroeder and colleagues (2001) reported that ornamental *Nicotiana alata* plants expressing the *SAG12-ipt* gene exhibited two to four times fewer senesced leaves, significantly longer in situ flower life, and greater shoot dry weight. The transgenic plants were shorter and produced more branches. On the other hand, the transgenic plants had 32 percent to 50 percent fewer flowers per branch.

A recent success in molecular cloning of cytokinin oxidase genes from maize and *Arabidopsis* makes it possible to reduce concentrations of endogenous cytokinins in transgenic plants (Houba-Herin et al., 1999; Werner et al., 2001). Cytokinin oxidase catalyzes the degradation of cytokinins bearing unsaturated isoprenoid side chains, resulting in reduced concentrations of active cytokinins (Armstrong, 1994). Because reducing endogenous cytokinin concentration in plant has previously not been possible, cytokinin oxidase genes provide an exciting possibility to explore the commercial potential of artificially lowering the cytokinin content in plants. Werner and colleagues (2001) found that overexpression of cytokinin oxidase genes in tobacco plants led to significant reductions in concentrations of various cytokinins. The transgenic plants developed stunted shoots with smaller apical meristems. The plastochrone was prolonged, and leaf cell production was only 3 to 4 percent of that found in wild-type plants. In contrast, root meristems of the transgenic plants were enlarged and elongation of roots was enhanced, and the transgenic plants developed more branched roots. Presumably, cytokinin oxidase genes could be used to promote lateral root development and to stimulate formation of adventitious roots in horticultural crops.

GIBBERELLINS

The effects of gibberellins on plant growth and development include seed dominance/germination, internode elongation, flower initiation, and fruit set (Crozier, 1983). Because gibberellin- or anti-gibberellin-type growth regulators are among the most widely used plant growth regulators in the horticultural industry, genes involved in GA production or tissue sensitivity to GA are of great use in the industry (Hedden and Kamiya, 1997; Lange, 1998). In higher plants, GA 20-oxidase, which catalyzes the sequential oxidation and elimination of C-20, is a regulatory enzyme for GA biosynthesis. Recently, genes encoding GA 20-oxidase were isolated from several species of higher plants (Lange et al., 1994; Phillips et al., 1995; Wu et al., 1996; Toyomasu et al., 1997). Transgenic *Arabidopsis* that overexpressed GA 20-oxidase genes exhibited a GA-overproduction phenotype with elongated hypocotyls and stems and early flowering (Huang et al., 1998; Coles et al., 1999). On the other hand, suppression of the stem-specific GA 20-oxidase gene expression by antisense technology resulted in reduced stem elongation and delayed flowering under short days (Coles et al., 1999).

In another study using *Solanum dulcamara,* in which GA_1 is the major biologically active GA, plants overexpressing the pumpkin 20-oxidase gene were semidwarfed, flowered earlier, and produced more fruits and seeds (Curtis et al., 2000). In transgenic *S. dulcamara* plants, the levels of the major biologically active GA_1 were reduced, whereas GA_4 was the same in stems or increased in the leaves. When lettuce was transformed with the pumpkin GA_{20}-oxidase gene under the control of the *CaMV 35S* promoter (Niki et al., 2001), endogenous levels of GA_1 and GA_4 were reduced, but large amounts of GA_{17} and GA_{25}, inactive GAs, were accumulated in the transgenic plants. Because GA_1 is also the major biologically active GA in lettuce, the transgenic plants were dwarfed. Also, the transgenic lettuce had reduced leaf size and stem height, and contained thinner roots and more lateral roots as compared to wild-type plants. Furthermore, flowering of the transgenic plants was delayed and seed production was reduced. These results suggest that overexpression of a functional pumpkin GA_{20}-oxidase in plants where GA_1 is the major biologically active GA can result in a diversion of the normal pathway of GA biosynthesis to inactive products and lead to a reduction in active GAs in plants. Also, expression of a single-chain antibody against GA_{24}/GA_{19} (Shimada et al., 1999) has been shown to reduce levels of GA_1 in transgenic plants, resulting in a dwarf phenotype. These findings suggest that desirable stature of many horticultural crops can be achieved by manipulating GA content using transgenic techniques.

Genes involved in GA signal transduction have been uncovered using GA-insensitive or constitutive GA response mutants of *Arabidopsis* (Sun, 2000). The GA-insensitive mutants are dwarfs, resembling GA-deficient mutants, but not rescued by exogenous GA applications. The GA-insensitive mutants typically produced dark green leaves and exhibited inhibition of seed germination, delay of flowering, and abnormal flower development. Of these genes, the most useful gene for horticultural biotechnology is probably a dominant or semidominant gain-of-function gibberellin insensitive *(gai-1)* gene (Koornneef et al., 1985). As shown in transgenic rice plants, expression of the dominant *Arabidopisis gai-1* gene resulted in dwarfing (Peng et al., 1999; Fu et al., 2001). Controlled expression of the *gai* gene could be used to produce compact traits in many horticultural and ornamental crops.

ETHYLENE

As early as in the beginning of the 1900s, prior to it being identified as a plant product, ethylene was known as a fruit-ripening inducer. During the era of gas lamps, leaking lamps along city streets often promoted leaf abscission. Today, grocery warehouses have ethylene rooms that are used for ripening most produce, which is shipped unripe. Synthetic ethylene-releasing compounds such as ethephon have a large number of commercial applications. For example, ethephon is used to ripen bananas, pineapples, melons, and tomatoes, and when applied as a preharvest spray it promotes uniform ripening of apples, cherries, and pineapple. Ethylene-releasing compounds such as ethephon have been used to increase the production of female flowers on cucumbers. On the other hand, inhibition of ethylene production and increase in ethylene insensitivity are commercially important ways to preserve fruits and vegetables and to extend shelf life of ornamentals and cut flowers.

In higher plants, ethylene is synthesized from *S*-adenosyl-L-methionine through the activity of the enzymes 1-aminocyclopropane-1-carboxylic acid (ACC) synthase and ACC oxidase. The rate-limiting step in the synthesis of ethylene is the formation of ACC. The gene encoding ACC synthase was initially cloned from zucchini (Sato and Theologis, 1989) and has been subsequently cloned from many different plant species. The ACC synthase gene has been used for manipulation of ethylene synthesis in plants. Overexpression of the ACC synthase gene caused plants to produce 100 times more ethylene than normal (Kosugi et al., 2000). The increase in ethylene production reduced internode length and lowered chlorophyll content. A bacterial ethylene-forming enzyme (EFE) that catalyzes oxygenation of 2-oxoglutarate to produce ethylene and carbon dioxide (in contrast to a plant enzyme that uses AAC as a substrate) has been cloned and characterized in transgenic plants (Araki et al., 2000). The EFE-overexpressing transgenic plants produced ethylene at consistently higher rates than untransformed plants. Dwarf morphology was observed in the transgenic tobacco that resembled the phenotype of a wild-type plant exposed to excess ethylene. Thus, developmental stage-specific expression of ethylene-producing genes could replace treatments of plants with ethylene-releasing compounds in horticultural practice.

Several strategies have been developed to suppress ethylene production in higher plants. Oeller and colleagues (1991) have shown that antisense suppression of the ACC synthase gene can reduce ethylene synthesis to 1 percent of that found in wild-type plants. As a result, the flower senescence of transgenic tomato plants was delayed and the ripening process was blocked both on and off the vine (Oeller et al., 1991). The ACC oxidase gene has also been used in antisense orientation for suppression of ACC oxidase gene expression in melon (Ayub et al., 1996), broccoli (Henzi et al., 1999), and the ornamental flower *Torenia fournieri* (Aida et al., 1998). In the case of *Torenia,* Aida and colleagues (1998) demonstrated that introducing a fragment of the ACC oxidase gene in sense or antisense orientation into *Torenia* resulted in significantly greater flower longevity than occurred in wild-type plants. Reducing ethylene levels in plants also was achieved using genes encoding enzymes to metabolize precursors of ethylene. Klee and colleagues (1991) showed that increasing degradation of ACC by overexpression of an ACC deaminase gene from *Pseudomonas* sp. resulted in a 90 to 95 percent reduction in ethylene synthesis in all tissues of tomato plants. ACC deaminase degrades ACC to α-ketobutyric acid, thus effectively preventing its conversion to ethylene. The fruit-ripening process of the ACC deaminase transgenic tomato fruits was dramatically delayed.

Changing sensitivity of plants to ethylene is a powerful method to block ethylene effects. Several genes involved in the ethylene-signaling pathway have been cloned and characterized. *ETR,* encoding a putative ethylene receptor, was isolated from *Arabidopisis* following identification of the ethylene-insensitive mutant *etr1* which failed to show the classical seedling "triple response" to ethylene (Bleecker et al., 1988). Wilkinson and colleagues (1997) used the *etr1* gene of

Arabidopsis, which confers dominant ethylene insensitivity, to extend shelf life of several crops. Expression of the etr1-1 gene caused significant delays in fruit ripening, flower senescence, and flower abscission when expressed in tomato and petunia plants (Wilkinson et al., 1997). Transgenic tomato plants exposed to ethylene exhibited a dramatic delay in fruit ripening and senescence compared with those on untransformed plants. Harvested tomato fruits retained their original golden yellow color even when stored for 100 days, while the regular tomato fruits soon turned red, became soft, and started to rot. Similarly, petunia flowers expressing the *etr-1* gene senesced slowly. When exposed to ethylene, the transgenic flowers stayed fresh for nine days in the vase while flowers from untransformed plants wilted within just three days.

ABSCISIC ACID

Abscisic acid, a "stress" hormone, promotes seed dormancy and enhances tolerance of plants to environmental stresses such as drought and low temperatures. ABA is synthesized by cleavage of violaxanthin or neoxanthin, and the cleavage of these xanthophylls is the rate-limiting step of stress-induced ABA biosynthesis (Walton and Li, 1995). Recently, the gene encoding the xanthophyll cleavage enzyme has been cloned from several species (Schwartz et al., 1997). Increases in ABA content in plants have been achieved using two different strategies. One strategy to enhance ABA production is to overexpress 9-*cis*-epoxycarotenoid dioxygenase genes, engineered for either constitutive or inducible expression in transgenic plants. The constitutive expression of 9-*cis*-epoxycarotenoid dioxygenase genes results in an increase in ABA. These plants showed activation of both drought- and ABA-inducible genes, a reduction in leaf transpiration rate (Iuchi et al., 2001), marked increases in tolerance to drought stress (Iuchi et al., 2001; Qin and Zeevaart, 2002), and increased seed dormancy (Thompson et al., 2000). In contrast, antisense suppression and disruption of *AtNCED3* gave a drought-sensitive phenotype. Alternatively, Frey and colleagues (1999) have shown that overexpression of a zeaxanthin epoxidase gene cloned from *Nicotiana plumbaginifolia* in both sense and antisense orientation in *N. plumbaginifolia* leads to alterations in ABA content. The seeds from overexpressing lines had increased ABA content and subsequently delayed germination, whereas those from plants expressing antisense *ABA2* had a reduced ABA content and germinated rapidly. Strauss and colleagues (2001) have shown that expression of a single-chain variable-fragment antibody against ABA in potato plants caused reduction in free ABA content and reduced growth. Transgenic plants produced smaller leaves than untransformed plants. Leaf stomatal conductivity of transgenic plants was increased due to larger stomatal pores. When expression of the antibody gene was restricted to seeds by using a seed-specific gene promoter from *Vicia faba,* the resulting transgenic tobacco plants were phenotypically similar to wild-type plants, but the embryo developed green cotyledons containing chloroplasts, accumulated photosynthetic pigments, and produced less seed storage protein and oil bodies (Phillips et al., 1997). The transgenic seeds germinated precociously if removed from seed capsules during development but were incapable of germination after drying. These studies demonstrate that it is possible to use transgenic technology to manipulate ABA levels to improve performance, such as stress tolerance, in horticultural crops.

CONCLUDING REMARKS

In comparison to exogenous applications of plant growth regulators, transgenic manipulation of either hormone concentration or sensitivity offers several advantages: (1) cost-effectiveness because once desirable transgenes are inserted into target plants, no additional manipulations are needed; (2) relatively low risk to human health and the environment compared to the exogenous application of synthetic growth regulators; (3) minimal side effects when the transgene is expressed only in target organs and at specific developmental stages; and (4) increased effectiveness. Although a large number of genes are now available for manipulating hormone concentration and sensitivity, relatively little has been accomplished in using these genes to improve horticultural crops. One reason is that we do not have a wide range of gene promoters to control the expression of genes that can alter hormone concentration or sensitivity in an organ- and stage-specific manner.

REFERENCES

Aida, R., T. Yoshida, K. Ichimura, R. Goto, and M. Shibata (1998). Extension of flower longevity in transgenic torenia plants incorporating ACC oxidase transgene. *Plant Science* 138:91-101.

Araki, S., M. Matsuoka, M. Tanaka, and T. Ogawa (2000). Ethylene formation and phenotypic analysis of transgenic tobacco plants expressing a bacterial ethylene-forming enzyme. *Plant and Cell Physiology* 41(3):327-334.

Armstrong, D.J. (1994) Chemistry, activity and function. In *Cytokinins* (pp. 139-154), Eds. Mok, D.W.S. and M.C. Mok. Boca Raton, FL: CRC Press.

Ayub, R.M., M. Guis, M. Ben Amor, L. Gillot, J.P. Roustan, A. Latché, M. Bouzayen, and J.C. Pech (1996). Expression of ACC oxidase antisense gene inhibits ripening of cantaloupe melon fruits. *Nature Biotechnology* 14:862-866.

Barg, R. and Y. Salts (1996). Method for the induction of genetic parthenocarpy in plants. Patent WO 97/30165.

Bleecker, A.B., M.A. Estelle, C. Somerville, and H. Kende (1988). Insensitivity to ethylene conferred by a dominant mutation in *Arabidopsis thaliana. Science* 241:1086-1089.

Chen, C. (1997). Cytokinin biosynthesis and interconversion. *Physiologia Plantarum* 101:665-673.

Chrispeels, M.J. and D.E. Sadava (1994). Plant genetic engineering: New genes in old crops. In *Plants, Genes, and Agriculture* (pp. 412-413), Eds. Chrispeels, M.J. and D.E. Sadava. Boston: Jones & Bartlett Publishers.

Coles, J.P., A.L. Phillips, S.J. Croker, R. García-Lepe, M.J. Lewis, and P. Hedden (1999). Identification of gibberellin production and plant development in *Arabidopsis* by sense and antisense expression of gibberellin 20-oxidase genes. *The Plant Journal* 17:547-556.

Crozier, A. (1983). *The Biochemistry and Physiology of Gibberellins*, Volume 2. New York: Praeger.

Curtis, I.S., D.A. Ward, S.G. Thomas, A.L. Phillips, M.R. Davey, J.B. Power, K.C. Lowe, S.J. Croker, M.J. Lewis, and S.L. Magness (2000). Induction of dwarfism in transgenic *Solanum dulcamara* by overexpression of a gibberellin 20-oxidase cDNA from pumpkin. *The Plant Journal* 23:329-338.

Estruch, J.J., E. Prinsen, H. Van Onckelen, J. Schell, and A. Spena (1991). Viviparous leaves produced by somatic activation of an inactive cytokinin-synthesizing gene. *Science* 254:1364-1367.

Frey, C., E. Audran, S.B. Marin, and A. Marion-Poll (1999). Engineering seed dormancy by the modification of zeaxanthin epoxidase gene expression. *Plant Molecular Biology* 39:1267-1274.

Fu, X., D. Sudhakar, J. Peng, D.E. Richards, P. Christou, and N.P. Harberd (2001). Expression of *Arabidopsis* GAI in transgenic rice represses multiple gibberellin responses. *The Plant Cell* 13:1791-1802.

Gan, S. and R.M. Amasino (1995). Inhibition of leaf senescence by autoregulated production of cytokinin. *Science* 270:1986-1988.

Gianfagna, T.J. (1995). Natural and synthetic growth regulators and their use in horticultural and agricultural crops. In *Plant Hormones: Physiology and Molecular Biology* (pp. 751-773), Ed. Davies, P.J. Boston: Kluwer Academic Publishers.

Guilfoyle, T.J., G. Hagen, Y. Li, T. Ulmasov, Z. Liu, T. Strabala, M.G. Gee, and G. Martin (1993). Auxin-regulation transcription. *Australian Journal of Plant Physiology* 20:489-506.

Hedden, P. and Y. Kamiya (1997). Gibberellin biosynthesis: Enzymes, genes and their regulation. *Annual Review of Plant Physiology and Plant Molecular Biology* 48:431-460.

Henzi, M.X., D.L. McNeil, M.C. Christey, and R.E. Lill (1999). A tomato antisense 1-aminocyclopropane-1-carboxylic acid oxidase gene causes reduced ethylene production in transgenic broccoli. *Australian Journal of Plant Physiology* 26:179-183.

Houba-Herin, N., C. Pethe, J. d'Alayer, and M. Laloue (1999). Cytokinin oxidase from *Zea mays:* Purification, cDNA cloning and expression in moss protoplasts. *The Plant Journal* 17(6):615-626.

Huang, S., A.S. Raman, J.E. Ream, H. Fujiwara, R.E. Cerny, and S.M. Brown (1998). Overexpression of 20-oxidase confers a gibberellin-overproduction phenotype in *Arabidopsis. Plant Physiology* 118:773-781.

Iuchi, S., M. Kobayashi, T. Taji, M. Naramoto, M. Seki, T. Kato, S. Tabata, Y. Kakubari, K. Yamaguchi-Shinozaki, and K. Shinozaki (2001). Regulation of drought tolerance by gene manipulation of 9-*cis*-epoxycarotenoid dioxygenase, a key enzyme in abscisic acid biosynthesis in *Arabidopsis. The Plant Journal* 27(4):325-333.

Klee, H.J., M.B. Hayford, K.A. Kretzmer, G.F. Barry, and G.M. Kishore (1991). Control of ethylene synthesis by expression of a bacterial enzyme in transgenic tomato plants. *The Plant Cell* 3:1187-1193.

Klee, H.J., R.B. Horsch, M.A. Hinchee, M.B. Hein, and N.L. Hoffmann (1987). The effects of overproduction of two *Agrobacterium tumefaciens* T-DNA auxin biosynthetic gene products in transgenic petunia plants. *Genes and Development* 1:86-96.

Klee, H.J. and M.B. Lanahan (1995). Transgenic plants in hormone biology. In *Plant Hormones: Physiology, Biochemistry and Molecular Biology* (pp. 340-353), Ed. Davies, P.J. Dordrecht, the Netherlands: Kluwer Academic Publishers.

Koornneef, M., A. Elgersma, C.J. Hanhart, E.P. van Loenen-Martinet, L. van Rign, and J.A.D. Zeevaart (1985). A gibberellin insensitive mutant of *Arabidopsis thaliana. Physiologia Plantarum* 65:33-39.

Kosugi, Y., K. Shibuya, N. Tsuruno, Y. Iwazaki, A. Mochizuki, T. Yoshioka, T. Hashiba, and S. Satoh (2000). Expression of genes responsible for ethylene production and wilting are differently regulated in carnation (*Dianthus caryophyllus* L.) petals. *Plant Science* 158(1-2):139-145.

Lange, T. (1998). Molecular biology of gibberellin synthesis. *Planta* 204:409-419.

Lange, T., P. Hedden, and J.E. Graebe (1994). Expression and cloning of a gibberellin 20-oxidase, a multifunctional enzyme involved in gibberellin biosynthesis. *Proceedings of National Academy of Sciences USA* 91:8552-8556.

Li, Y. (1997). Transgenic seedless fruit and methods. Patent WO 98/49888.

Li, Y., G. Hagen, and T.J. Guilfoyle (1992). Altered morphology in transgenic tobacco plants that overproduce cytokinins in specific tissues and organs. *Developmental Biology* 153:386-395.

Ludford, P.M. (1995). Postharvest hormone changes. In *Plant Hormones: Physiology, Biochemistry and Molecular Biology* (pp. 725-750), Ed. Davies, P.J. Boston: Kluwer Academic Publishers.

McCabe, M.S., L.C. Garratt, F. Schepers, W.J. Jordi, G.M. Stoopen, E. Davelaar, J.H. van Rhijn, J.B. Power, and M.R. Davey (2001). Effects of *P(SAG12)-IPT* gene expression on development and senescence in transgenic lettuce. *Plant Physiology* 127(2):505-516.

McGaw, B.A. and L.R. Burch (1995). Cytokinin biosynthesis and metabolism. In *Plant Hormones: Physiology, Biochemistry and Molecular Biology* (pp. 98-117), Ed. Davies, P.J. Dordrecht, the Netherlands: Kluwer Academic Publishers.

Niki, T., T. Nishijima, M. Nakayama, T. Hisamatsu, N. Oyama-Okubo, H. Yamazaki, P. Hedden, T. Lange, L.N. Mander, and M. Koshioka (2001). Production of dwarf lettuce by overexpressing a pumpkin gibberellin 20-oxidase gene. *Plant Physiology* 126:965-972.

Norcini, J.G., W.G. Hudson, M.P. Garber, R.K. Jones, A.R. Chase, and K. Bondari (1996). Pest management in the U.S. greenhouse and nursery industry: III. Plant growth regulation. *HortTechnology* 6(3):207-210.

Oeller, P.W., M.W. Lu, L.P. Taylor, D.A. Pike, and A. Theologis (1991). Reversible inhibition of tomato fruit senescence by antisense RNA. *Science* 254:437-439.

Ortiz (1998). Critical role of plant biotechnology for the genetic improvement of food crops: Perspectives for the next millennium. *Electronic Journal of Biotechnology* 1(3): <http://www.ejbiotechnology.info/content/vol1/issue3/full/7/index.html>.

Peng, J., D.E. Richards, N.M. Hartley, G.P. Murphy, K.M. Devos, J.E. Flintham, J. Beales, L.J. Fish, A.J. Worland, F. Pelica, et al. (1999). "Green revolution" genes encode mutant gibberellin response modulators. *Nature* 400:256-261.

Phillips, A.L., D.A. Ward, S. Uknes, N.E.J. Appleford, T. Lange, A.K. Huttly, P. Gaskin, J.E. Graebe, and P. Hedden (1995). Isolation and expression of three gibberellin 20-oxidase cDNA clones from *Arabidopsis. Plant Physiology* 108:1049-1057.

Phillips, J., O. Artsaenko, U. Fiedler, C. Horstmann, H.P. Mock, K. Muntz, and U. Conrad (1997). Seed-specific immunomodulation of abscisic acid activity induces a developmental switch. *EMBO Journal* 16(15):4489-4496.

Qin, X. and J.A. Zeevaart (2002). Overexpression of a 9-*cis*-epoxycarotenoid dioxygenase gene in *Nicotiana plumbaginifolia* increases abscisic acid and phaseic acid levels and enhances drought tolerance. *Plant Physiology* 128(2):544-551.

Romano, C.P., M.B. Hein, and H. Klee (1991). Inactivation of auxin in tobacco transformed with the indoleacetic acid-lysine synthetase gene of *Pseudomonas savastanoi. Genes and Development* 5(3):438-446.

Romano, C.P., P.R. Robson, H. Smith, M. Estelle, and H. Klee (1995). Transgene-mediated auxin overproduction in *Arabidopsis:* Hypocotyl elongation phenotype and interactions with the hy6-1 hypocotyl elongation and axr1 auxin-resistant mutants. *Plant Molecular Biology* 27(6):1071-1083.

Rotino, G.L., E. Perri, M. Zottini, H. Sommer, and A. Spena (1997). Genetic engineering of parthenocarpic plants. *Nature Biotechnology* 15(13):1398-1401.

Sato, T. and A. Theologis (1989). Cloning the mRNA encoding 1-aminocyclopropane-1-carboxylate synthase, the key enzyme for ethylene biosynthesis in plants. *Proceedings of National Academy of Sciences USA* 86(17):6621-6625.

Schroeder, K.R., D.P. Stimart, and E.V. Nordheim (2001). Response of *Nicotiana alata* to insertion of an autoregulated senescence-inhibition gene. *Journal of the American Society for Horticultural Science* 126 (5):523-530.

Schwartz, S.H., B.C. Tan, D.A. Gage, J.A.D. Zeevaart, and D.R. McCarty (1997). Specific oxidative cleavage of carotenoids by VP14 of maize. *Science* 276:1872-1874.

Shimada, N., Y. Suzuki, M. Nakajima, U. Conrad, N. Murofushi, and I. Yamaguchi (1999). Expression of a functional single-chain antibody against GA24/19 in transgenic tobacco. *Bioscience, Biotechnology Biochemistry* 63:779-783.

Smart, C.M., S.R. Scofield, M.W. Bevan, and T.A. Dyer (1991). Delayed leaf senescence in tobacco plants transformed with *tmr,* a gene for cytokinin production in *Agrobacterium. The Plant Cell* 3:647-656.

Smigocki, A., J.W. Neal Jr., I. McCanna, and L. Douglass (1993). Cytokinin-mediated insect resistance in *Nicotiana* plants transformed with the *ipt* gene. *Plant Molecular Biology* 23(2):325-335.

Spena, A. (1998). Methods for producing parthenocarpic or female sterile transgenic plants. Patent WO 98/28430.

Strauss, M., F. Kauder, M. Peisker, U. Sonnewald, U. Conrad, and D. Heineke (2001) Expression of an abscisic acid-binding single-chain antibody influences the subcellular distribution of abscisic acid and leads to developmental changes in transgenic potato plants. *Planta* 213:361-369.

Sun, T. (2000). Gibberellin signal transduction. *Current Opinion in Plant Biology* 3:374-380.

Thompson, A.J., A.C. Jackson, R.C. Symonds, B.J. Mulholland, A.R. Dadswell, P.S. Blake, A. Burbidge, and I.B. Taylor (2000). Ectopic expression of a tomato 9-*cis*-epoxycarotenoid dioxygenase gene causes over-production of abscisic acid. *The Plant Journal* 23:363-374.

Toyomasu, T., H. Kawaide, C. Sekimoto, C. von Numers, A.L. Phillips, P. Hedden, and Y. Kamiya (1997). Cloning and characterization of a cDNA encoding gibberellin 20-oxidase from rice (*Oryza sativa* L.) seedlings. *Plant Physiology* 99:111-118.

Varoquaux, F., R. Blanvillain, M. Delseny, and P. Gallois (2002). Less is better: New approaches for seedless fruit production. *Trends in Biotechnology* 18:233-242.

Walton, D.C. and Y. Li (1995). Abscisic acid biosynthesis and metabolism. In *Plant Hormones Physiology, Biochemistry and Molecular Biology,* Second Edition (pp. 140-157), Ed. Davies, P.J. Dordrecht, the Netherlands: Kluwer Academic Publishers.

Weaver, R.J. (1972). *Plant Growth Substance in Agriculture.* San Francisco: W.H. Ferrman and Company.

Werner, T., V. Motyka, M. Strnad, and T. Schmulling (2001). Regulation of plant growth by cytokinin. *Proceedings of National Academy of Sciences USA* 98:10487-10492.

Wilkinson, J.Q., M.B. Lanahan, D.G. Clark, A.B. Bleecker, C. Chang, E.M. Meyerowitz, and H.J. Klee (1997). A dominant mutant receptor from *Arabidopsis* confers ethylene insensitivity in heterologous plants. *Nature Biotechnology* 15:444-447.

Wu, K., L. Li, D.A. Gage, and J.A.D. Zeevaart (1996). Molecular cloning and photoperiod-regulated expression of gibberellin 20-oxidase from the long-day plant spinach. *Plant Physiology* 110:547-554.

Zhang, J., T. Van Toai, L. Huynh, and J. Preiszner (2000). Development of flooding-tolerant *Arabidopsis thaliana* by autoregulated cytokinin production. *Molecular Breeding* 6:135-144.

Chapter 5

Avidin: An Egg-Citing Insecticidal Protein in Transgenic Corn

Karl J. Kramer

INTRODUCTION

With all of the controversy surrounding the safety of transgenes and their encoded proteins in foods or feeds, one attempt to find a solution was to develop a protein that is already being consumed in human and animal diets as a biopesticide in transgenic grain. This type of development might result in greater public acceptance of the use of transgenes in foods or feeds for humans and animals than has occurred up to now. That goal was the motivation for a collaborative research project between biochemists, molecular biologists, and entomologists at the U.S. Department of Agriculture (USDA) Agricultural Research Service (ARS) Grain Marketing and Production Research Center (http://www.usgmrl.ksu.edu) in Manhattan, Kansas, and two agricultural biotechnology companies, Pioneer Hi-Bred International, Inc. (http://www.pioneer.com), in Johnston, Iowa, and ProdiGene, Inc. (http://www.prodigene.com), in College Station, Texas. Results from the project eventually led to the development of a commercial industrial protein, avidin produced from *Zea mays,* a value-added grain containing the chicken egg white protein avidin, which is used in re-

I am grateful to Thomas Morgan, James Throne, Floyd Dowell, Michelle Bailey, John Howard, and the late Thomas Czapla for excellent collaboration in this research, and to Marc Harper and Craig Roseland for critical comments about an early draft of this chapter. Mention of a proprietary product does not constitute a recommendation by the USDA. The Agricultural Research Service, USDA, is an equal employment opportunity/affirmative action employer, and all agency services are available without discrimination.

search and medical diagnostic testing procedures. Another end result of the project was the potential for a commodity that is very resistant to attack by stored-product insect pests.

MECHANISM OF ACTION

Historically, the insecticidal activity of chicken avidin has been known since 1959 when Levinson and his associates in Germany first reported that the protein was toxic to the housefly, *Musca domestica,* when administered to larvae in the diet (Levinson and Bergmann, 1959). Our project actually began about thirty years later when members of my research group in Kansas and Thomas Czapla of Pioneer Hi-Bred in Iowa discovered that avidin or streptavidin, a homologous protein produced by the bacterium *Streptomyces avidinii,* when administered in semiartificial diets at approximately 100 ppm, caused in the European corn borer, *Ostrinia nubilalis,* and several species of stored-product beetles and moths a deficiency of the vitamin biotin, a biheterocyclic nitrogen-sulfur compound with a valeric acid side chain (Morgan et al., 1993). That vitamin deficit in turn led to stunted growth and mortality of those species.

At that time and even now, no biopesticides have been commercially developed in transgenic plants for insect pest control other than several endotoxins from the bacterium *Bacillus thuringiensis* (Bt), which, unlike avidin, are not human or animal dietary proteins. ARS and Pioneer were interested in prospecting for pesticidal proteins that not only were already naturally found in foods or feeds but also would have a broader spectrum of activity and be as safe as or exhibit greater safety than did the Bt-types of proteins as well as employ an alternative mode of action as compared to Bt.

Until avidin, no known single biopesticidal proteins exhibited a broad enough spectrum of activity for controlling the many primary and secondary lepidopteran and coleopteran insect pests encountered in stored products. Recently, some Bt corn kernels were evaluated for stored-product insect resistance (Sedlacek et al., 2001). Two of five hybrids tested containing the Cry1Ab endotoxin exhibited reduced laboratory populations of Indian meal moth, *Plodia interpunctella,* and angoumois grain moth, *Sitotroga cerealella,* relative to populations that developed on non-Bt-transformed isolines. Those results were not unexpected because lepidopterans are the targets of Cry1A proteins.

In another related study, development, survival, sex ratio, and adult body length of *P. interpunctella* were negatively impacted by several Bt corn hybrids (Giles et al., 2000). However, the overall results of those studies did not demonstrate sufficient efficacy of Bt corn for economic control of those lepidopteran species or any stored-product Coleoptera. Thus, a need remains to increase the expression levels of the toxins in the grain or to identify proteins more efficacious than Bt-derived ones for controlling stored-grain insect pests. Other kinds of proteins that have been tested for oral toxicity include digestive enzyme inhibitors, hydrolytic enzymes, and carbohydrate-binding proteins, but none of these have been found to be as efficacious as either avidin or Bt endotoxins.

Avidin and streptavidin have a very strong affinity for the vitamin biotin and have a long history of use in a variety of biochemical and medical diagnostic procedures (Wilchek and Bayer, 1990). The former protein has an apparent molecular mass of 17 kDa, is glycosylated, and binds very tightly only one biotin molecule with a dissociation constant $Kd = 10^{-15} \cdot M^{-1}$ (Nardone et al., 1998; Freitag et al., 1999). Biotin is a coenzyme required for enzymes that catalyze carboxylation, decarboxylation, and transcarboxylation reactions in all forms of life. An example of a ubiquitous biotin-containing protein in insects is pyruvate decarboxylase, which synthesizes oxaloacetate from pyruvate and bicarbonate for gluconeogenesis and serves to replenish intermediates of the citric acid cycle (Ziegler et al., 1995). What makes avidin particularly unique as a biopesticide is that it not only is a common dietary protein, but also has a naturally occurring antidote, biotin, which can be used as a supplement in diets to prevent toxicity or to rescue potential nontarget victims from adverse side effects. Many biopesticides do not have any kind of antidote.

HISTORY OF COMMERCIALIZATION

Our encouraging initial results (Morgan et al., 1993) led to the next phase of the project, which was the creation of a variety of transgenic avidin maize by scientists at both Pioneer and ProdiGene. John Howard, who at that time led Pioneer's Protein Products Group and later formed ProdiGene in Texas in 1996, and members of his group at Pioneer were developing transgenic cultivars of maize for commercial production of industrial proteins. The first transcription unit attempted, a chicken avidin cDNA with its codons optimized for the preferred maize codon usage pattern, the *Streptomyces Bar* (bialaphos-resistance) gene for selection, and the potato proteinase inhibitor II transcription terminator region, was not successful. The recombinant protein product was not secreted from the plant's cells, which caused high concentrations and toxic symptoms in the plant (Hood et al., 1997; Ginzberg and Kapulnik, 2000). In plants, biotin is an essential cofactor for a small number of enzymes involved mainly in the transfer of CO_2 during bicarbonate-dependent carboxylation reactions, which apparently are disrupted when avidin is present in the cytoplasm (Alban et al., 2000). When DNA encoding a barley alpha-amylase signal peptide was also included in the construct, however, the avidin protein was targeted to the secretory pathway and was secreted from the cytoplasm to the cell wall and beyond, which made the plants much more viable. Eventually, utilization of this second construct allowed avidin corn to became the first transgenic variety of corn commercialized by ProdiGene.

Recombinant avidin is now sold for use as a research chemical and diagnostic reagent (Hood et al., 1997). The protein is produced from corn at only a fraction of the cost (approximately tenfold less expensive) than it can be produced from chicken eggs, where it is found in egg white. ProdiGene and Sigma Chemical Co., St. Louis, Missouri, began marketing avidin produced from transgenic maize in 1997 (Hood et al., 1999; Fischer and Emans, 2000). The first generation of avidin corn exhibited a level of only about 100 ppm avidin. Recently, best line selections after seven generations in the field have resulted in seeds containing as much as 3,000 ppm avidin. Egg white contains only ~500 ppm. Thus, corn has become a high-yielding factory for avidin production. In the corn kernels, avidin is found primarily in the germ and endosperm (77 percent of the total), but it is found in other milling fractions as well. In addition, the recombinant protein is very stable in the kernel, in contrast to avidin's instability in hen egg white.

DEMONSTRATION OF HOST PLANT RESISTANCE AND SPECTRUM OF ACTIVITY

At about the same time that avidin corn became a commercial success, the Kansas research group began testing it for use in a second application, as a host plant resistance factor for stored-product insect pests. When present in maize kernels at levels of approximately 100 ppm or higher, avidin was found to be toxic to and to prevent development of many internally and externally feeding insect pests that damage grains during storage, including the rice weevil, *Sitophilus oryzae;* lesser grain borer, *Rhyzopertha dominica;* angoumois grain moth; warehouse beetle, *Trogoderma variabile;* sawtoothed grain beetle, *Oryzaephilus surinamensis;* flat grain beetle, *Cryptolestes pusillus;* red and confused flour beetles, *Tribolium castaneum* and *T. confusum;* Indian meal moth; and Mediterranean flour moth, *Anagasta kuehniella* (Kramer et al., 2000). That level of efficacy was approximately the same as had been reported earlier when bioassays were conducted using semiartificial diets (Morgan et al., 1993). Other pest species also observed to be susceptible to avidin toxicity include the house fly (Levinson and Bergmann, 1959); hide beetle, *Dermestes maculatus* (Levinson et al., 1967); fruit fly, *Drosophila melanogaster* (Bruins et al., 1991); olive fruit fly, *Dacus oleae* (Tsiropoulos, 1985); flour mite, *Acarus siro* (Levinson et al., 1992); tobacco hornworm, *Manduca sexta* (Du and Nickerson, 1996); tobacco budworm, *Heliothis virescens;* black cutworm, *Agrotis ipsilon;* sunflower moth, *Homoeosoma electellum;* beet armyworm, *Spodoptera exigua;* cotton bollworm, *Helicoverpa zea* (Czapla et al., unpublished data); light brown apple moth, *Epiphyas postvittana;* greenheaded leafroller, *Planotortrix octo;* brownheaded leafroller, *Ctenopseustis obliquana;* and potato tuber moth, *Phthorimaea operculella* (Markwick et al., 2001).

The only species identified to date that is not susceptible to avidin toxicity is the larger grain borer, *Prostephanus truncatus* (Kramer et al., 2000). This species is occasionally found in southern Texas, but it generally is not a significant pest in the United States. It has been a serious pest, however, in Mexico, Central America, northern South America, and Africa. Why and how the larger grain borer tolerates high levels of avidin is unknown, but that question will be addressed in a future study. Perhaps avidin is degraded by borer gut proteinases or by symbiotic microorganisms, who might also produce an endogenous level of biotin which is sufficient to saturate the exogenous avidin. Another possibility is that the biotin-containing enzymes in the gut tissues of the larger grain borer are somehow protected from interaction with dietary avidin.

The existence of an avidin-tolerant species, however, indicates the very real possibility of the development of insect resistance to avidin if resistance management is not conducted properly if and when avidin is utilized as an insect resistance factor in the field. Nonetheless, avidin does act as a potent biopesticide in transgenic maize against many pest species with a specific toxicity comparable to Bt toxins, and the spectrum of activity of the former protein is much broader than the latter. Out of 28 pest species tested to date, 27 (96 percent), including 10 species of Coleoptera, 13 of Lepidoptera, and 3 of Diptera, were found to be susceptible to avidin toxicity (Kramer et al., unpublished data). Currently, many other species of insect pests are being screened for susceptibility to avidin toxicity including several cockroach species and various pests of cowpea, field rice, and cotton.

ALLERGENICITY

The immune response is a potential impediment to the consumption by humans and livestock of any exogenous or recombinant proteins in transgenic plants (Metcalfe et al., 1996). However, it should not be an impediment for the development of avidin as a biopesticide because the protein is not highly allergenic (Subramanian and Adiga, 1997; Breiteneder and Ebner, 2000). It is absent from the Official List of Allergens maintained by the International Allergen Nomenclature Subcommittee of the International Union of Immunological Societies (Larsen and Lowenstein, 2001). It also is interesting to note that corn kernels contain two potent allergens on the list, a lipid transfer protein and a bifunctional amylase-trypsin inhibitory protein, and chicken egg white has four allergenic proteins listed, including ovomucoid, ovalbumin, conalbumin and lysozyme (Hoffman, 1983; Langeland, 1983). Nonetheless, for avidin corn to be approved for use as a human food or animal feed, it would have to be subjected to the FAO/WHO Decision Tree assessment procedure for evaluation of potential allergenicity for a food containing a gene derived from a source known to be allergenic to some people, such as the chicken egg (Food and Agriculture Organization of the United Nations, 2001). The procedure involves amino acid sequence homology searching, specific serum screening, targeted serum screening, pepsin resistance screening, and animal model screening.

Allergenic residues, domains, or motifs in proteins now can be eliminated by genetic modification. Recent studies have demonstrated that the antibody response to streptavidin, a bacterial homologue of avidin, can be greatly reduced through site-directed mutagenesis (Meyer et al., 2001). Streptavidin is a protein used medically as an effective receptor for biotinylated tumoricidal molecules, such as radionuclides, when conjugated to an antitumor antibody and administered systemically. Because treatment of cancer patients may require repeated administration, so as to maximize the antitumor effect, the researchers attempted to reduce the antigenicity of streptavidin by mutating surface residues capable of forming ionic or hydrophobic interactions that would facilitate an immune response (Subramanian and Adiga, 1997). Surface residues with no known impact on biotin binding and protein folding were identified. Some of those residues, when mutated, reduced the immunoreactivity with patient antisera to <10 percent that of wild-type streptavidin. Another form with ten amino acids mutated was only about 20 percent as antigenic as streptavidin. Also, the molecule's ability to elicit an immune response in rabbits was reduced substantially when some charged, aromatic, or large hydrophobic residues located on the protein's surface were substituted with smaller neutral amino acid residues.

SAFETY

Avidin corn demonstrated excellent resistance to storage insect pests when ground into a meal (Kramer et al., 2000). The average concentration of avidin in the meal needs to be approximately 100 ppm avidin, which, though substantially lower than the level present in chicken egg white, is sufficient for protection from insects. Humans can suffer from megadoses of avidin or "egg white injury," but only after they consume exceedingly large quantities of avidin, such as by eating dozens of raw eggs a day for several months. Avidin maize is not toxic to mice when administered as the sole component of their diet for three weeks (Kramer et al., 2000). One of the reasons why avidin is an insect-selective biocide is because it is very stable and functional in the insect's gut where the pH is near neutrality or alkaline. In the human stomach, however, where the pH is very acidic, avidin is much less stable and more susceptible to degradation by proteinases. Another reason avidin is much more potent against insects is the fact that insects eat a rather limited diet such as corn kernels, whereas humans consume a more diverse and nutritionally balanced diet which supplies a better variety of vitamins including biotin.

Following a thorough and satisfactory safety risk assessment of avidin corn, its utilization as a food or feed grain would be an exciting development that could impact upon postharvest losses caused by stored-product insect pests. Even if there were a safety concern, avidin maize might be processed prior to consumption to include supplementation with biotin or treatment with heat to neutralize or inactivate avidin. The latter process would denature the avidin as well as the avidin-biotin complexes and release most of the biotin for use by the consumer. Another use of avidin maize could be as an insect-resistant background host plant germplasm for farming of other valuable biopharmaceutical or industrial proteins such as enzymes, sweeteners, adhesives, and oral vaccines. Those types of value-added grains could then be stored in warehouses where they would be self-protected from insect pest contamination. Whatever utilization of avidin corn does occur in the future, it is at least a proof of the efficacy of a common human food protein for use as a biopesticide in transgenic plants.

MALE STERILITY AND NONUNIFORM EXPRESSION

Because the ubiquitin promoter controls expression of avidin in avidin corn, expression of the protein occurs in all of the tissues, including the anthers (Hood et al., 1997). This localization causes male sterility and nonuniform protein expression in many of the plants. There was a 97.5 percent correlation between the presence of the avidin gene and the male sterile/limited fertility phenotype. This phenotype, however, does express the recombinant protein in the grain and other portions of the plant, and the male sterility trait can be used for scoring of the transformed plants. Disruption of normal pollen production can be averted by timely application of biotin in a foliar spray. The avidin transgene system, therefore, has utility in plant biotechnology for hybrid seed production and research on the regulation of plant growth and development.

Because of male sterility, avidin corn produces little if any functional pollen that could drift outside of the cornfield (Hood et al., 1997). Thus, genetically modified organism (GMO) pollen pollution is highly unlikely with the male-sterile varieties of avidin corn because there would be no outcrossing of transgenes to wild-type species or to neighboring fields with the same species. Commercial Bt corn hybrids, on the other hand, have viable pollen that may contaminate the environment, which may have occurred in the field with StarLink corn.

Only about half of the individual kernels from harvested lots of avidin corn have inadequate insect resistance and little or no avidin due to male sterility. That result led to the development of a nondestructive method utilizing near-infrared spectroscopy to screen for avidin content in individual kernels and to separate low- from high-avidin kernels prior to conducting bioassays of intact kernels (Kramer et al., 2000; Xie et al., 2000). However, for a biopesticide resistance management program designed for application in a bin of grain, it might be fortuitous that all of the corn kernels do not contain ≥ 100 ppm avidin (i.e., microrefuges) so that pest resistance to avidin would not develop so quickly. Nevertheless, a future goal of this research project will be to increase the proportion of avidin-containing, insect-resistant kernels by performing new transformation events involving promoters designed to avoid expression of the transgene in the anthers and to yield kernels containing more uniform and effective levels of avidin. The use of a seed-specific promoter, for example, should allow recovery of homozygous plants since male sterility would presumably be eliminated, thus increasing expression levels per individual kernel and doubling the amount of positive seeds.

CONCLUDING REMARKS

Corn originated in Mexico where it was grown as a food crop as early as 5000 B.C. (Armstrong et al., 2000; Pope et al., 2001). Efforts to genetically transform corn using directed recombination methods began in only the 1980s and fertile transgenic maize was recently described only in 1990. Since then transformation technology has been rapidly applied for the genetic improvement of corn, leading to the first commercial transgenic hybrids in 1996. Soon thereafter, hybrids carrying Bt and/or herbicide tolerance genes were adopted enthusiastically by growers. Subsequently, transgenic maize was developed as a biofactory for high-value proteins, which led to the commercialization of avidin corn for such specialty protein production, a grain that fortuitously possessed superior resistance to insects in the field and in storage (Kramer et al., 2000 and unpublished data). Avidin is a noteworthy protein because it has many positive attributes for use as a biopesticide. It has an insecticidal potency comparable to Bt endotoxins, but it has a much broader spectrum of toxicity than Bt. It also is insect specific at a dose of only 10 to 100 ppm in the diet. It can be heat inactivated and has a naturally occurring antidote if toxicity happens to be a problem for nontarget organisms. It is not allergenic and also is a common dietary protein for humans and other animals. Overall, it has many properties that are beneficial to the environment, producer, and consumer. Current work is underway to transfer avidin biotechnology to other grains and commodities for insect resistance purposes, which should lead to other types of developments similar to avidin corn in the future.

REFERENCES

Alban, C., D. Job, and R. Douce (2000). Biotin metabolism in plants. *Annual Review of Plant Physiology and Plant Molecular Biology* 51:17-47.

Armstrong, C.L., T.M. Spencer, M.A. Stephens, and S.M. Brown (2000). Transgenic maize. In *Transgenic Cereals* (pp. 115-152), Eds. O'Brien, L. and R.J. Henry. St. Paul, MN: American Association of Cereal Chemists.

Breiteneder, H. and C. Ebner (2000). Molecular and biochemical classification of plant-derived food allergens. *Journal of Allergy and Clinical Immunology* 106:27-36.

Bruins, B.G., W. Scharloo, and G.E.W. Thorig (1991). The harmful effect of light on *Drosophila* is diet dependent. *Insect Biochemistry* 21:535-539.

Du, C. and K.W. Nickerson (1996). The *Bacillus thuringiensis* toxin binds biotin-containing proteins. *Applied and Environmental Microbiology* 62:2932-2939.

Fischer, R. and N. Emans (2000). Molecular farming of pharmaceutical proteins. *Transgenic Research* 9:279-299.

Food and Agriculture Organization of the United Nations (2001). Evaluation of allergenicity of genetically modified foods. Report of a Joint FAO/WHO Expert Consultation on Allergenicity of Foods Derived from Biotechnology. Rome, Italy. (For the most recent information about the subject of this document, see the WHO Web site at <http://www.who.int/foodsafety/publications/biotech/ec_jan2001/en/>).

Freitag, S., I.L. Trong, L.A. Klumb, P.S. Stayton, and R.E. Stenkamp (1999). Atomic resolution structure of biotin-free Tyr43Phe streptavidin: What is in the binding site? *Acta Crystallographica* D55:1118-1126.

Giles, K.L., R.L. Hellmich, C.T. Iverson, and L.L. Lewis (2000). Effects of transgenic *Bacillus thuringiensis* maize grain on B. thuringiensis susceptible *Plodia interpunctella* (Lepidoptera: Pyralidae). *Journal of Economic Entomology* 93: 346-349.

Ginzberg, I. and Y. Kapulnik (2000). Expression of streptavidin in plants: A novel approach to control plant development. *Biomolecular Engineering* 16:169.

Hoffman, D.R. (1983). Immunochemical identification of the allergens in egg white. *Journal of Allergy and Clinical Immunology* 71:481-486.

Hood, E.E., A. Kusnadi, Z. Nikolov, and J.A. Howard (1999). Molecular farming of industrial proteins from transgenic maize. *Advances in Experimental Medicine and Biology* 464:127-147.

Hood, E.E., D.R. Witcher, S. Maddock, T. Meyer, C. Baszczynski, M. Bailey, P. Flynn, J. Register, L. Marshall, D. Bond, et al. (1997). Commercial production of avidin from transgenic maize: Characterization of transformant, production, processing, extraction and purification. *Molecular Breeding* 3:291-306.

Kramer, K.J., T.D. Morgan, J.E. Throne, F.E. Dowell, M. Bailey, and J.A. Howard (2000). Transgenic maize expressing avidin is resistant to storage insect pests. *Nature Biotechnology* 18:670-674.

Langeland, T. (1983). A clinical and immunological study of allergy to hen's egg white: IV. Specific IgE antibodies to individual allergens in hen's egg white related to clinical and immunological parameters in egg-allergic patients. *Allergy* 38:493-500.

Larsen, J.N. and H. Lowenstein (2001). Official List of Allergens. See the ftp site at <ftp://biobase.dk/pub/who-iuis/allergen.list>.

Levinson, H.Z., J. Barelkovsky, and A.R. Bar Ilan (1967). Nutritional effects of vitamin omission and antivitamin administration on development and longevity of the hide beetle *Dermestes maculatus* (Coleoptera: Dermestidae). *Journal of Stored Product Research* 3:345-352.

Levinson, H.Z. and E.D. Bergmann (1959). Vitamin deficiencies in the housefly produced by antivitamins. *Journal of Insect Physiology* 3:293-305.

Levinson, H.Z., A.R. Levinson, and M. Offenberger (1992). Effect of dietary antagonists and corresponding nutrients on growth and reproduction of the flour mite (*Acaris siro* L.). *Experientia* 48:721-729.

Markwick, N.P., J.T. Christeller, L.C. Dochterty, and C.M. Lilley (2001). Insecticidal activity of avidin and streptavidin against four species of pest Lepidoptera. *Entomologia Experimentalis Applicata* 98:59-66.

Metcalfe, D.D., J.D. Astwood, R. Townsend, H.A. Sampson, S.L. Taylor, and R.L. Fuchs (1996). Assessment of the allergenic potential of foods derived from genetically engineered crop plants. *Critical Reviews in Food Science and Nutrition* 36:S165-S186.

Meyer, D.L., J. Schultz, Y. Lin, A. Henry, J. Sanderson, J.M. Jackson, S. Goshorn, A.R. Rees, and S.S. Graves (2001). Reduced antibody response to streptavidin through site directed mutagenesis. *Protein Science* 10:491-503.

Morgan, T.D., B. Oppert, T.H. Czapla, and K.J. Kramer (1993). Avidin and streptavidin as insecticidal and growth inhibiting dietary proteins. *Entomologia Experimentalis Applicata* 69:97-108.

Nardone, E., C. Rosano, P. Santambrogio, F. Curnis, A. Corti, F. Magni, A.G. Siccardi, G. Paganelli, R. Losso, B. Apreda, M. Bolognesi, A. Sidoli, and P. Arosio (1998). Biochemical characterization and crystal structure of a recombinant hen avidin and its acidic mutant expressed in *Escherichia coli. European Journal of Biochemistry* 256:453-460.

Pope, K.O., M.E.D. Pohl, J.G. Jones, D.L. Lentz, C. von Nagy, F.J. Vega, and I.R. Quitmyer (2001). Origin and environmental setting of ancient agriculture in the lowlands of Mesoamerica. *Science* 292:1370-1373.

Sedlacek, J.D., S.R. Komaravalli, A.M. Hanley, B.D. Price, and P.M. Davis (2001). Life history attributes of Indian meal moth (Lepidoptera: Pyralidae) and angoumois grain moth (Lepidoptera: Gelechiidae) reared on transgenic corn kernels. *Journal of Economic Entomology* 94:586-592.

Subramanian, N. and P.R. Adiga (1997). Mapping the common antigenic determinants in avidin and streptavidin. *Biochemistry and Molecular Biology International* 43:375-382.

Tsiropoulos, G.R. (1985). Dietary administration of antivitamins affected the survival and reproduction of *Dacus oleae. Zeitschrift fur Angewandte Entomologie* 100:35-39.

Wilchek, M. and E.A. Bayer (1990). *Avidin-Biotin Technology.* Methods in Enzymology, Volume 184. New York: Academic Press.

Xie, F., E. Maghirang, T. Pearson, D. Wicklow, K. Kramer, T. Morgan, and F. Dowell (2000). NIRS applied to detecting single corn kernel characteristics. Abstracts of the 85th Annual Meeting of the American Association of Cereal Chemists, Kansas City, Missouri, p. 195.

Ziegler, R., D.L. Engler, and N.T. Davis (1995). Biotin-containing proteins of the insect nervous system, a potential source of interference with immunocytochemical localization procedures. *Insect Biochemistry and Molecular Biology* 25:569-574.

Chapter 6

Genetic Engineering of Wheat: Protocols and Use to Enhance Stress Tolerance

Tom Clemente
Amitava Mitra

INTRODUCTION

Two components are associated with plant genetic engineering protocols: first, a cell must be competent to receive and integrate foreign DNA, and, second, the lineage of that cell must differentiate to form the germline. With the exception of the floral dip protocol described by Bechtold and colleagues (1993) for *Arabidopsis,* in which the integration event occurs in the egg cell (Desfeux et al., 2000), recovering germline lineage of a transgenic cell requires an in vitro regeneration regime.

Wheat genetic engineering studies were begun in the early 1980s. Initial attempts targeted wheat protoplasts coupled with the direct-DNA delivery methods of electroporation (Hauptmann et al., 1987) or polyethylene glycol (PEG)-mediated transformation (Lörz et al., 1985). Although these early studies demonstrated the feasibility of expressing foreign genes in wheat cells, reliable protocols for the in vitro regeneration from wheat protoplast impeded further progress in recovery of whole plant transformants via this route. A number of alternative wheat explant sources are competent for in vitro regeneration (reviewed by Bajaj, 1990; Vasil and Vasil, 1999); however, wheat transformation efforts were subsequently primarily focused toward implementing embryo explants as the target tissue for transformation due to their reported prolific in vitro culture response (Ozias-Akins and Vasil, 1982, 1983). The explants targeted were immature embryos derived from wheat stock plants in which spikes are harvested 12 to 15 days postanthesis. Embryos harvested at this point generally range in size from 0.5 to 2.0 mm. Culturing of the immature embryos in the dark on a Murashige and Skoog (MS) (Murashige and Skoog, 1962) based medium supplemented with 1 to 2 mg·L^{-1} 2,4-dichlorophenoxyacetic acid (2,4-D) induces somatic embryogenesis about the scutellar region (Vasil et al., 1992, 1993; Perl et al., 1992), an observation first reported by Gosch-Wackerle and colleagues (1979). Under these conditions callus induction from immature embryos is induced in approximately three to four days and somatic embryos formed within seven days. Modulating the growth regulator regime and transferring the cultures to the light will bring about conversion of the somatic embryos. The shoots derived from converted embryos develop into plantlets following total withdrawal of exogenous growth regulators, although some reported protocols supplement the medium with a low level of auxin for plantlet development (Fennell et al., 1996).

Vasil and colleagues (1992) were the first to exploit the prolific somatic embryogenic response of wheat immature embryos as targets for genetic engineering. Starting with established embryogenic cultures and employing microprojectile bombardment (Klein et al., 1987) as the gene delivery approach, wheat transformants were recovered. The prototype wheat transformation protocol (Vasil et al., 1992, 1993) required a protracted time in culture, seven to nine months, to recover transformed lines. The longer plant cells remain in culture, the greater the probability of inducing a tissue culture-derived mutation. To reduce the culture time, and thus limit the potential for mutations arising, genetic engineering protocols subsequently targeted immature embryos for microprojectile-mediated transformation following a brief preculture period (Weeks et al., 1993; Nehra et al., 1994) or immediate bombardment following isolation of the explants (Becker et al., 1994). Wheat transformation using immature embryos as the explant coupled with microprojectile-mediated DNA delivery has since been reported from a number of independent laboratories (Zhou et al., 1995; Ortiz et al., 1996; De Block et al., 1997; Witrzens et al., 1998; Zhang et al., 2000; Campbell et al., 2000).

A modification of the protocol described by Zhou and colleagues (1995) has some positive attributes including short culture time and reliability. In this procedure wheat spikes are harvested 12 to 15 days postanthesis. The quality of the stock plants will have a significant impact on the culture response of the embryos, which has been observed by Altpeter and colleagues (1996), with stock plants maintained under growth chamber conditions producing higher-quality explant material than those produced under field or greenhouse environments. Moreover, pesticide application to stock plants generally has a negative impact on culture performance of the immature embryo explant. The husked seeds are sterilized by 1 min wash in 70 percent ethanol, followed by 10 min wash in 0.5 percent sodium hypochlorite solution and subsequent washes in sterile water. The embryos should be approximately 1.5 to 2.0 mm long and slightly opaque. The isolated embryos are cultured on callus induction medium consisting of a MS-based medium supplemented with 0.5 $mg \cdot L^{-1}$ 2,4-D, 2.2 $mg \cdot L^{-1}$ picloram, 40 $g \cdot L^{-1}$ maltose, 20 mM 2-(*N*-morpholino)-ethanesulfonic acid (MES) (pH 5.8), and 100 $mg \cdot L^{-1}$ ascorbic acid. The medium is solidified with 2 $g \cdot L^{-1}$ phytagel. The use of picloram as an alternative auxin during the callus induction phase has been shown to positively impact transformation frequencies of wheat (Barro et al., 1998). The embryos are precultured in the dark at 24°C for four days on this medium prior to being subjected to microprojectile bombardment. Approximately four to six hours prior to microprojectile bombardment the tissue is transferred to a high osmotic medium consisting of the callus induction medium supplemented with 74.3 $g \cdot L^{-1}$ raffinose and 22.8 $g \cdot L^{-1}$ mannitol. An osmotic pretreatment of tissue prior to bombardment was reported to enhance transformation frequencies in maize tissue (Vain et al., 1993); an analogous observation was also reported for wheat (Altpeter et al., 1996). The osmotic pulse impacts transformation frequency by reducing tissue damage induced by the bombardment process and, in turn, augments tissue survival and ultimately enhances the in vitro culture response.

An array of particle preparation protocols have been employed to deliver gene(s) of interest into plant cells. Rasco-Gaunt and colleagues (1999) investigated a series of parameters associated with the bombardment process, including DNA load, particle size, chamber vacuum, gap distance, bursting plate pressure, and number of bombardments per plate. The monitoring of transient expression across a series of experiments resulted in a fine-tuned procedure that enhanced transformation frequencies from five- to tenfold in elite wheat genotypes (Rasco-Gaunt et al., 1999).

Following the direct DNA delivery step the explants remain on the high osmotic medium for a period of 16 to 20 hours in the dark at 24°C. The tissue is subsequently placed on callus induction medium supplemented with the appropriate selection agent. Witrzens and colleagues (1998) evaluated four selection agents for use in wheat transformation employing immature embryos as the target tissue. The selection agents tested included bialaphos, hygromycin B, and geneticin or paromomycin. These were used in combination with the selectable marker genes, *bar, hpt,* and *nptII,* respectively. The use of geneticin combined with the *nptII* gene resulted in an increase in the recovery of transgenic wheat lines with fewer escapes (nontransformed wheat lines) (Witrzens et al., 1998). The alternative aminoglycosides, kanamycin and paromomycin, show significantly less effectiveness in preventing callus growth, and hence relatively higher concentration must be used to prevent escapes as compared to geneticin. In practice, 25 $mg{\cdot}L^{-1}$ of geneticin (G418) (Campbell et al., 2000), 0.5 to 5 $mg{\cdot}L^{-1}$ bialaphos (Weeks et al., 1993; Becker et al. 1994), or 25 $mg{\cdot}L^{-1}$ hygromycin (Ortiz et al., 1996) have all been successful in recovering transgenic wheat lines in microprojectile-derived protocols. The initial selection step is conducted in the dark at 24°C for a period of two weeks.

At the end of the two-week callus induction step, the tissue is subsequently transferred to maturation medium composed of a modified MS-based medium (Zhou et al., 1995) supplemented with 0.2 $mg{\cdot}L^{-1}$ 2,4-D and 40 $g{\cdot}L^{-1}$ maltose. The selection pressure is maintained during the maturation step. One important aspect at this transfer point is breakup of the callus tissue into small pieces (2 to 3 mm) to ensure efficient selection of transformed lines. The callus tissue should be identity preserved with respect to initial embryo explant, because multiple plant lines derived from the initial immature embryo are generally clones. The embryo maturation step is conducted for two weeks under a 16-hour light regime at 24°C. At the end of this step, embryo conversion and greening of tissue can be observed. Following the maturation step, developing green tissue is subcultured to elongation medium consisting of maturation medium devoid of 2,4-D. The elongation step should be conducted in a sundae cup* or equivalent vessel to permit plantlet development, while maintenance of selection pressure to prevent escapes and identity preservation of the culture lineage documented to accurately record independent transformed lines also should be carried on during this step.

The elongation step is conducted for four to six weeks with a transfer to fresh medium every two weeks until plantlets are fully developed and ready for acclimation to soil. The protocol and modifications outlined have been successful for the genetic engineering of a number of wheat genotypes, albeit a strong genotype variation is encountered as is true of all transformation protocols.

Although the use of microprojectile-mediated transformation for the delivery of transgenes into wheat has been effectively implemented in a number of independent laboratories, the patterns of integration of the foreign DNA element(s) tend to be highly complex (Pawlowski and Somers, 1998). The complexity of transgene integration is reflected in the hybridization patterns observed from a Southern blot analysis on transgenic wheat lines derived from microprojectile-mediated transformation (Figure 6.1). Complex integration patterns will have a higher probability of unstable expression of the desired phenotype over multiple generations due to an increased risk of manifesting induction of the gene-silencing cascade (Matzke et al., 1996). Moreover, in most transformation protocols employing microprojectile bombardment, whole plasmids are delivered; hence extraneous plasmid DNA sequences are integrated along with the gene(s) of interest. Therefore, although genetically engineered wheat lines derived from microprojectile bombardment are being generated routinely in a number of laboratories, the lines are generally of low quality with respect to stable expression patterns. Hence, additional labor, greenhouse space, and ultimately cost in subsequent screening of the transformed lines are required to identify a desirable line.

Agrobacterium-mediated transformation has become the most desirable method for the introduction of foreign genes into plant cells

---

*Available in the catalog of Sweetheart Cup Co., Owings Mills, MD; lid #LDS58; base #DSD8x.

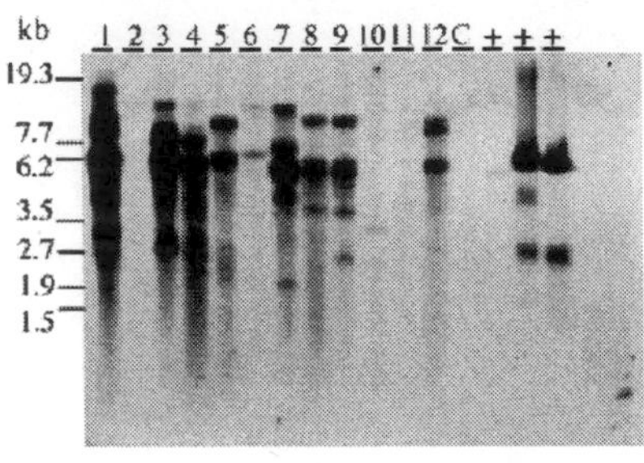

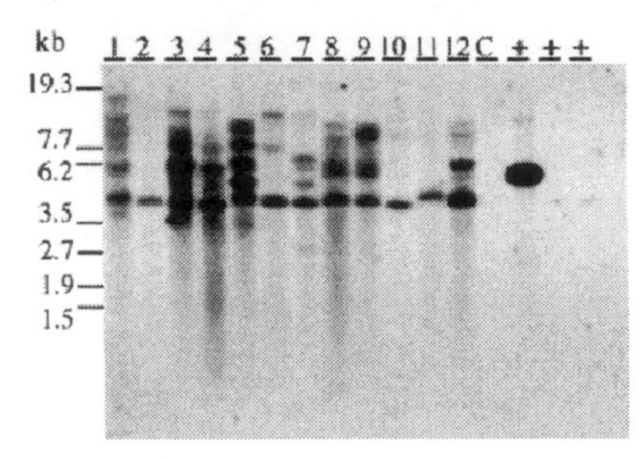

FIGURE 6.1. Southern blot analysis on microprojectile-derived wheat transformants. Transformations were conducted using a cotransformation approach by delivering two plasmids. Plasmid one harbored a *GUS* cassette, while plasmid two carried an *nptII* cassette. Lanes 1 through 12 were loaded with ten µg total genomic DNA from transgenic wheat lines derived from microprojectile bombardment. Lane C indicates negative-control wild-type wheat DNA. Three + lanes refer to the positive control lanes loaded with 50 pg of either *nptII* vector or *GUS* vectors (last two + lanes). The blots were hybridized with either *GUS* (L) or *nptII* (R).

and the subsequent regeneration of transgenic plants. *Agrobacterium* carries out a precise transfer and stable integration of foreign genes into the genome of host plants. Monocot plants were initially considered not amenable to *Agrobacterium*-mediated genetic engineering until recent development of methodologies for rice (Hiei et al., 1994; Cheng et al., 1998), sugarcane (Enriquez-Obregon, 1998), maize (Ishida et al., 1996), sorghum (Zhao et al., 2000), and banana (May et al., 1995). Woolston and colleagues (1988) demonstrated the susceptibility of wheat to infection by *Agrobacterium tumefaciens* by injecting the nopaline strain C58 that harbored the wheat dwarf virus (WDV) genome directly into the basal portion of one- to four-day-old wheat seedlings. Virus replication and subsequent symptom expression suggested successful transmission of the WDV genome clone from *A. tumefaciens* to wheat cells. The first report on successful *Agrobacterium*-mediated transformation of wheat was by Cheng and colleagues (1997).

This protocol employs a standard binary vector system coupled with a nopaline strain of *A. tumefaciens,* ABI (Cheng et al., 1997).

The system merely incorporates the in vitro regeneration regime outlined for the microprojectile-mediated method and couples it with *A. tumefaciens* inoculation preparation and inoculation steps. We have successfully used *A. tumefaciens* strain C58C1 coupled with the Ti helper plasmid pMP90 (Koncz and Schell, 1986) for *Agrobacterium*-mediated transformation of wheat (Clemente, unpublished). In this protocol, derivatives of the pPZP family of binary vectors (Hajdukiewicz et al., 1994) are used that harbor the standard CaMV 35S *nptII* cassette as the selectable marker gene. C58C1 transconjugants carrying the binary vector of interest are cultured in yeast extract-phosphate (YEP) medium supplemented with the appropriate antibiotics. The culture is grown to an OD_{650} of 0.8 and harvested by centrifugation at 3,500 rpm. The bacterial pellet is suspended to an OD_{650} of 0.6 in 1/2 MS salts, full-strength vitamins supplemented with 0.5 $mg{\cdot}L^{-1}$ 2,4-D, 2.2 $mg{\cdot}L^{-1}$ picloram, 1 percent glucose, 200 μM acetosyringone, and 20 mM MES (pH 5.4). Immature wheat embryos are harvested and precultured on callus induction medium as described for the microprojectile-mediated transformation system. Following the four-day embryo preculture, the explants are immersed for 30 min in freshly prepared *Agrobacterium* inoculum. The embryos are subsequently cocultured for two days in the dark at 24°C on the inoculation medium solidified with 2 $g{\cdot}L^{-1}$ phytagel. The explants are transferred to selection following the two-day cocultivation step. The selection plates are composed of the callus induction medium supplemented with 10 $mg{\cdot}L^{-1}$ G418, 50 $mg{\cdot}L^{-1}$ ticarcillin, 50 $mg{\cdot}L^{-1}$ cefotaxime, and 50 $mg{\cdot}L^{-1}$ vancomycin. The last three antibiotics are used to counter select *A. tumefaciens* cells. The cultures are incubated in the dark at 24°C for two weeks. Following the two-week callus induction step callus tissue is broken into 2 mm pieces and subcultured to maturation medium (described earlier) supplemented with 25 $mg{\cdot}L^{-1}$ G418 and the antibiotic regime used to counter select *A. tumefaciens* cells. The cultures are incubated at 24°C under a 16-hour light regime. Following a two-week maturation period, differentiating segments are subcultured to elongation medium (described earlier) supplemented with 25 $mg{\cdot}L^{-1}$ G418 and the antibiotic regime. The cultures are incubated under the same environmental conditions described for the maturation step. Rapidly elongating shoots are subsequently moved to sundae cups containing the elongation medium without the antibiotic regime used to counter select *A. tumefaciens,* but maintaining the 25 mg·L^{-1} G418 selection pressure. The removal of the antibiotic regime greatly facilitates rooting while overgrowth of the bacterium at this stage is rarely observed. Images of the various steps in the *Agrobacterium*-mediated wheat transformation protocol are displayed in Figure 6.2. Although we have not performed a wide genotype screen with this protocol, in our hands it has been successful in a number of wheat genotypes, namely 'Bobwhite', 'Sakha 206', 'Yangmai 10', and 'PM97034', with transformation frequen-

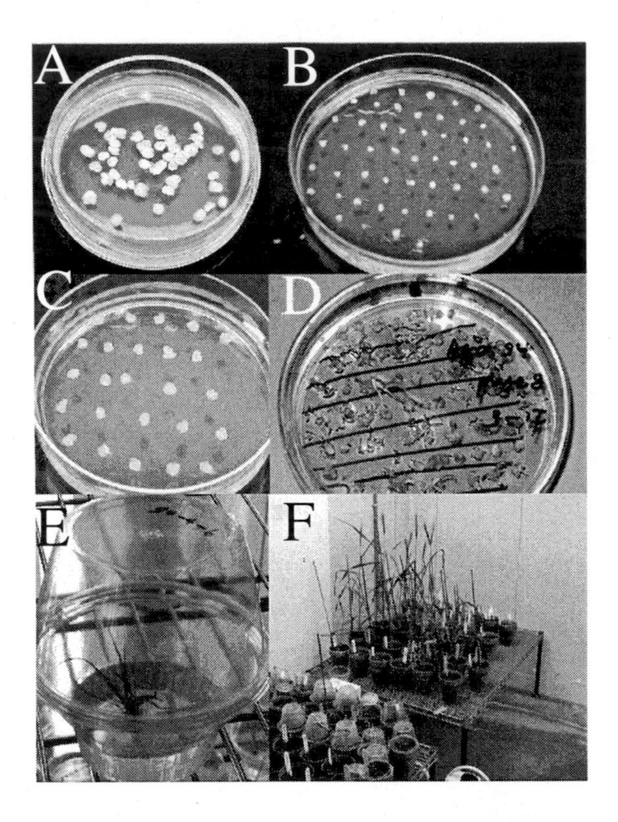

FIGURE 6.2. Steps in *Agrobacterium*-mediated transformation of wheat. Panel 1A: Inoculation step. 1B: Cocultivation step. 1C: Tissue following two-week callus induction step on G418. 1D: Developing shoots on maturation medium supplemented with G418. 1E: Elongating shoot in sundae cup on G418 supplemented medium. 1F: Transgenic wheat lines derived through *Agrobacterium*-mediated transformation developing in the growth chamber.

cies ranging from 1 percent to 3 percent on an independent Southern blot–positive plant in soil per explant basis.

The patterns of integration of foreign genes are often drastically different in *Agrobacterium*-mediated transformants from those patterns observed when using direct DNA delivery techniques such as microprojectile bombardment (Chilton, 1993). The relatively simple patterns of integration produced from the *Agrobacterium*-mediated transformation method in wheat are reflected from data derived from Southern blot analysis (Figure 6.3). The differences between direct DNA and *Agrobacterium*-mediated methodologies can be seen in reduction of integrated copies, fewer DNA rearrangements, and less fragmented copies in transformants derived from the latter approach. Hence, a higher-quality transformant is produced from *Agrobac-*

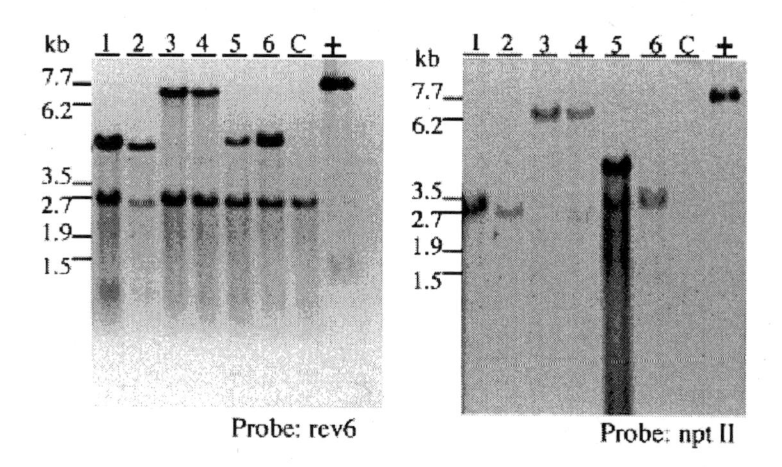

FIGURE 6.3. Southern blot analysis on *Agrobacterium*-mediated wheat transformants. Transformations were conducted using strain C58C1 carrying a binary vector that harbored within its T-DNA element an *nptII* cassette and *rev-6* (a maize mutant ADP-glucose pyrophosphorylase) cassette. Lanes 1 through 6 were loaded with 10 μg of total genomic DNA from transgenic lines derived from *Agrobacterium*-mediated transformation. Lane C indicates negative-control wild-type wheat DNA. Lane + refers to the positive control lane loaded with 50 pg of the binary plasmid used in the transformations. The blots were hybridized with either the *rev-6* gene (L) or the *nptII* gene (R). *Note:* The approximate 2.7 kb signal on the *rev-6* blot represents the endogenous *rev-6* wheat homolog. Lanes 1 and 2 and lanes 3 and 4 represent two plants derived from the same initial immature embryo explant (i.e., clones).

terium-mediated transformation technology verses direct DNA delivery techniques. Additional benefits that can be realized with *Agrobacterium*-mediated transformation protocols include the ability to transfer intact, large segments of DNA of up to 150 kb (Hamilton et al., 1996) and a reliable strategy to produce marker-free transformants (Komari et al., 1996; Daley et al., 1998; Xing et al., 2000).

SELECTION AND VISUAL MARKER GENES FOR TRANSFORMATION

Gene transfer protocols for plant species require the use of visual or selectable marker genes to allow for the efficient identification of the relatively small number of cells in which integration of the foreign DNA actually occurs. The choice of selectable marker gene and selective agent can significantly impact transformation efficiency. Wheat transformations initially relied upon the use of the *bar* gene (Thompson et al., 1987) coupled with the herbicidal agent bialaphos as the selection strategy. The *bar* gene, from *Streptomyces hygroscopicus,* codes for phosphinothricin acetyltransferase which detoxifies the herbicidal agent phosphinothricin. The bacterial antibiotic resistance gene neomycin phosphotransferase II *(nptII)* (Fraley et al., 1994) has been widely used in plant genetic engineering. As with other marker genes, multiple selective agents can be employed. For example, *nptII* can detoxify a number of aminoglycosides, including kanamycin, paromomycin, and geneticin (G418), while the *bar* gene can be used to inactivate the herbicidal activity of the two phosphinothricin derivatives bialaphos and glufosinate. Although each of the aminoglycosides and phosphinothricin agents possess similar modes of action (i.e., aminoglycosides inhibit translation and phosphinothricin derivatives target glutamine synthetase), levels used and effectiveness vary depending on the transformation system. For example, with respect to wheat, levels of G418 ranging from 10 $mg \cdot L^{-1}$ to 50 $mg \cdot L^{-1}$ are sufficient to impede development of nontransgenic tissue, while significantly higher levels of either paromomycin (50 to 100 $mg \cdot L^{-1}$) or kanamycin (>200 $mg \cdot L^{-1}$) are required to effectively control proliferation of nontransformed wheat tissue. Moreover, at the levels of either paromomycin or kanamycin required to impede growth of nontransformed wheat tissue, the transformation frequencies are significantly lower than those realized with coupling G418 with the *nptII* marker gene. A strong genotype variation is observed with regard to levels of G418 required for retarding growth of nontransgenic wheat tissue, and in some culture condition/genotype combinations paromomycin has been more effective than G418 in recovering transgenic wheat lines (Witrzens et al., 1998). Hence, when evaluating a selectable marker gene for wheat transformation, parameters will always need adjusting.

The antibiotic resistance gene hygromycin phosphotransferase *(hph)* (Waldron et al., 1985) has been used successfully as a selectable marker gene for wheat transformation (Ortiz et al., 1996). Transformation frequencies observed with a 25 $mg \cdot L^{-1}$ hygromycin selection pressure ranged from 4.7 percent to 6.0 percent, as compared to 2.3 percent to 2.9 percent when incorporating glufosinate within the system at 1 $mg \cdot L^{-1}$ (Ortiz et al., 1996). Contrary to this, Witrzens and colleagues (1998) found paromomycin, G418, and bialaphos superior to hygromycin for selection of wheat transgenics. This discrepancy may be explained by the different genotypes used and/or the slight variation in culture conditions.

Other herbicidal molecules have also been incorporated as selection agents in wheat transformation protocols, including glyphosate (Zhou et al., 1995) and cyanamide (Weeks et al., 2000). Glyphosate, the active ingredient of the herbicide Roundup, induces cell death by blocking the production of aromatic amino acids in plants by specifically impeding the activity of the plastid localized enzyme 5-enolpyruvyl shikimic acid-3-phosphate (EPSP) synthase (Steinrücken and Amrhein, 1980). Glyphosate tolerance has been achieved via two approaches, a detoxification strategy and the identification of the glyphosate-tolerant EPSP synthase gene. The gene *gox,* derived from *Achromobacter* sp. strain LBAA, codes for a glyphosate oxidoreductase which in turn metabolizes glyphosate to the nonherbicidal molecule aminomethylphosphonic acid (AMPA) (Barry et al., 1992). A glyphosate-tolerant EPSP synthase gene designated *cp4* was isolated from *Agrobacterium* strain CP4. A glyphosate-tolerant phenotype is imparted to plant cells by targeting either one of these genes to the plastids of higher plants (Barry et al., 1992). Zhou and colleagues (1995) effectively incorporated glyphosate as a selection agent in the microprojectile-mediated wheat transformation system. A reliable and tight selection for transgenic wheat recovery was obtained by utilizing both glyphosate marker genes, *cp4* and *gox,* coupled with a selection pressure of 2 mM glyphosate for nine to 12 weeks during the callus induction phase and systematically reducing the selection pressure during the maturation and elongation steps (Zhou et al., 1995).

Calcium cyanamide was introduced as an industrial nitrogen fertilizer in 1905. Unlike most nitrogen fertilizers, cyanamide first must be metabolized to become an available nitrogen source for plants. In the presence of moisture, Ca-cyanamide is converted to hydrogen cyanamide which possesses herbicidal activity. Cyanamide hydratase (*Cah*) gene converts cyanamide to urea which, in turn, plant cells metabolize to NH^+, effectively circumventing the herbicidal activity of the compound. The *Cah* gene was isolated from the soil organism *Myrothecium verrucaria* (Maier-Greiner et al., 1991). Weeks and colleagues (2000) utilized the *Cah* gene as a selectable marker to recover stable transgenic lines of wheat via microprojectile-mediated gene transfer by implementing a selection pressure of 37 $mg \cdot L^{-1}$ throughout the in vitro regeneration process.

The *Escherichia coli manA* (phosphomannose isomerase) gene (*pmi*) can be utilized as a positive selection system for the identification of transgenic plant cells using mannose-supplemented culture medium (Joersbo et al., 1999). Mannose is phosphorylated to mannose-6-phosphate, which is not metabolized further in plant cells, and accumulation of this metabolite is lethal (Goldsworthy and Street, 1965). The *pmi* gene converts mannose-6-phosphate to fructose-6-phosphate, which is a nontoxic metabolite. Wright and colleagues (2001) implemented the *pmi* positive selection strategy in the wheat microprojectile-mediated gene transfer protocol by permitting callus induction to proceed in the absence of mannose and subsequently subjecting the embryogenic tissue to a carbohydrate regime composed of 10 $g \cdot L^{-1}$ mannose:5 $g \cdot L^{-1}$ sucrose and 5 $g \cdot L^{-1}$ mannose:10 $g \cdot L^{-1}$ sucrose during the maturation and elongation steps, respectively, with the final elongation step using mannose as the sole carbon source at 15 $g \cdot L^{-1}$.

Visual marker genes are a valuable resource for use in optimization of transformation parameters. The *E. coli* β-glucuronidase *(gus)* gene was extremely beneficial for this purpose in the early development of wheat transformation protocols (Vasil et al., 1992; Weeks et al., 1993; Becker et al., 1994). Green fluorescent protein (GFP) and luciferase enzyme are two nondestructive visual marker genes that have been used in wheat transformation protocols as a strategy to rapidly monitor transformation events. A spectrally modified synthetic version of GFP from *Aequorea victoria* was tested as a way to detect transformed cells and tissues in wheat (Pang et al., 1996). Multicellular calli clusters and shoots emitting green fluorescence could be observed as early as 14 days and 21 days, respectively, after bombardment (Jordan, 2000). The fire fly luciferase gene has also been used in stable transformation of wheat (Lonsdale et al., 1998). Both GFP and luciferase permit real-time *in planta* analysis of transgene activity. However, unlike GFP, luciferase requires a substrate to be sprayed onto the tissue or taken up through the root system.

FUTURE NEEDS FOR WHEAT TRANSFORMATION

The embryogenic regeneration regime initiated from immature wheat embryos has proven to be a reliable target for the genetic engineering of wheat via both *Agrobacterium*-mediated and microprojectile bombardment methodologies. However, as stated earlier, the quality of the stock plants used to obtain the immature embryos will have a significant impact on the in vitro regeneration response from the explant. Moreover, maintenance of the stock plants adds significant labor and cost to the transformation system. Hence, it would be extremely desirable to have an alternative explant in which to initiate target tissue for genetic engineering. To this end, one attractive choice would be mature seed-derived embryogenic tissue (Delporte et al., 2001; Mendoza and Kaeppler, 2002).

Another limitation in wheat genetic engineering, as in all plant transformation systems, is the lack of availability of a large repository of useful promoter elements. A number of constitutive promoters are repeatedly used for wheat transformation studies. When introducing multiple foreign genes into a plant cell it is desirable to link each of the respective genes to a different promoter element in order to reduce the risk of manifesting gene silencing (Matzke and Matzke, 1995). The majority of regulator elements that have been used to control transgene expression in wheat include the constitutive promoter region of the maize alcohol dehydrogenase gene *(adh1)*, which was first used to monitor transient expression of GUS in wheat cells (Wang et al., 1988). In addition, the CaMV 35S promoter (Kay et al., 1987) derived from cauliflower mosaic virus was demonstrated to effectively express transgenes both stably and transiently in transgenic wheat cells (Chibbar et al., 1991; Vasil et al., 1991). The maize Ubiquitin1 promoter (Christensen and Quail, 1996) and the rice actin promoter (McElroy et al., 1990) also are commonly used in wheat transformation studies.

A few additional viral-derived promoter elements have been shown to direct transgene expression in monocotyledonous plants. Elements within an approximate 1.4 kb unit of the sugarcane bacilliform virus (ScBV) genome were shown to direct expression of visual marker gene *gus* in most cell types of oat and maize (Tzafrir et al., 1998; Schenk et al., 1999). Schenk and colleagues (2001) recently identified two promoter elements from banana streak badnavirus (BSV), designated My and Cv which were isolated from BSV strains BSV-Mys and BSV-Cav, respectively. The My and Cv promoter elements are approximately 2.3 kb and 1.5 kb in size, respectively. These regulatory elements directed high levels of expression of marker genes in both sugercane and banana. A synthetic promoter, designated Emu, has also been shown to direct high levels of transgene expression in wheat and rice (Chamberlain et al., 1994; Li et al., 1997).

Although strong constitutive promoters are of great value for plant genetic engineering, in many cases, tissue-specific expression of a desired trait is preferred. Lamacchia and colleagues (2001) characterized the endosperm-specific promoter element of the wheat high-molecular-weight glutenin *Glu-1D-1* gene. Wheat transformants that harbored a chimeric *gus* gene under the control of this approximate 1.3 kb 5' untranslated region (UTR) of the *Glu-1D-1* gene displayed specific expression in the starchy endosperm in a temporal fashion only during the middevelopment stage (Lamacchia et al., 2001). Anderson and colleagues (1998) described a set of promoter sequences from the high-molecular-weight glutenin locus *Glu-B1-1*. In an early study it was shown that a hybrid vector composed of the promoter region from the *Dy10* gene coupled with the 5'UTR and first 145 codons of the *Dy10* subunit fused to the 3' open reading frame (ORF) of *Dx5* coupled with the endogenous 3' UTR of *Dx5* specifically expressed the hybrid *Dy10-Dx5* high-molecular-weight glutenin subunit in a seed-specific fashion (Blechl and Anderson, 1996), revealing that the promoter elements of these subunits can be used in wheat transformation studies to specifically drive expression of desired transgenes in an endosperm-specific fashion in wheat.

Targeting a specific trait such as cold acclimation can have a profound impact on the ability to broaden the acreage of a crop plant. Due to the unpredictable potential negative impact on crop production that may result from constitutive expression of a normally stress-induced transgene, a valuable addition to a promoter repository would be the availability of inducible promoter elements. With respect to cold acclimation, the use of a low temperature inducible promoter to regulate expression of transgene(s) associated with cold tolerance acclimation in plants would be prudent. Ouellet and colleagues (1998) demonstrated the cold induction of the *wcs120* promoter. The *wcs120* gene was previously shown to be a cold-specific inducible gene in wheat (Vazquez-Tello et al., 1998). The enormous amount of research emphasis in the area of plant genomics promises to expand our understanding of the major crop plants including wheat (Lagudah et al., 2001). These efforts undoubtedly will lead to greater numbers of useful promoter elements and novel genes that will complement wheat genetic engineering studies which ultimately can be exploited to enhance the genetics of wheat to combat both biotic and abiotic stresses.

Transgenes with Potential to Combat Biotic and Abiotic Stresses

Biotic and abiotic stresses annually plague crop plants and drastically limit production by curtailing the yield potential of the respective plant species. Wheat, similar to all major crop plants, is negatively impacted by a number of stresses. The average reduction in wheat production over a 33-year period attributed to pathogen attack was estimated at 11.5 percent, while production was reduced by winter survival alone over the same period by 7 percent (Patterson et al., 1990). In addition to the direct impact on yield, the various stresses also may reduce grain quality. Current strategies of integrated pest management coupled with deployment of elite wheat genotypes developed through conventional breeding practices have significantly contributed to limiting the wheat production losses associated with pathogen attack and various environmental stresses. The world population is estimated to grow by more than 8 billion people within the next 30 years. With the advent of the tools of biotechnology leading to the identification and functionality of an array of plant genes involved in protecting plants from disease and adverse environmental impacts, wheat-breeding efforts will be drastically augmented via plant genetic engineering methodology in the future, providing additional avenues to combat yield loss by enhancing tolerance to stresses.

POTENTIAL TRANSGENES FOR RESISTANCE TO PATHOGENS

Viral Resistance Strategies

The development of plant gene transfer systems has allowed for the introgression of alien genes into plant genomes, thus providing a mechanism for broadening the genetic resources available to plant breeders. This technology has enabled plant scientists to introduce novel disease control strategies into plant species, targeting a variety of pathogens. The use of a transgene to control plant disease was demonstrated in the mid-1980s. The introduction of the coat protein gene of tobacco mosaic virus (TMV) into tobacco led to a delay in viral pathogenesis upon challenging the tobacco transformants with TMV virions (Powell et al., 1986). This discovery led to a plethora of investigations targeting virus protection by introducing various components of viral genomes into plants (Prins and Goldbach, 1996). This technology, referred to as pathogen-derived resistance (PDR), has been successful against a multitude of plant pathogenic viruses. PDR has been demonstrated in wheat. Karunaratne and colleagues (1996) introduced the coat protein gene from barley yellow mosaic virus (BaYMV) into wheat and showed expression of the capsid protein of the virus in the subsequent generation. BaYMV does not infect wheat, and PDR generally is effective only against the specific virus in which the gene was derived or closely related strains thereof. Nonetheless, the authors justified the introduction of the capsid protein of BaYMV as a potential strategy to impart resistance toward wheat spindle streak virus and wheat yellow mosaic virus because of their significant homology of the respective capsid proteins with BaYMV (Karunaratne et al., 1996). However, protection toward the wheat viral agents has not been demonstrated with this approach to date. Nonetheless, this work was the first demonstration of expression of a phytopathogenic virus gene in wheat.

Wheat streak mosaic virus (WSMV) causes significant losses in wheat throughout the production areas of the world. Sivamani, Brey, and colleagues (2000) introduced the replicase gene of WSMV, *NIb,* into the wheat genome under the control of the maize *Ubiquitin1* promoter coupled with its first intron. Transgenic wheat lines carrying the *NIb* cassette displayed varying degrees of resistance when challenged with the WSMV, although none of the lines were completely immune to infection (Sivamani, Brey, et al., 2000).

As mentioned earlier, the PDR strategy is limited to the specific virus from which the gene was derived or a closely related strain of that virus. Therefore a transgene(s) that potentially imparts a more broad-spectrum virus resistance would be more appealing for wheat breeding programs. Most plant viruses will undergo a double-stranded RNA (dsRNA) intermediate during their replication cycle. The dsRNA can be specifically targeted for degradation and thus block viral replication and halt disease progress. Sano and colleagues (1997) evaluated this strategy to control potato spindle tuber viroid (a viruslike, self-replicating RNA) by expressing the yeast dsRNA specific ribonuclease gene, *pac1,* in potatoes. Transgenic potato lines expressing the *pac1* gene displayed resistance toward the infectious RNA, demonstrating the feasibility of this approach (Sano et al., 1997). Zhang and colleagues (2001) introduced into wheat the *E. coli* mutant *rnc70* gene, which encodes an RNase III, a dsRNA-specific ribonuclease. Unlike the wild-type *rnc70* gene product, the mutant retains binding capacity but cannot cleave the substrate. Previous work in tobacco demonstrated that both the wild type and mutant *rnc70* genes were capable of imparting a virus resistance phenotype *in planta* (Langenberg et al., 1997). Expression of mutant *rnc70* gene in wheat, under the control of the maize *Ubiquitin1* promoter, was detected via Western blot analysis and had no obvious negative phenotype in the wheat lines generated. Upon challenging the transgenic wheat lines expressing the mutant *rnc70* gene with barley stripe mosaic virus (BSMV) a significant reduction in BSMV virus accumulation was observed along with a concomitant lack of typical virus symptom development (Zhang et al., 2001).

Fungal Protection Approaches

Plant genetic engineering has provided for novel stress resistance traits including control of lepidopterous and coleopterous insects via expression of endotoxins derived from *Bacillus thuringiensis* (Bt) in transgenic plants (Vaeck et al., 1987; Perlak et al., 1993) and immunity toward numerous viral agents in various vegetables, potatoes, and papaya via PDR (Prins and Goldbach, 1996). The products currently on the market that carry these respective resistance traits have proven to be very effective, resulting in significant reduction of pesticide applications without compromising productivity, food quality, or safety. However, with respect to fungal pathogens, clearly the most problematic of all the plant pathogens, no resistance trait to date introduced via genetic engineering is currently on the marketplace, indicating that developing durable field resistance toward a fungal agent by genetic engineering approaches may require a multiprong strategy (i.e., more than a single gene). To this end, a number of studies have been conducted that reflect the ability of single transgene approach to impart a partial resistance phenotype in a number of host-fungal pathogen interactions.

Plants respond to fungal attack by inducing a myriad of defense genes. These include enhancing structural barriers to block pathogen ingress and impeding hyphal growth through production of phytoalexins or pathogenesis-related (PR) proteins (Dixon and Harrison, 1990). The latter group include the hydrolytic enzymes β-1,3-glucanases and chitinases, along with ribosome-inactivating proteins (RIP). Bliffeld and colleagues (1999) designed a set of plasmids for wheat transformation, each harboring dual monocot expression cassettes consisting of either β-1,3-glucanase and class II chitinase or β-1,3-glucanase and RIP. The expression of the chitinase gene within the respective plasmids was regulated by the maize *Ubiquitin1* promoter while expression of both the RIP and β-1,3-glucanase genes were controlled by the rice actin 1 promoter. However, only chitinase expression was detected; coexpression of either the RIP or β-1,3-glucanase was not observed in the transgenic wheat lines generated. In the lines in which chitinase was overexpressed an enhanced resistance phenotype was observed when the plants were challenged with the fungal pathogen *Erysiphe graminis,* the causal agent of powdery mildew (Bliffeld et al., 1999). However, the level of resistance observed in these lines is not sufficient to impart a desirable level of field control toward the pathogen. In a related strategy Oldach and colleagues (2001) designed a set of plasmids for monocot expression that harbored either an antifungal protein (afp) derived from *Aspergillus giganteus* (Wnendt et al., 1994), a class II chitinase, or RIP. All of the potential antifungal transgenes were under the control of the maize *Ubitquitin 1* promoter. Coexpression of the transgenes was verified by Northern blot analysis over two subsequent generations. Upon challenging the respective transgenic wheat lines with either *E. graminis* or the leaf rust pathogen, *Puccinia recondita* f. sp. *tritici,* significant reduction in colonization was observed in the chitinase and afp expressing lines, while no potential resistance phenotype was displayed with the RIP transgenics (Oldach et al., 2001). The reduction in colonization observed in the chitinase and afp wheat lines, however, was inoculum dose dependent (i.e., under higher inoculum densities the reduction in colonization was not observed). Nonetheless, the data are rather promising and perhaps pyramiding these two transgenes will provide an additive effect as was previously observed in transgenic tobacco lines (Jach et al., 1995).

Modulating expression of thaumatin-like PR proteins (TLP) has been shown to enhance fungal resistance in transgenic plants (Datta et al., 1999). Chen and colleagues (1999) evaluated a rice TLP along with rice chitinase gene as a mechanism to enhance resistance toward fusarium head blight (FHB) in wheat. FHB is a devastating disease of both barley and wheat. The primary causal agent of FHB is *Fusarium graminearum.* In addition to direct impact on yield, the fungus also produces a number of mycotoxins that severely reduce grain quality. Wheat transgenics harboring both the rice TLP and chitinase gene were developed, but expression of only the former was observed (Chen et al., 1999). Upon challenging developing wheat spikes with conidia of *Fusarium* and monitoring for infection spread within the head, the TLP transgenics displayed a significantly lower number of infected spikelets (Chen et al., 1999). Thus, an enhanced resistance phenotype toward scab due to an impediment of pathogen spread within the spike was a result of the TLP transgene. Clearly the overriding objective of the work was to evalute TLP for scab resistance in wheat. However, the authors also mapped the transgenic locus in one of the trangenic lines expressing the TLP transgene via fluorescent in situ hybridization (FISH) to chromosome 6A (Chen et al., 1999). The authors correctly point out that such a map position will reduce the probability of outcrossing to the related weed *Aegilops cylindrica,* but it brings up a more important point that knowledge of the linkage relationship of a desirable transgene is valuable information when incorporating the new germplasm into a breeding program.

Production of antimicrobial compounds called phytoalexins is another strategy plants employ to combat pathogen attack. Stilbene synthase carries out the enzymatic step to produce *trans*-resveratrol, a phytoalexin, utilizing substrates found in most plant cells. Expression of the grape stilbene synthase gene *(vst1)* (Hain et al., 1993) in rice resulted in reduced lesion area on leaves inoculated with the rice blast fungus *Pyricularia oryzae* (Stark-Lorenzen et al., 1997). Leckband and Lörz (1998) extended this approach to barley and wheat. Plant expression cassettes carrying the *vst1* gene under the control of its endogenous promoter element were transformed into wheat and barley. The regulation of the *vst1* promoter in barley and wheat mirrored that observed in grapevine; moreover, in barley lines carrying the *vst1* cassette, stilbene synthase activity was demonstrated and data from a detached leaf assay suggested that accumulation of *trans*-resveratrol may enhance resistance in a monocot host-fungal interaction (Leckband and Lörz, 1998).

A novel source of potential antifungal transgenes exploited by Clausen and colleagues (2000) provided enhanced resistance toward stinking bunt disease in wheat. Viruslike particles associated with certain strains of *Ustilago maydis* encode for antifungal proteins that have been termed killing proteins (KPs) (Hankin and Puhalla, 1971). The KPs are excreted from the host cell, which itself is immune to KPs, but are able to induce lethality in other *U. maydis* strains. The KPs excreted from the viruslike infected *U. maydis* cells have antifungal properties against many of the so-called smut fungi including the causal agent of loose smut of wheat, *U. tritici* (Koltin and Day,

1975). A KP gene designated *KP4* was fused to the maize *ubiquitin1* promoter and delivered into wheat via microprojectile-mediated transformation. Wheat seed expressing the *KP4* gene inhibited growth of known sensitive *U. maydis* strains in plate assays, while hyphal growth of a nonsensitive *KP4* producing *U. maydis* strain was not impacted (Clausen et al., 2000). These data suggest that the specificity of the protein produced *in planta* is equivalent to that excreted from the virally infected *U. maydis* cells. Moreover, a wheat line challenged with the stinking bunt pathogen, *Tilletia tritici,* displayed a significant reduction of diseased spikes (Clausen et al., 2000).

Genetic Engineering for Control of Insect Pests

The Bt technology engineered into maize, cotton, and potatoes has been extremely successful in controlling the respective target pests. Importantly, significant reductions in insecticide sprays have been realized as a result of this single gene trait, which in turn has augmented producer profitability while reducing the potential negative environmental impacts associated with agrochemical applications. A diversity of insect pests infest crop plants on an annual basis which directly and indirectly impact yield through feeding and vectoring of plant pathogens. The biotechnology tools have provided and will continue to provide additional avenues to enhance the genetics of crop plants that can be incorporated into integrated pest management regimes to combat these pests. Hilder and Boulter (1999) have comprehensively reviewed the various strategies that have been tested to enhance insect pest resistance in plants. Therefore, the topic will not be overviewed herein.

Wheat production is not immune to insect attack. Two economically important insect pests within the Great Plains are the Russian wheat aphid *(Diuraphis noxia)* and greenbug *(Schizaphis graminum)* (Stoger et al., 1999). Carbohydrate binding proteins, referred to as lectins, possess multiple binding sites and agglutinate animal cells. In some cases a lectin may bind two sugars with equal capacity, but in most cases one sugar is more tightly bound while other sugars have loose affinity. When fed to various insect pests, purified lectins can influence proper development and/or survival. Hilder and colleagues (1995) demonstrated that expression of a mannose-specific lectin from snowdrops *(Galanthus nivalis)* in tobacco imparted resistance toward peach potato aphid *(Myzus persicae).* Stoger and colleagues (1999) evaluated transgenic wheat lines expressing the snowdrop lectin, designated *GNA,* for tolerance to grain aphid *(Sitobion avenae)* predation. The rice sucrose synthase promoter (Wang et al., 1992) or the maize *ubiquitin1* promoter were used to regulate *GNA* expression in transgenic wheat. Wheat lines expressing a range of *GNA* were subsequently challenged with neonate aphids. Data were collected on aphid survival and fecundity. The results reveal that *in planta* expression obtained in wheat of *GNA* did not alter aphid survival but had a negative impact on fecundity (Stoger et al., 1999). These data suggest that implementing the *GNA* gene in a wheat-breeding program may complement existent aphid management programs by impacting population growth of the insect.

An alternative strategy toward enhancing insect resistance in plants is to block essential steps in the organism's metabolism. To this end, obstruction of protein digestion via expression of various protease inhibitors has been the focus of extensive research (reviewed by Hilder et al., 1992). A barley trypsin inhibitor, designated *CMe,* accumulates in the endosperm of the grain and has been shown to possess activity toward trypsinlike proteases found in the gut of the fall armyworm *(Spodoptera frugiperda)* (Alfonso et al., 1997). Altpeter and colleagues (1999) characterized transgenic wheat lines expressing the *CMe* gene under the control of the maize *Ubiquitin1* promoter in feeding studies with both *Melanoplus sanguinipes* (grasshopper) and *Sitotroga cerealella* (angoumois grain moth). The former is a significant pest of grains, while the latter is a postharvest pest of stored bread wheat. The data revealed a significant reduction in grain moth larvae survival reared on homozygous seed derived from wheat lines expressing the *CMe* gene. However, expression of the *CMe* transgene in wheat showed no effect on survival or development of grasshoppers (Altpeter et al., 1999). This work demonstrated the potential usefulness of this approach in controlling a postharvest pest of wheat. However, as the authors point out in this study, protease inhibitors have been associated with reduction in nutritional quality. Therefore, this and other novel traits derived through biotechnology need to be demonstrated to be substantially equivalent or superior to their conventional counterpart prior to entry on the marketplace.

POTENTIAL TRANSGENES FOR ENHANCED TOLERANCE TOWARD ABIOTIC STRESS

Crop productivity is often limited by various environmental factors such as drought, salinity, and temperature fluctuations. Plants respond to these environmental factors through induction of a series of stress-related genes. These gene products are primarily categorized into two broad classes, those that have a direct impact on the particular stress and those involved in modulating signal transduction as a consequence of the stress (Shinozaki and Yamaguchi-Shinozaki, 1997). Conventional plant breeding approaches have had limited success in developing resistance toward abiotic stresses. This is primarily due to the complexity underlying the genetic control, limited stress-tolerance phenotypic variation within the respective germplasm, and insufficient rapid techniques for identifying an abiotic resistance phenotype *in planta* (Cushman and Bohnert, 2000).

Drought is clearly the most significant abiotic stress that impacts crop productivity. Plant breeders will monitor for specific parameters to identify germplasm with potential enhanced drought stress. These include accumulation of biomass and water use efficiency (Blum, 1993). Sivamani, Bahieldin, and colleagues (2000) developed transgenic wheat lines that expressed the barley *HVA1* gene and evaluated the transgenic lines performance under water stress conditions. HVA1 is a late embryogenesis abundant (LEA) protein. LEA proteins accumulate during the late maturation and desiccation phases of embryogenesis and are induced in vegetative tissues upon exposure to various environmental stresses including drought, salinity, and cold (Ingram and Bartels, 1996). The *HVA1* gene, a plant stress inducible gene that falls in the direct impact class of stress response genes, was previously shown to enhance tolerance toward both salinity and water stress in transgenic rice (Xu et al., 1996). Wheat lines that constitutively expressed the *HVA1* gene developed phenotypically normal under well-watered conditions. However, under the water stress treatment a subset of the transformants displayed an increase in a number of parameters over the controls, including root fresh and dry weights, shoot fresh and dry weights, and water use efficiency (Sivamani, Bahieldin, et al., 2000). These data are in agreement with previous work suggesting the role of the LEA proteins, specifically HVA1, as a direct protectant against water stress.

The work by Sivamani, Bahieldin, and colleagues (2000) demonstrated the positive impact a single gene could have toward enhancing abiotic stress in wheat. However, it is unlikely that expression of a single stress gene, with a direct effect mode of action, will be able to impart field tolerance for drought or other abiotic stresses. With the current state of technology, coordinating expression of multiple transgenes is rather difficult, hence imparting a multigene trait via genetic engineering will be cumbersome. One potential strategy that may circumvent the need to introduce multiple transgenes to provide for a multigene trait is to deliver a transcription factor that plays a role in the signal transduction pathway activated upon the plant's response to abiotic stress. Kasuga and colleagues (1999) demonstrated the feasibility of this approach by expressing the transcription factor *DREBA1A* under the control of the stress inducible promoter element *rd29A* in *Arabidopsis*. *DREBA1A* binds to the regulatory element designated DRE, dehydration response element, which is found in the *rd29* promoter and other regulatory elements controlling expression of genes induced in response to various abiotic stresses (Yamaguchi-Shinozaki and Shinozaki, 1994). The binding of the transcription factor, *DREBA1*, to the DRE element will activate gene expression. Therefore expression of a single transgene will coordinately activate the endogenous stress response pathway. When *DREB1A* was constitutively expressed in *Arabidopsis* under the control of the CaMV 35S promoter, the transformants suffered growth abnormalities (Kasuga et al., 1999). Replacing the CaMV 35S with the *rd29A* promoter, which harbors the *cis*-element DRE, resulted in low-level constitutive *DREBA1A* expression, with strong expression upon exposing the plants to a stress signal, thereby negating the growth abnormality phenotype associated with strong constitutive expression (Kasuga et al., 1999). Importantly, the *rd29-DREBA1A* lines displayed dramatic enhancement of tolerance toward freezing, drought, and salinity (Kasuga et al., 1999).

CONCLUDING REMARKS

Acreage of wheat sown ranks first among the cereals and, based on production, holds a leading position in human nutrition in the world. Therefore, a sustained and increased wheat production is vital to ensure a plentiful food supply. Wheat breeders will increasingly exploit genetic engineering tools to enhance diversity within wheat germplasm. The ongoing research programs in the areas of genomics and functional genomics will augment such efforts. This in turn will greatly expand our knowledge on the biology underlying a plant's response to the environment. This information will enable scientists to design effective strategies to combat both biotic and abiotic stresses, which will ultimately expedite the release of elite wheat cultivars.

REFERENCES

Alfonso, J., F. Ortego, R. Sanchez-Monge, G. Garcia-Casado, I. Pujol, P. Castanera, and G. Salcedo (1997). Wheat and barley inhibitors active toward α-amylase and trypsin-like activities from *Spodoptera frugiperda*. *Journal of Chemical Ecology* 23:1729-1741.

Altpeter, F., I. Diaz, H. McAuslane, K. Gaddour, P. Carbonero, and I.K. Vasil (1999). Increased insect resistance in transgenic wheat stably expressing trypsin inhibitor CMe. *Molecular Breeding* 5:53-63.

Altpeter, F., V. Vasil, V. Srivastava, E. Stöger, and I.K. Vasil (1996). Accelerated production of transgenic wheat (*Triticum aestivum* L.) plants. *Plant Cell Reports* 16:12-17.

Anderson, O.D., F.A. Abraham-Pierce, and A. Tam (1998). Conservation in wheat high-molecular-weight glutenin gene promoter sequences: Comparisons among loci and among alleles of the GLU-B1 locus. *Theoretical and Applied Genetics* 96:568-576.

Bajaj, Y.P.S. (1990). Biotechnology in wheat breeding. In *Biotechnology in Agriculture and Forestry Wheat,* Volume 3 (pp. 3-23), Ed. Baja, Y.P.S. Berlin: Springer-Verlag.

Barro, F., M.E. Cannell, P.A. Lazzeri, and P. Barcelo (1998). The influence of auxins on transformation of wheat and tritordeum and analysis of transgene integration patterns in transformants. *Theoretical and Applied Genetics* 97:684-695.

Barry, G., G. Kishore, S. Padgette, M. Taylor, K. Kolacz, M. Weldon, D. Re, D. Eichholtz, K. Fincher, and L. Hallas (1992). Inhibitors of amino acid biosynthesis: Strategies for imparting glyphosate tolerance to crop plants. In *Biosynthesis and Molecular Regulation of Amino Acids in Plants* (pp. 139-145), Eds. Singh, B.K., H.E. Flores, and J.C. Shannon. Rockville, MD: American Society of Plant Physiologists.

Bechtold, N., J. Ellis, and G. Pelletier (1993). *In planta Agrobacterium* mediated gene transfer by infiltration of adult *Arabidopsis thaliana* plants. *Comptes Rendue Academy of Science Paris, Sciences de la vie/ Life Science* 316:1194-1199.

Becker, D., R. Brettschneider, and H. Lörz (1994). Fertile transgenic wheat from microprojectile bombardment of scutellar tissue. *The Plant Journal* 5:299-307.

Blechl, A.E. and O.D. Anderson (1996). Expression of a novel high-molecular-weight glutenin subunit gene in transgenic wheat. *Nature Biotechnology* 14: 875-879.

Bliffeld, M., J. Mundy, I. Potrykus, and J. Fütterer (1999). Genetic engineering of wheat for increased resistance to powdery mildew disease. *Theoretical and Applied Genetics* 98:1079-1086.

Blum, A. (1993). Selection for sustained production in water-deficit environments. In *International Crop Science I* (pp. 343-347), Ed. Buxton, D. III. Madison, WI: Crop Science Society of America.

Campbell, B.T., P.S. Baenziger, A. Mitra, S. Sato, and T. Clemente (2000). Inheritance of multiple transgenes in wheat. *Crop Science* 40:1133-1141.

Chamberlain, D.A., R.I.S. Brettell, D.I. Last, B. Witrzens, D. McElroy, R. Dolferus, and E.S. Dennis (1994). The use of the Emu promoter with antibiotic and herbicide resistance genes for the selection of transgenic wheat callus and rice plants. *Australian Journal of Plant Physiology* 21:95-112.

Chen, W.P., P.D. Chen, D.J. Liu, R. Kynast, B. Friebe, R. Velazhahan, S. Muthukrishnan, and B.S. Gill (1999). Development of wheat scab symptoms is delayed in transgenic wheat plants that constitutively express a rice thaumatin-like protein gene. *Theoretical and Applied Genetics* 99:755-760.

Cheng, M., J.E. Fry, S. Pang, H. Zhou, C.M. Hironaka, D.R. Duncan, T.W. Conner, and Y. Wan (1997). Genetic transformation of wheat mediated by *Agrobacterium tumefaciens*. *Plant Physiology* 115:971-980.

Cheng, X., R. Sardana, H. Kaplan, and I. Altosaar (1998). *Agrobacterium*-transformed rice plants expressing synthetic cryIA(b) and cryLA(c) genes are highly toxic to striped stem borer and yellow stem borer. *Proceedings of the National Academy of Sciences USA* 95:2767-2772.

Chibbar, R.N., K.K. Kartha, N. Leung, J. Qureshi, and K. Caswell (1991). Transient expression of marker genes in immature zygotic embryos of spring wheat through microprojectile bombardment. *Genome* 34:453-460.

Chilton, M.-D. (1993). *Agrobacterium* gene transfer: Progress on a "poor man's vector" for maize. *Proceedings of the National Academy of Sciences USA* 90:3119-3120.

Christensen, A.H. and P.H. Quail (1996). Ubiquitin promoter-based vectors for high-level expression of selectable and/or screenable marker genes in monocotyledonous plants. *Transgenic Research* 5:213-218.

Clausen, M., R. Kräuter, G. Schachermayr, I. Potrykus, and C. Sautter (2000). Antifungal activity of a virally encoded gene in transgenic wheat. *Nature Biotechnology* 18:446-449.

Cushman, J.C. and H. J. Bohnert (2000). Genomic approaches to plant stress tolerance. *Current Opinion in Plant Biology* 3:117-124.

Daley, M., V.C. Knauf, K.R. Summerfelt, and J.C. Turner (1998). Co-transformation with one *Agrobacterium tumefaciens* strain containing two binary plasmids as a method for producing marker-free transgenic plants. *Plant Cell Reports* 17:489-496.

Datta, K., R. Velazhahan, N. Oliva, I. Ona, T. Mew, G.S. Khush, S. Muthukrishnan, and S.K. Datta (1999). Over expression of the cloned rice thaumatin-like protein

(PR-5) gene in transgenic rice plants enhances environmental friendly resistance to *Rhizoctonia solani* causing sheath blight disease. *Theoretical and Applied Genetics* 98:1138-1145.

DeBlock, D., D. Debrouwer, and T. Moens (1997). The development of a nuclear sterility system in wheat: Expression of the *barnase* gene under the control of tapetum specific promoters. *Theoretical and Applied Genetics* 95:125-131.

Delporte, F., O. Mostade, and J.M. Jacquemin (2001). Plant regeneration through callus initiation from thin mature embryo fragments of wheat. *Plant Cell Tissue and Organ Culture* 67:73-80.

Desfeux, C., S.J. Clough, and A.F. Bent (2000). Female reproductive tissues are the primary target of *Agrobacterium*-mediated transformation by *Arabidopsis* floral-dip method. *Plant Physiology* 123:895-904.

Dixon, R.A. and M. Harrison (1990). Activation, structure and organization of genes involved in microbial defense in plants. *Advances in Genetics* 28:165-234.

Enriquez-Obregon, G.A., R.I. Vazquez-Padron, D.L. Prieto-Samsonov, G.A. Dela Riva, and G. Selman-Housein (1998). Herbicide-resistant sugarcane (*Saccharum officinarum* L.) plants by *Agrobacterium*-mediated transformation. *Planta* 206:20-27.

Fennell, S., N. Bohorova, M. van Ginkel, J. Crossa, and D. Hoisington (1996). Plant regeneration from immature embryos of 48 elite CIMMYT bread wheats. *Theoretical and Applied Genetics* 92:163-169.

Fraley, R.T., R.B. Horsch, and S.G. Rodgers (1994). Chimeric genes for transforming plant cells using viral promoters. U.S. Patent No. 5,352,605.

Goldsworthy, A. and H.E. Street (1965). The carbohydrate nutrition of tomato roots: VIII. The mechanism of the inhibition by D-mannose of the respiration of excised roots. *Annals of Botany* 29:45-58.

Gosch-Wackerle, G., L. Avivi, and E. Galun (1979). Induction, culture and differentiation of callus from immature rachises, seeds and embryos of *Triticum*. *Zeitschrift für Pflanzenphysiologie* 91:267-278.

Hain, R., H.J. Reif, E. Krause, R. Langebartels, H. Kindl, B. Vornam, W. Wiese, E. Schmelzer, P.H. Schreier, and K. Stenzel (1993). Disease resistance results from foreign phytoalexin expression in a novel plant. *Nature* 361:153-156.

Hajdukiewicz, P., Z. Svab, and P. Maliga (1994). The small versatile pPZP family of *Agrobacterium* binary vectors for plant transformation. *Plant Molecular Biology* 25:989-994.

Hamilton, C.M., A. Frary, C. Lewis, and S.D. Tanksley (1996). Stable transfer of intact high molecular weight DNA into plant chromosomes. *The Proceedings of the National Academy of Sciences USA* 93:9975-9979.

Hankin, L. and J.E. Puhalla (1971). Nature of a factor causing interstrain lethality in *Ustilago maydis*. *Phytopathology* 61:50-53.

Hauptmann, R.M., P. Ozias-Akins, V. Vasil, Z. Tabaeizadeh, S.G. Rogers, R.B. Horsch, I.K. Vasil, and R.T. Fraley (1987). Transient expression of electroporated DNA in monocotyledonous and dicotyledonous species. *Plant Cell Reports* 6:265-270.

Hiei, Y., S. Ohta, T. Komari, and T. Kumashiro (1994). Efficient transformation of rice (*Oryza sativa* L.) mediated by *Agrobacterium* and sequence analysis of the boundaries of the T-DNA. *The Plant Journal* 6:271-282.

Hilder, V.A. and D. Boulter (1999). Genetic engineering of crop plants for insect resistance—A critical review. *Crop Protection* 18:177-191.

Hilder, V.A., A.M.R. Gatehouse, and D. Boulter (1992). Transgenic plants conferring insect tolerance: Protease inhibitor approach. In *Transgenic Plants* (pp. 317-338), Eds. Kung, S. and R. Wu. New York: Academic Press.

Hilder, V.A., K.S. Powell, A.M.R. Gatehouse, J.A. Gatehouse, L.N. Gatehouse, Y. Shi, W.D.O. Hamilton, A. Merryweather, C.A. Newell, J.C. Timans, et al. (1995). Expression of snowdrop lectin in transgenic tobacco plants results in added protection against aphids. *Transgenic Research* 4:18-25.

Ingram, J. and D. Bartels (1996). The molecular basis of dehydration tolerance in plants. *Annual Review of Plant Physiology* and *Plant Molecular Biology* 47:377-403.

Ishida, Y., H. Saito, S. Ohta, Y. Hiei, T. Komari, and T. Kumashiro (1996). High efficiency transformation of maize (*Zea mays* L.) mediated by *Agrobacterium tumefaciens*. *Bio/Technology* 14:745-750.

Jach, G., B. Görnhardt, J. Mundy, J. Logemann, E. Pinsdorf, R. Leah, J. Schell, and C. Maas (1995). Enhanced quantitative resistance against fungal diseases by combinatorial expression of different barley antifungal proteins in transgenic tobacco. *The Plant Journal* 8:97-109.

Joersbo, M., S.G. Petersen, and F.T. Okkels (1999). Parameters interacting with mannose selection employed for the production of transgenic sugar beet. *Physiologia Plantarum* 105:109-115.

Jordan, M.C. (2000). Green fluorescent protein as a visual marker for wheat transformation. *Plant Cell Reports* 19:1069-1075.

Karunaratne, S., A. John, A. Mouradov, J. Scott, H.H. Steinbiss, and K.J. Scott (1996). Transformation of wheat with the gene encoding the coat protein of barley yellow mosaic virus. *Australian Journal of Plant Physiology* 23:429-435.

Kasuga, M., Q. Liu, S. Miura, K. Yamaguchi-Shinozaki, and K. Shinozaki (1999). Improving plant drought, salt and freezing tolerance by gene transfer of a single stress-inducible transcription factor. *Nature Biotechnology* 17:287-291.

Kay, R., A. Chan, M. Daly, and J. McPherson (1987). Duplication of CaMV 35S promoter sequences creates a strong enhancer for plant genes. *Science* 236:1299-1302.

Klein, T.M., E.D. Wolf, R. Wu, and J.C. Sanford (1987). High velocity microprojectiles for delivering nucleic acids into living cells. *Nature* 327:70-73.

Koltin, Y. and P.R. Day (1975). Specificity of *Ustilago maydis* killer proteins. *Applied Microbiology* 30:694-696.

Komari, T., Y. Hiei, Y. Saito, N. Murai, and T. Kumashiro (1996). Vectors carrying two separate T-DNAs for co-transformation of higher plants mediated by *Agrobacterium tumefaciens* and segregation of transformants free from selection markers. *The Plant Journal* 10:165-174.

Koncz, C. and J. Schell (1986). The promoter of T_L-DNA gene 5 controls the tissue specific expression of chimaeric genes carried out by a novel type *Agrobacterium* binary vector. *Molecular and General Genetics* 204:383-396.

Lagudah, E. S., J. Dubcovsky, and W. Powell (2001). Wheat genomics. *Plant Physiology and Biochemistry* 39:335-344.

Lamacchia, C., P.R. Shewry, N. Di Fonzo, J.L. Forsyth, N. Harris, P.A. Lazzeri, J.A. Napier, N.G. Halford, and P. Barcelo (2001). Endosperm-specific activity of a storage protein gene promoter in transgenic wheat seed. *Journal of Experimental Botany* 52:243-250.

Langenberg, W.G., L. Zhang, D.L. Court, L. Gjunchedi, and A. Mitra (1997). Transgenic tobacco plants expressing the bacterial *rnc* gene resist virus infection. *Molecular Breeding* 3:391-399.

Leckband, G. and H. Lörz (1998). Transformation and expression of a stilbene synthase gene of *Vitis vinifera* L. in barley and wheat for increased fungal resistance. *Theoretical and Applied Genetics* 96:1004-1012.

Li, Z., M.N. Upadhyaya, S. Meena, A.J. Gibbs, and P.M. Waterhouse (1997). Comparison of promoters and selectable marker for use in indica rice transformation. *Molecular Breeding* 3:1-14.

Lonsdale, D.M., S. Lindup, L.J. Moisan, and A.J. Harvey (1998). Using firefly luciferase to identify the transition from transient to stable expression in bombarded wheat scutellar tissue. *Physiologia Plantarum* 102:447-452.

Lörz, H., B. Baker, and J. Schell (1985). Gene transfer to cereal cells mediated by protoplast transformation. *Molecular and General Genetics* 199:178-182.

Maier-Greiner, U.H., B.M.M. Obermaier-Skrobranek, L.M. Estermaier, W. Kammerloher, C. Freund, C. Wülfing, U.I. Burkert, D.H. Matern, M. Breuer, M. Eulitz, et al. (1991). Isolation and properties of a nitrile hydratase from the soil fungus *Myrothecium verrucaria* that is highly specific for the fertilizer cyanamide and cloning of its gene. *Proceedings of the National Academy of Sciences USA* 88:4260-4264.

Matzke, M.A. and A.J.M. Matzke (1995). How and why do plants inactivate homologous (trans) genes? *Plant Physiology* 107:679-685.

Matzke, M.A., A.J.M. Matzke, and W.B. Eggleston (1996). Paramutation and transgene silencing: A common response to invasive DNA? *Trends in Plant Science* 1:382-388.

May, G.D., R. Afza, H.S. Mason, A. Wiecko, F.J. Novak, and C.J. Arntzen (1995). Generation of transgenic banana *(Musa acuminata)* plants via *Agrobacterium*-mediated transformation. *Bio/Technology* 13:486-492.

McElroy, D., W. Zhang, J. Cao, and R. Wu (1990). Isolation of an efficient actin promoter for use in rice transformation. *Plant Cell* 2:163-171.

Mendoza, M.C. and H.F. Kaeppler (2002). Auxin and sugar effects on callus induction and plant regeneration frequencies from mature embryos of wheat (*Triticum aestivum* L.). *In Vitro Cellular and Developmental Biology—Plant* 38:39-45.

Murashige, T. and F. Skoog (1962). A revised medium for rapid growth and bioassays with tobacco tissue cultures. *Physiologia Plantarum* 15:473-497.

Nehra, N.S., R.N. Chibbar, N. Leung, K. Caswell, C. Mallard, L. Steinhaur, M. Bága, and K. Kartha (1994). Self-fertile transgenic wheat plants regenerated from isolated scutellar tissues following microprojectile bombardment with two distinct gene constructs. *The Plant Journal* 5:285-297.

Oldach, K.H., D. Becker, and H. Lörz (2001). Heterologous expression of genes mediating enhanced fungal resistance in transgenic wheat. *Molecular Plant-Microbe Interactions* 14:832-838.

Ortiz, J.P.A., M.I. Reggiardo, R.A. Ravizzini, S.G. Altabe, G.D.L. Cervigni, M.A. Spitteler, M.M. Morata, F.E. Elías, and R.H. Vallejos (1996). Hygromycin, resistance as an efficient selectable marker for wheat stable transformation. *Plant Cell Reports* 15:877-881.

Ouellet, F., A. Vazquez-Tello, and F. Sarhan (1998). The wheat *wcs120* promoter is cold-inducible in both monocotyledonous and dicotyledonous species. *FEBS Letters* 423:324-328.

Ozias-Akins, P. and I.K. Vasil (1982). Plant regeneration from cultured immature embryos and inflorescences of *Triticum aestivum* L. (wheat): Evidence for somatic embryogenesis. *Protoplasma* 110:95-105.

Ozias-Akins, P. and I.K. Vasil (1983). Improved efficiency and normalization of somatic embryogenesis in *Triticum aestivum* (wheat). *Protoplasma* 116:40-44.

Pang, S.Z., D.L. Deboer, Y. Wan, G. Ye, J.G. Layton, M.K. Neher, C.L. Armstrong, J.E. Fry, M.A.W. Hinchee, and M.E. Fromm (1996). An improved green fluorescent protein gene as a vital marker in plants. *Plant Physiology* 112:893-900.

Patterson, F.L., G.E. Shaner, H.W. Ohm, and J.E. Foster (1990). A historical perspective for the establishment of research goals for wheat improvement. *Journal of Production Agriculture* 3:30-38.

Pawlowski, W.P. and D.A. Somers (1998). Transgenic DNA integrated into the oat genome is frequently interspersed by the host DNA. *Proceedings of the National Academy of Sciences USA* 95:12106-12110.

Perl, A., H. Kless, A. Blumenthal, G. Galili, and E. Galun (1992). Improvement of plant regeneration and GUS expression in scutellar wheat calli by optimization of cuture condition and DNA-microprojectile delivery procedures. *Molecular and General Genetics* 235:279-284.

Perlak, F., T. Stone, Y. Muskopf, L. Petersen, G. Barker, S. McPherson, J. Wyman, S. Love, G. Reed, D. Biever, and D. Fischoff (1993). Genetically improved potatoes: Protection from damage by Colorado potato beetles. *Plant Molecular Biology* 22:313-321.

Powell, P., R. Nelson, D. Re, N. Hoffmann, S. Rogers, R. Fraley, and R. Beachy (1986). Delay of disease development in transgenic plants that express the tobacco mosaic virus coat protein gene. *Science* 232:738-743.

Prins, M. and R. Goldbach (1996). RNA mediated virus resistance in transgenic plants. *Archives in Virology* 141:2259-2276.

Rasco-Gaunt, S., A. Riley, P. Barcelo, and P.A. Lazzeri (1999). Analysis of particle bombardment parameters to optimise DNA delivery into wheat tissues. *Plant Cell Reports* 19:118-127.

Sano, T., A. Nagayama, T. Ogawa, I. Ishida, and Y. Okada (1997). Transgenic potato expressing a double-stranded RNA-specific ribonuclease is resistant to potato spindle tuber viroid. *Nature Biotechnology* 15:1290-1294.

Schenk, P.M., L. Sági, A.R. Elliott, R.G. Dietzgen, R. Swennen, P.R. Ebert, C.P.L. Grof, and J.M. Manners (2001). Promoters from pregenomic RNA of banana streak badnavirus are active for transgene expression in monocot and dicot plants. *Plant Molecular Biology* 47:399-412.

Schenk, P.M., L. Sági, T. Remans, R.G. Dietzgen, M.J. Bernard, M.W. Graham, and J.M. Manners (1999). A promoter from sugarcane bacilliform badnavirus drives transgene expression in banana and other monocot and dicot plants. *Plant Molecular Biology* 39:1221-1230.

Shinozaki, K. and K. Yamaguchi-Shinozaki (1997). Gene expression and signal transduction in water-stress response. *Plant Physiology* 115:327-334.

Sivamani, E., A. Bahieldin, J.M. Wraith, T. Al-Niemi, W.E. Dyer, T-H.D. Ho, and R. Qu (2000). Improved biomass productivity and water use efficiency under water deficit conditions in transgenic wheat constitutively expressing the barley *HVA1* gene. *Plant Science* 155:1-9.

Sivamani, E., C.W. Brey, W.E. Dyer, L.E. Talbert, and R. Qu (2000). Resistance to wheat streak mosaic virus in transgenic wheat expressing the viral replicase (NIb) gene. *Molecular Breeding* 6:469-477.

Stark-Lorenzen, P., B. Nelke, G. Hänssler, H.P. Mühlbach, and J.E. Thomzik (1997). Transfer of a grapevine stilbene synthase gene to rice (*Oryza sativa* L.). *Plant Cell Reports* 16:668-673.

Steinrücken, H.C. and N. Amrhein (1980). The herbicide glyphosate is a potent inhibitor of 5-enolpyruvylshikimic acid-3-phosphate synthase. *Biochemical and Biophysical Research Communications* 94:1207-1212.

Stoger, E., S. Williams, P. Christou, R.E. Down, and J.A. Gatehouse (1999). Expression of the insecticidal lectin from snowdrop (*Galanthus nivalis* agglutinin; GNA) in transgenic wheat plants: Effects on predation by the grain aphid *Sitobion avenae*. *Molecular Breeding* 5:65-73.

Thompson, C.J., N.R. Movva, R. Tizard, R. Crameri, J.E. Davies, M. Lauwereys, and J. Botterman (1987). Characterization of the herbicide resistance gene *bar* from *Streptomyces hygroscopicus*. *The European Molecular Biology Organization Journal* 6:2519-2523.

Tzafrir, I., K.A. Torbert, B.E.L. Lockhart, D.A. Somers, and N.E. Olszewski (1998). The sugarcane bacilliform badnavirus promoter is active in both monocots and dicots. *Plant Molecular Biology* 38:347-356.

Vaeck, M., A. Reynaerts, H. Hofte, S. Jansens, M. De Beukeleer, C. Dean, M. Zabeau, M. Van Montagu, and J. Leemans (1987). Transgenic plants protected from insect attack. *Nature* 328:33-37.

Vain, P., M.D. McMullen, and J.J. Finer (1993). Osmotic treatment enhances particle bombardment mediated transient and stable transformation of maize. *Plant Cell Reports* 12:84-88.

Vasil, I.K. and V. Vasil (1999). Transgenic cereals: *Triticum aestivum* (wheat). In *Molecular Improvement of Cereal Crops* (pp. 137-147), Ed. Vasil, I.K. Dordrecht, the Netherlands: Kluwer Academic Publishers.

Vasil, V., S.M. Brown, D. Re, M.E. Fromm, and I.K. Vasil (1991). Stably transformed callus lines from microprojectile bombardment of cell suspension cultures of wheat. *Bio/Technology* 9:743-747.

Vasil, V., A.M. Castillo, M.E. Fromm, and I.K. Vasil (1992). Herbicide-resistant fertile transgenic wheat plants obtained by microprojectile bombardment of regenerable embryogenic callus. *Bio/Technology* 10:662-674.

Vasil, V., V. Srivastava, A.M. Castillo, M.E. Fromm, and I.K. Vasil (1993). Rapid production of transgenic wheat plants by direct bombardment of cultured immature embryos. *Bio/Technology* 11:1553-1558.

Vazquez-Tello, A., F. Ouellet, and F. Sarhan (1998). Low temperature-stimulated phosphorylation regulates the binding of nuclear factors to the promoter of *Wcs120,* a cold-specific gene in wheat. *Molecular and General Genetics* 257:157-166.

Waldron, C., E.B. Murphy, J.L. Roberts, G.D. Gustafson, S.L. Armour, and S.K. Malcolm (1985). Resistance to hygromycin B a new marker for plant transformation studies. *Plant Molecular Biology* 5:103-108.

Wang, M.B., D. Boulter, and J.A. Gatehouse (1992). A complete sequence of the rice sucrose synthase-1 *(Rss1)* gene. *Plant Molecular Biology* 19:881-885.

Wang, Y.C., T.M. Klein, M. Fromm, J. Cao, J.C. Sanford, and R. Wu (1988). Transient expression of foreign genes in rice, wheat and soybean cells following particle bombardment. *Plant Molecular Biology* 11:433-439.

Weeks, J.T., O.D. Anderson, and A.E. Blechl (1993). Rapid production of multiple independent lines of fertile transgenic wheat *(Triticum aestivum). Plant Physiology* 102:1077-1084.

Weeks, J.T., K.Y. Koshiyama, U. Maier-Greiner, T. Schaeeffner, and O.D. Anderson (2000). Wheat transformation using cyanamide as a new selective agent. *Crop Science* 40:1749-1754.

Witrzens, B., R.I.S. Brettell, F.R. Murray, D. McElroy, Z. Li, and E.S. Dennis (1998). Comparison of three selectable marker genes for transformation of wheat by microprojectile bombardment. *Australian Journal of Plant Physiology* 25:39-44.

Wnendt, S., N. Ulbrich, and U. Stahl (1994). Molecular cloning, sequence analysis and expression of the gene encoding an antifungal-protein from *Aspergillus giganteus. Current Genetics* 25:519-523.

Woolston, C.J., R. Barker, H. Gunn, M.I. Boulton, and P.M. Mullineaux (1988). Agroinfection and nucleotide sequence of cloned wheat dwarf virus DNA. *Plant Molecular Biology* 11:35-44.

Wright, M., J. Dawson, E. Dunder, J. Suttie, J. Reed, C. Kramer, Y. Chang, R. Novitzky, H. Wang, and L. Artim-Moore (2001). Efficient biolistic transformation of maize (*Zea mays* L.) and wheat (*Triticum aestivum* L.) using phosphomannose isomerase, pmi, as the selectable marker. *Plant Cell Reports* 20:429-436.

Xing, A., Z. Zhang, S. Sato, P. Staswick, and T. Clemente (2000). The use of the two T-DNA binary system to derive marker-free transgenic soybeans. *In Vitro Cellular and Developmental Biology—Plant* 36:456-463.

Xu, D., X. Duan, B. Wang, B. Hong, T.-H.D. Ho, and R. Wu (1996). Expression of a late embryogenesis abundant protein gene, HVA1, from barley confers tolerance to water deficit and salt stress in transgenic rice. *Plant Physiology* 110:249-257.

Yamaguchi-Shinozaki, K. and K. Shinozaki (1994). A novel *cis*-acting element in an *Arabidopsis* gene is involved in responsiveness to drought, low-temperature, or high-salt stress. *Plant Cell* 6:251-264.

Zhang, L., R. French, W.G. Langenberg, and A. Mitra (2001). Accumulation of barley stripe virus is significantly reduced in transgenic wheat plants expressing a bacterial ribonuclease. *Transgenic Research* 10:13-19.

Zhang, L., J.J. Rybczynski, W.G. Langenberg, A. Mitra, and R. French (2000). An efficient wheat transformation procedure: Transformed calli with long-term morphogenic potential for plant regeneration. *Plant Cell Reports* 19:241-250.

Zhao, Z.-Y., T. Cai, L. Tagliani, M. Miller, N. Wang, H. Pang, M. Rudert, S. Schroeder, D. Hondred, J. Seltzer, and D. Pierce (2000). *Agrobacterium*-mediated sorghum transformation. *Plant Molecular Biology* 44:789-798.

Zhou, H., J.W. Arowsmith, M.E. Fromm, C.M. Hironaka, M.L. Taylor, D. Rodriguez, M.E. Pajeau, S.M. Brown, C.G. Santino, and J.E. Fry (1995). Glyphosate-tolerant CP4 and GOX genes as a selectable marker in wheat transformation. *Plant Cell Reports* 15:159-163.

Chapter 7

Development and Utilization of Transformation in *Medicago* Species

Deborah A. Samac
Stephen J. Temple

INTRODUCTION

The genus *Medicago* includes both perennial and annual species that are important forage crops for livestock. Alfalfa *(Medicago sativa)* is a deep-rooted perennial forage legume grown extensively throughout the world. It is one of the most important crops in the United States, ranking fourth in terms of acreage and economic value following corn, soybeans, and wheat (Barnes et al., 1988). A number of the annual species are used in pastures, as cover crops, and for weed control. One of the annual species, *M. truncatula,* has become a model in legume biology and is the focus of a number of genome projects (Cook, 1999; Frugoli and Harris, 2001).

In contrast to many legumes and other forages, alfalfa and several of the annual *Medicago* species can be regenerated relatively easily from tissue culture and are thus amenable to transformation by *Agrobacterium.* This chapter will review the methods used to transform *Medicago* species, the expression patterns of promoters tested, and traits introduced for crop improvement, as well as provide a case study for posttransformation breeding required to develop a commercial transgenic variety of alfalfa.

PLANT REGENERATION

Alfalfa Tissue Culture Systems

Alfalfa regeneration in tissue culture occurs via somatic embryogenesis either indirectly from callus cells or directly from leaf explants. Indirect somatic embryogenesis in alfalfa was first described by Saunders and Bingham (1972) and shown to be genotype dependent. Because alfalfa is an open-pollinated plant, each variety is a highly heterogeneous and heterozygous population. Only a small percentage of plants within most varieties are able to form somatic embryos in tissue culture (Bingham et al., 1988). In a study of 35 alfalfa varieties that encompass the spectrum of alfalfa germplasm, Mitten and colleagues (1984) found regenerable genotypes in seven of nine germplasm sources. The varieties with the greatest number of individuals capable of regenerating in culture were from the germplasm sources 'Ladak' and 'Turkestan'. In a study of 76 varieties, Brown and Atanassov (1985) found that approximately 34 percent of the varieties had genotypes with some degree of somatic embryogenesis. The varieties with greater proportions of regenerating plants had strong genetic contributions from *M. sativa* ssp. *falcata,* the germplasm source 'Ladak', and *M. sativa* ssp.×*varia*. Several studies have shown that regenerating individuals can be obtained from most varieties by screening large numbers of plants (Desgagnés et al., 1995; Fuentes et al., 1993; Matheson et al., 1990; Saunders and Bingham, 1972). In fact, the highly regenerable variety 'Regen-S', in which approximately 67 percent of the genotypes will regenerate in culture, was developed by recurrent selection for regeneration from plants selected from the varieties 'DuPuits' and 'Saranac', which have relatively low proportions of regenerating genotypes (Bingham et al., 1975). Although direct regeneration can be a more rapid technique than indirect regeneration (Denchev et al., 1991), most transformation systems have utilized indirect embryogenesis protocols. Only recently a protocol was described for transformation using direct embryogenesis (Shao et al., 2000).

Several studies concluded that the ability of alfalfa genotypes to regenerate in culture is controlled by two independent dominant and complementary loci. Reisch and Bingham (1980) proposed that embryo formation in a diploid alfalfa was controlled by two dominant genes. Similarly, embryogenesis was shown to be under the control of two dominant genes with complementary effects in tetraploid alfalfa genotypes selected from 'Ladak' and 'Lahontan' (Wan et al., 1988). These conclusions were confirmed with regenerating genotypes selected from 'Rangelander' (Hernández-Fernández and Christie, 1989) and 'Rangelander' and 'Regen-S' (Kielly and Bowley, 1992). Cloning and characterization of the loci controlling regeneration would provide significant insight into the nature of totipotency in alfalfa and other plants. A random amplified polymorphic DNA (RAPD) marker has been identified that is associated with somatic embryogenesis in alfalfa (Yu and Pauls, 1993). The development of chromosome maps and genomics tools should make it possible to map and clone the genes controlling regeneration and embryo formation.

Although certain varieties have a high regeneration capacity, the genotypes that regenerate in culture may not necessarily have all the desired agronomic traits. The genotypes most frequently reported in transformation experiments are derived from 'Rangelander', 'Regen-S', and 'Regen-SY' (Bingham, 1991). If these genotypes are used for transformation, backcrossing to other varieties is required to alter dormancy and increase yield, persistence, and disease resistance. Micallef and colleagues (1995) showed rapid improvements in yield in two to three backcrosses of transgenic lines to elite alfalfa lines. Alternatively, because the regeneration trait is highly heritable (Bingham et al., 1975), it can be introgressed into breeding lines and those plants used as stock plants in transformation experiments (Bowley et al., 1993). Transformation and regeneration of individuals selected from commercial breeding lines has also been reported (Desgagnés et al., 1995; Matheson et al., 1990; Tabe et al., 1995) and may facilitate development of improved transgenic alfalfa varieties.

Regeneration is highly influenced by the developmental state of explants, environmental conditions, and media components. Regeneration systems use either a two- or three-step procedure. In the two-step procedure, callus is initiated on a basal medium with 2,4-dichlorophenooxyacetic acid (2,4-D) and kinetin followed by transfer of calli to hormone-free medium for initiation and differentiation of embryos (Austin et al., 1995; Desgagnés et al., 1995; Hernández-

Fernández and Christie, 1989; Saunders and Bingham, 1972; Wan et al., 1988). In the three-step process, callus is initiated on a medium with either 2,4-D or napthaleneacetic acid (NAA) and kinetin, transferred to a medium with a high 2,4-D content for a short period, then transferred to hormone-free medium (Brown and Atanassov, 1985; Skokut et al., 1985; Walker et al., 1978).

The optimal medium for regeneration has been shown to vary with genotype, and several amendments to the basal medium increase embryo induction from regenerable genotypes. In all genotypes, incorporation of 2,4-D in the induction medium is critical for initiation of embryogenesis (Finstad et al., 1993). Optimal regeneration occurs on a medium without growth regulators containing 12.5 to 100 mM NH_4^+ (Walker and Sato, 1981) or amended with 30 mM alanine or proline (Skokut et al., 1985). Once embryos form, they are removed from the callus, usually at the cotyledonary stage, and placed on a hormone-free medium for conversion to plantlets. Some genotypes were shown to have a higher rate of conversion when exposed to a conditioning treatment that included abascisic acid (ABA), glutamine, and sucrose (Hindson et al., 1998). In transformation protocols, optimization of embryogenesis and conversion is of particular concern if the genotypes being used for production of transgenic plants have a relatively low efficiency of regeneration. In addition, cocultivation with *Agrobacterium* may negatively effect the efficiency of embryo formation (Desgagnés et al., 1995).

Wenzel and Brown (1991) concluded that alfalfa somatic embryos arise from epidermal and subepidermal cells of petiole explants. This suggests that the cells developing into somatic embryos can be easily accessed by *Agrobacterium* for T-DNA transfer during a cocultivation step. Embryos developing after cocultivation are most likely the result of a single transformation event since somatic embryos probably arise from a single cell. However, many genotypes are capable of secondary embryogenesis, as was first noted by Saunders and Bingham (1972). Thus, cultures should be managed to limit repetitive embryogenesis and plants evaluated to ensure that they are derived from independent transformation events. Alternatively, somatic embryos can be used as the explant material for cocultivation (Ninkovic et al., 1995).

Regeneration of Annual Medicago *Species*

Somatic embryogenesis has been described for several annual *Medicago* species including *M. lupulina* (Li and Demarly, 1995), *M. polymorpha* (Scarpa et al., 1993), *M. suffruticosa* (Li and Demarly, 1996), and *M. truncatula* (Nolan et al., 1989); however, transformation systems have been described only for *M. truncatula*. Although the annual *Medicago* species are closely related to *M. sativa,* the protocols for somatic embryogenesis have distinct differences. Alfalfa regeneration protocols do not promote regeneration of *M. truncatula* (Nolan et al., 1989) or *M. lupulina* (Li and Demarly, 1995), and amendments to media that increase embryogenesis in alfalfa, such as addition of proline or 50 mM NH_4^+, increase embryo mortality of *M. truncatula* (das Neves et al., 1999). In contrast to alfalfa, in which auxin is necessary for induction of embryos, in the annual species, cytokinins are required for embryoid formation (Li and Demarly, 1995).

Medicago truncatula can be regenerated by indirect somatic embryogenesis or by direct shoot organogenesis. Embryogenesis is genotype specific, and most research has focused on optimizing regeneration from a few highly regenerable seed lines (Hoffmann et al., 1997; Trinh et al., 1998; Rose et al., 1999). Embryogenic callus of *M. truncatula* is initiated from explants on a medium containing NAA and benzylaminopurine (BAP) (Nolan et al., 1989), 2,4-D and BAP (Chabaud et al., 1996), or 2,4-D alone (Trinh et al., 1998) in darkness. Transfer of calli to a hormone-free medium stimulates embryogenesis (Nolan and Rose, 1998). Embryos emerging from calli are removed to fresh medium for plantlet conversion. The highly embryogenic lines of *M. truncatula* can form prolific secondary embryos that prevent conversion of the primary embryo. Efficient conversion of embryos has been the greatest challenge for optimizing regeneration systems in *M. truncatula*. If embryos develop in darkness, the rate of primary embryo conversion was found to be higher (Nolan and Rose, 1998). Furthermore, if embryos are removed from the callus at the torpedo stage and placed on fresh medium, secondary embryogenesis is decreased (das Neves et al., 1999). Production of plants from somatic embryos takes three to ten months depending on genotype. In contrast, shoot organogenesis is more rapid, requiring approximately 2.5 months of culture to produce plantlets, and is highly efficient, avoiding the problem of formation of secondary embryos. Shoot organogenesis from cotyledonary explants has been demonstrated for plants from the cultivar Jemalong and can be adapted to *Agrobacterium*-mediated transformation (Trieu and Harrison, 1996).

PLANT TRANSFORMATION

Transformation of Alfalfa

Agrobacterium tumefaciens-*Mediated Transformation*

Alfalfa has been transformed using a number of different methods; however, the most common procedure involves cocultivation of explants with *Agrobacterium tumefaciens*. Early experiments focused on the parameters for efficient and rapid production of transgenic alfalfa plants. For many alfalfa genotypes, efficient transformation was found to depend significantly on the strain of *A. tumefaciens* used. The first report of transformation of *M. sativa* involved cocultivation of stem sections from plants of the variety CUF101 with the disarmed *A. tumefaciens* strain LBA4404 (Shahin et al., 1986). The first report of transformation of *M. sativa* ssp.×*varia* involved cocultivation of alfalfa stem explants with *A. tumefaciens* A281, an armed strain that retains the tumor-inducing genes (Deak et al., 1986). Chabaud and colleagues (1988) also transformed *M. sativa* ssp.×*varia* using either A281 or the disarmed strain LBA4404 and found that strain A281 increased the number of calli giving rise to transgenic embryos twofold compared to LBA4404. Du and colleagues (1994) evaluated four strains of *A. tumefaciens,* three armed strains (A281, C58, and C58-R1000) and one disarmed strain (GV3101) with three genotypes of alfalfa. Although each genotype was previously selected for efficient regeneration, cocultivation substantially reduced embryogenesis and only one strain-genotype combination resulted in transgenic plants. A strong strain-genotype interaction was observed in a similar study by Desgangés and colleagues (1995). All three genotypes tested produced transgenic embryos when transformed with strain LBA4404, two genotypes produced transgenic embryos after cocultivation with strain A281, while only one genotype produced transgenic embryos with strain C58. Clearly, the optimal genotype-strain combination is a primary factor to consider in strain selection for a high transformation frequency. Cocultivation of explants above a feeder culture, a layer of alfalfa suspension culture cells on the agar medium covered by sterile filter paper moistened with liquid culture medium, has been used in alfalfa transformation and may reduce the negative responses of explants to *Agrobacterium* infection.

The length of the cocultivation period has also been shown to affect transformation frequency (Austin et al., 1995; Samac, 1995). Longer periods of cocultivation, up to eight days, may increase the frequency of transformation due to the delay in applying selection pressure. However, the optimal cocultivation period is most likely genotype dependent, and regrowth of some *Agrobacterium* strains may be more difficult to control following longer cocultivations.

The most widespread selectable marker in alfalfa transformation is the neomycin phosphotransferase *(nptII)* gene that confers kanamycin resistance; however, transgenic alfalfa with resistance to hygromycin or the herbicide phosphinothricin (PPT) has also been produced (D'Hulluin et al., 1990; Tabe et al., 1995). A concentration of 10 to 50 $mg{\cdot}L^{-1}$ kanamycin was found to prevent callus formation from explants that did not undergo transformation (Chabaud et al., 1988; Desgagnés et al., 1995), and most transformation systems use 25 to 100 $mg{\cdot}L^{-1}$ kanamycin in media for callus growth, embryo formation, and plantlet development. High amounts of kanamycin have been observed to inhibit transgenic callus growth and embryo production (Desgagnés et al., 1995) and may inhibit root formation (Austin et al., 1995). Transformation of alfalfa with the *bar* gene encoding phophinothricin acetyl transferase and direct selection on medium containing phophinothricin increases the frequency of transformation by practically eliminating nontransformed escapes (D'Hulluin et al., 1990; Tabe et al., 1995).

Although an array of variables must be evaluated to optimize *Agrobacterium*-mediated transformation, systems have been developed for rapid and efficient production of substantial numbers of transgenic alfalfa plants. Over the past seven years, scientists in the USDA-ARS-Plant Science Unit and collaborators have expressed more than 70 different constructs in alfalfa and produced thousands of transgenic plants using a relatively simple system based on that of Austin and colleagues (1995). Using this system, more than 75 percent of the callus pieces produce multiple transgenic embryos, and transgenic plantlets can be generated within nine to 12 weeks. This system uses leaf explants from a highly embryogenic plant selected from 'Regen-SY' (Bingham, 1991). Although stem and petiole explants can be used, leaf explants give a more uniform and rapid callusing response and more rapid production of embryos. The stock plants are vegetatively propagated in vitro on Murashige and Skoog (MS) medium without hormones (Murashige and Skoog, 1962) and cultured at 25°C with a light intensity of approximately 100 $\mu mol \cdot m^{-2} \cdot sec^{-1}$. For production of explants, plants are transplanted to a soil mix and maintained in a growth chamber under conditions to maximize leaf production: 24°C/19°C day/night temperature, 16 hours of light per day with a light intensity of approximately 300 $\mu mol \cdot m^{-2} \cdot sec^{-1}$. Plants are fertilized weekly with a complete fertilizer and watered daily. Leaves from the top second to fifth node are surface disinfested by a brief rinse in 70 percent ethanol followed by gentle agitation in a 20 percent bleach solution for 90 seconds. After three rinses in sterile water, leaflet margins are trimmed away with a scalpel and the leaflet piece is cut in half. In vitro grown plants can be used for explant material, however the tissues are fragile and more sensitive to handling and cocultivation. Leaflet pieces are placed in 12 ml of liquid SH medium without hormones (Schenk and Hildebrandt [SH], 1972). After sufficient pieces have been prepared, 3 ml of an overnight culture of *A. tumefaciens* strain LBA4404 containing the vector of interest is added to the leaf pieces and incubated at room temperature for 10 to 15 minutes. The leaf pieces are removed from the inoculum, blotted briefly on sterile filter paper, and placed on the surface of a callus-inducing medium. The medium contains B5 salts and vitamins (Gamborg et al., 1968), 30 $g \cdot L^{-1}$ sucrose, 0.5 $g \cdot L^{-1}$ KNO_3, 0.25 $g \cdot L^{-1}$ $MgSO_4$-$7H_2O$, 0.5 $g \cdot L^{-1}$ proline, 798 $mg \cdot L^{-1}$ L-glutamine, 99.6 $mg \cdot L^{-1}$ serine, 0.48 $mg \cdot L^{-1}$ adenine, 9.6 $mg \cdot L^{-1}$ glutathione, 1 $mg \cdot L^{-1}$ 2,4-D, 0.1 $mg \cdot L^{-1}$ kinetin, and 7 $g \cdot L^{-1}$ Phytagar (Gibco), pH 5.7. The plates are sealed with gas-permeable tape (#394 3M Venting Tape) and placed in an incubator at 24°C with a light intensity of approximately 100 $\mu mol \cdot m^{-2} \cdot sec^{-1}$. After seven days of cocultivation the leaf pieces are removed from the plate, rinsed three to four times in sterile distilled water, and placed on the surface of fresh callus-inducing plates containing 25 $mg \cdot L^{-1}$ kanamycin and 100 $mg \cdot L^{-1}$ ticarcillin. Plates are sealed as before and returned to the incubator. After two to three weeks callus clumps are transferred to a modified hormone-free MS medium (MMS) containing MS salts, 30 $g \cdot L^{-1}$ sucrose, vitamins as described by Nitsch and Nitsch (1969), 100 $mg \cdot L^{-1}$ myo-inositol, 7 $g \cdot L^{-1}$ Phytagar, plus the same antibiotics and cultured for an additional two to three weeks. As cotyledonary stage embryos arise from the calli, they are removed and transferred to fresh MMS medium with antibiotics for conversion to plantlets. When the plantlets have formed a primary leaf, they are moved to a Magenta vessel containing fresh MMS with ticarcillin but without kanamycin for further root and shoot development. Rooted plants are removed from the medium after two to five weeks and the roots placed in a test tube of water on the lab bench for 24 to 48 hours to condition the plants. The plants are then planted in a potting medium and placed in a growth chamber. The conditioning step reduces leaf desiccation, although some leaf loss occurs after transplanting.

Transformation by Particle Bombardment

Due to the ease of *Agrobacterium*-mediated transformation of alfalfa, biolistic transformation (particle bombardment) has received less attention than in crops that are recalcitrant to *Agrobacterium* transformation. Nonetheless, direct delivery of DNA into alfalfa cells has several important applications. One potential avenue to circumvent tissue culture steps and decrease the time and labor involved in generating transgenic plants is particle bombardment of pollen. Ramaiah and Skinner (1997) described a method of bombarding alfalfa pollen that was subsequently used to pollinate male sterile flowers. Approximately 27 percent of the plants obtained showed integration of the *gus* gene by DNA blot hybridization. Blotting results indicated that some plants contained multiple inserts as well as truncated copies of the *gus* gene, a frequent consequence of particle bombardment. Unexpectedly, after vegetative propagation, some lines appeared to lose the integrated *gus* gene and in others the copy number decreased.

Transformation of chloroplasts has been accomplished in several model systems using particle bombardment of leaf explants and has several advantages over nuclear transformation (Bogorad, 2000). Foreign proteins can accumulate to very high amounts because multiple copies of the gene can be introduced. Integration into the chloroplast genome occurs by homologous recombination, which eliminates the "position effect" frequently observed in nuclear transformants. In many crops, plastid transformation also addresses concerns about gene escape through pollen or gene expression in pollen. Alfalfa would not have this advantage because plastid inheritance in alfalfa is biparental with a strong paternal bias (Smith et al., 1986). Nevertheless, chloroplast transformation would greatly facilitate projects in which expression of a large amount of a target protein is desired, such as for production of industrial enzymes or biodegradable plastics. Pereira and Erickson (1995) described stable nuclear transformation of alfalfa by particle bombardment, indicating that chloroplast transformation of alfalfa can be achieved.

***Transformation of* Medicago truncatula**

Transformation by Agrobacterium *Cocultivation*

The first report of transformation of *M. truncatula* showed that transformation could be achieved using either *A. tumefaciens* or *A. rhizogenes* (Thomas et al., 1992). However, transformation frequencies using *A. rhizogenes* were low and most embryos did not develop past the globular stage, suggesting that the genes on the *A. rhizogenes* plasmid (pRi) have a negative effect on somatic embryogenesis. Transformation of several genotypes has been achieved by cocultivation of leaf explants with *A. tumefaciens* (Chabaud et al., 1996; Trinh et al., 1998; Wang et al., 1996). Trinh and colleagues (1998) investigated different growth regulators to develop a rapid and highly efficient transformation protocol. A suspension of the disarmed *A. tumefaciens* strain EHA101 or GV3101 is vacuum infiltrated into explants of the highly regenerable line, R108-1(c3), and cocultivated on agar medium with the virulence gene inducer acetosyringone in the dark for three days. Explants are cultured for five weeks in darkness for callus and embryoid induction, then transferred to medium in the light for embryo development. With this protocol, large numbers of somatic embryos can be obtained from leaf (Trinh et al., 1998) and floral explants (Kamaté et al., 2000) and up to 80 percent of the embryos regenerate into plants three to four months after culture initiation. This highly efficient transformation method has enabled initiation of a T-DNA insertional mutagenesis program for *Medicago truncatula* (Scholte et al., 2002).

Rapid transformation of *M. truncatula* has been described based on cocultivation of *A. tumefaciens* with cotyledonary explants followed by culture to induce shoot organogenesis (Trieu and Harrison, 1996). Multiple shoots developed from the explants; however, not all shoots were derived from independent transformation events, and the transformation efficiency was low. Use of vectors containing the *bar* gene conferring phosphinothricin resistance coupled with culture of explants on media containing PPT was found to be more efficient for selection of transformants than selection for kanamycin resistance. Transgenic plantlets were produced with the system in 2.5 months.

Two methods have been described that eliminate tissue culture steps in *M. truncatula* transformation. Trieu and colleagues (2000) described a method based on infiltration of flowers with *Agrobacterium,* similar to the *Arabidopsis* flower infiltration protocol (Clough and Bent, 1998), and a method based on infiltration of seedlings. After infiltration of flowers, seed production was fairly low but transformation frequency was relatively high, from 21 to 76 percent of the seeds recovered. The seedling infiltration method allows for the simultaneous transformation of a large number of individuals. A high frequency of transformation was reported, with 9.4 percent of resulting seedlings showing resistance to PPT. Although promising, these results have not been repeated or further extended by this group and they have yet to be corroborated by other laboratories.

PROMOTERS FOR CONSTITUTIVE AND TISSUE-SPECIFIC EXPRESSION OF TRANSGENES IN MEDICAGO

Transformation in *Medicago* species has been used to characterize genes from a range of legumes and for crop improvement. For some traits, the expression pattern desired is constitutive, whole-plant expression. For other traits, expression limited to specific organs or upon certain environmental cues may be required. A wide variety of promoters have been described from crop species, but relatively few have been tested for use in alfalfa.

Constitutive Promoters

Undoubtedly the most frequently used promoter in genetically modified plants is the 35S promoter from the cauliflower mosaic virus (CaMV). Early studies showed that in virus-infected plant cells, this promoter directs high constitutive expression of a viral gene encoding a 35S RNA without requiring additional viral products (Guilley et al., 1982). When 343 bp of sequence upstream of the protein coding region of the 35S gene is fused to a marker gene such as β-glucuronidase *(gus)* and inserted into a plant chromosome, the gene is strongly expressed in nominally all cells. In contrast to the strong activity in other plant species, several reports suggest that the 35S promoter may have less activity in alfalfa (Khoudi et al., 1997; Narváez-Vásquez et al., 1992; Tabe et al., 1995). In leaves of tobacco plants containing a 35S::*gus* gene, GUS activity is reported to range from 5,000 to 200,000 pmol 4-MU·min^{-1}·mg^{-1} protein. As shown in Figure 7.1, the amounts of GUS activity in alfalfa leaves with a 35S::*gus* gene are considerably lower than the GUS activity in tobacco, even though alfalfa leaves contain approximately five to ten times more soluble protein per unit fresh weight than tobacco leaves. In addition, not all alfalfa cells express the 35S::*gus* gene. In stems,

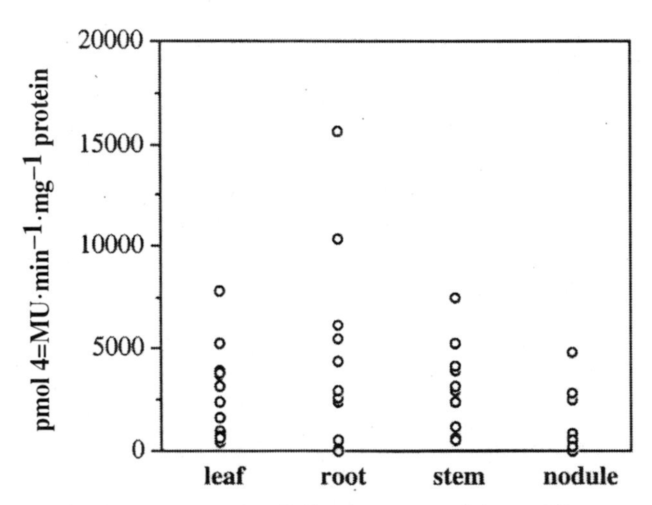

FIGURE 7.1. GUS expression in alfalfa plants containing a 35s::*gus* transgene. Each circle represents the mean value obtained from an individual transgenic plant.

activity is found in the epidermis, chlorenchyma, phloem, and cambium but rarely in xylem or pith cells (Figure 7.2A). In roots and nodules, activity is limited in most plants to vascular tissues. The Mac promoter, a chimeric promoter with elements from the 35S promoter and mannopine synthetase promoter, has been shown to have greater activity than the 35S promoter in model systems (Comai et al., 1990). It has been used for production of proteins that accumulate to high levels in alfalfa (Austin-Phillips and Ziegelhoffer, 2001). Although differences in expression may be dependent on the gene product, additional constitutive promoters should be investigated for expression of genes in alfalfa for biotechnological applications.

Expression of Heterologous Promoters in Alfalfa

Tissue-specific and inducible promoters from other plant species must be tested empirically in alfalfa as the expression pattern may differ from that in the original plant species. Table 7.1 lists the ex-

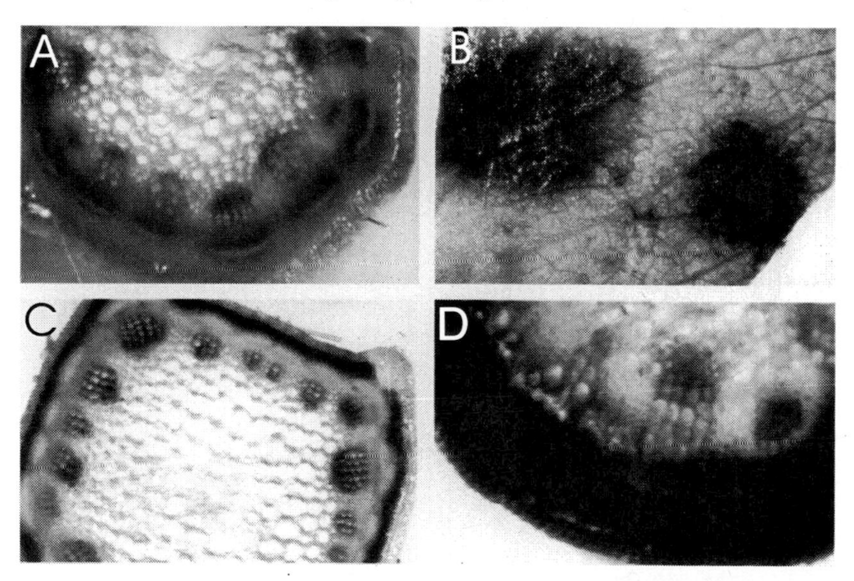

FIGURE 7.2. Expression patterns of three promoters in alfalfa: (A) constitutive expression of the 35S::*gus* gene in alfalfa stem; (B) constitutive expression of the *Arabidopsis* class III chitinase promoter in alfalfa stem; (C) induction of the *Arabidopsis* class III chitinase promoter in alfalfa leaves by infection with *Phoma medicaginis;* (D) constitutive expression of the PEPC-4 promoter in alfalfa stem.

pression patterns of heterologous promoters in alfalfa. The Blec4, PsUGT-1, PAL2, and TA29 promoters had expression patterns similar to those in the original plant species. For example, the TA29 promoter when fused to the *bar* gene resulted in production of barnase in alfalfa anthers, leading to male sterility (Rosellini et al., 2001). In contrast, the *Arabidopsis* class III chitinase promoter and potato *pin2* promoter had unexpected expression patterns in alfalfa. In *Arabidopsis,* the class III chitinase promoter has enhanced expression in roots and expression is induced by fungal pathogens (Samac and Shah, 1991). When the *Arabidopsis* class III chitinase promoter::*gus* gene is expressed in alfalfa, GUS activity is limited to vascular tissue, particularly phloem (Figure 7.2B-C). Interestingly, upon infection by the foliar fungal pathogen *Phoma medicaginis,* GUS activity occurs around lesions. The potato protease inhibitor II *(pin2)* gene promoter is expressed constitutively in root tips of potato. After wounding leaves strong expression occurs in vascular tissue throughout the plant (Keil et al., 1989). In alfalfa plants containing the *pin2-gus* gene, low constitutive activity occurred in leaf and root vascular tissues and root tips or plants had low constitutive activity in leaf mesophyll and root tips. Weak gene induction was seen 24 hours after wounding leaves (Samac and Smigocki, unpublished).

TABLE 7.1. Expression of heterologous promoters in alfalfa

Gene promoter	Origin	Expression pattern	Reference
Blec4	*Pisum*	Epidermal cells	Mandaci and Dobres, 1997
PsUGT-1	*Pisum*	Root meristems	Woo et al., 1999
PAL2	*Phaseolus*	Vascular tissue	Guo, Chen, Inoue, et al., 2001
TA29	*Nicotiana*	Anther tapetum	Rosellini et al., 2001
Class III chitinase	*Arabidopsis*	Vascular tissue	This chapter
pin2	*Solanum*	Vascular tissue and mesophyll	Samac and Smigocki, unpublished

Expression of Homologous Promoters in Alfalfa

Several promoters have been isolated and characterized from alfalfa, primarily from nodule-enhanced genes central to nitrogen and carbon cycling (Table 7.2). The expression patterns of the *gus* gene fusions of aspartate aminotransferase (AAT-1, AAT-2), phosphoenolpyruvate (PEP) carboxylase (PEPC), asparagine synthetase (AS), and NADH-glutamate synthase (GOGAT) were identical or highly similar to mRNA accumulation as determined by RNA blotting and in situ hybridization. Although the expression of these genes is enhanced in root nodules, expression is also observed in other organs and tissues. For example, the full-length promoter (1277 bp) of the PEPC nodule-enhanced gene directs expression in the nitrogen-fixing zone of nodules as well as in root tips, leaf pulvini, and pollen (Pathirana et al., 1997). In experiments to identify promoter elements

TABLE 7.2. Expression of promoters from alfalfa

Gene promoter	Primary expression pattern	Reference
Rubisco	Light inducible, leaf tissue	Khoudi et al., 1997
IFR	Root meristem and cortex, nodules; fungal induced	Oomen et al., 1994
AAT-1	Uninfected nodule cells	Yoshioka et al., 1999
AAT-2	Infected nodule cells	Yoshioka et al., 1999
PEPC	Infected nodule cells, pulvinal cells	Pathirana et al., 1997
PEPC-4	Vascular tissue, xylem cells	Pathirana et al., 1997
AS	Infected and uninfected nodule cells	Shi et al., 1997
GOGAT	Infected nodule cells	Trepp et al., 1999
MAPK	Meristematic cells, glandular hairs	Schoenbeck et al., 1999

involved in nodule-enhanced expression, a 536 bp sequence of the PEPC promoter (PEPC-4) was found to be a strong promoter for expression in vascular tissue, particularly xylem (Figure 7.2D). Strong xylem-specific expression is relatively unusual and could be exploited for expression of specific genes in alfalfa for improving nutritional quality of stems for ruminant animals.

A mitogen-activated protein kinase (MAP kinase) gene promoter from alfalfa also has relatively specific activity. MAP kinases are expressed in dividing cells but are also associated with response to stress responses such as wounding, drought, cold, and fungal elicitors. The alfalfa nodule-enhanced MAP kinase promoter directed strong GUS expression in the meristematic cells of the root nodule, the root tip, and in glandular trichomes (Schoenbeck et al., 1999). The gene appears to be especially active in developing nodules and roots. Expression was induced in leaves by mechanical wounding and pathogen infection. With fairly specific and strong root tip expression, this promoter may also be useful for alfalfa improvement.

UTILIZATION OF TRANSFORMATION FOR CROP IMPROVEMENT

Crop Plants As "Bioreactors"

Genetic engineering has the potential to improve the economic value of alfalfa through introducing genes for value-added products such as industrial raw materials. Alfalfa is particularly well suited as a "bioreactor." Once established, alfalfa requires very few agronomic inputs, is well adapted to many environmental conditions, and is highly productive over many years. In addition, processes for fractionation are well established and do not interfere with subsequent use of the residue as an animal food. For a number of industrial materials, production in alfalfa would be a cost-effective alternative to bacterial fermentation. Austin and colleagues (1995) demonstrated expression of two industrial enzymes in alfalfa, α-amylase from *Bacillus licheniformis,* which can be used in starch degradation, and manganese-dependent lignin peroxidase (Mn-P), an enzyme with potential in biopulping processes, from the fungus *Phanerochaete chrysosporium.* Two bacterial cellulase genes, with potential for use in conversion of plant biomass to ethanol, have also been produced in alfalfa (Ziegelhoffer et al., 1999). These enzymes were produced at relatively low levels, from 0.01 percent to 0.5 percent of total soluble protein. However, when the gene encoding phytase from *Aspergillus niger* was expressed in alfalfa, a high amount of active enzyme, from 0.1 to 1.5 percent of total soluble protein, was recovered from fresh alfalfa juice and dried plant material (Austin-Phillips and Ziegelhoffer, 2001). Most phosphorus (P) in plant seeds, stored as phytic acid, is unavailable to monogastric animals such as poultry and swine. Phytase releases P from phytic acid and can be added to animal feeds to increase P availability. The enzyme can be delivered inexpensively as a crude alfalfa juice extract or as leaf meal. By eliminating the need to add P to animal diets and utilizing the endogenous phytic acid, the amount of P entering the waste stream is greatly reduced, providing both economic and environmental benefits.

Alfalfa plants have also been developed to produce nonenzymatic, high-value compounds. Wigdorovitz and colleagues (1999) expressed the structural protein VP1 of foot and mouth disease virus (FMDV) in alfalfa. When mice were fed freshly harvested alfalfa leaves from transgenic plants, or were immunized parenterally with leaf extracts, they developed a virus-specific immune response and were protected against challenge with the virus. Production of animal vaccines in a forage plant would be a convenient means to deliver the antigens.

Antihuman IgG is a widely used reagent in blood banks for phenotyping and cross-matching red blood cells. It is usually produced by large-scale culture of murine hybridomas. To develop a less costly source of this reagent, Khoudi and colleagues (1999) expressed full-length cDNAs for the heavy and light chains separately and then intercrossed plants. In plants expressing both chains, the antibody accumulated from 0.13 to 1.0 percent of total soluble protein. In diagnostic assays the antibody produced in alfalfa and the hybridoma-derived antibody reacted similarly. Under field conditions the antibody accumulated to 140 $\mu g \cdot g^{-1}$ dry hay and was stable in dried material for at least 12 weeks after harvest.

An additional nonenzymatic industrial material produced in alfalfa is the biodegradable plastic polymer polyhydroxybutyrate (PHB) (Saruul et al., 2002). This compound is made naturally by many bacteria under nutrient-limiting conditions and can be produced commercially by fermentation. The three genes required for PHB synthesis, *phba, phbb,* and *phbc,* have been expressed in several model plant species, and PHB accumulates to high levels if the gene products are targeted to chloroplasts. In alfalfa the three genes were placed in the same vector, each controlled by the 35S promoter, and the gene products targeted to the chloroplast. Granules of PHB were observed in chloroplasts and PHB accumulated in leaves to 2 $g \cdot kg^{-1}$ dry weight (Saruul et al., 2002). Production of bioplastics in alfalfa leaves could be combined with use of alfalfa for feed and energy. In such a system, alfalfa leaves could be harvested to produce PHB, leaf by-product could be processed into a feed, and stems could be used for producing either electricity by burning or ethanol by fermentation.

Enhancing Nutritional Qualities

Alfalfa is often called the "queen of forages" because of its excellent nutritional properties for ruminant animals. However, the protein in alfalfa leaves can be digested too rapidly in the rumen and the nitrogen lost by excretion as urea. While leaf tissues are easily broken down, lignin retards digestion of alfalfa stems, limiting feed intake value and negatively impacting milk and meat production. Additional problems include bloating of animals fed fresh alfalfa and insufficient digestible carbohydrates in alfalfa hay to balance the amount of available protein. One approach to improving protein quality is overexpression of genes for proteins resistant to digestion by rumen fluids. Schroeder and colleagues (1991) overexpressed the cDNA for chicken ovalbumin, a rumen-stable, sulfur amino acid rich protein in alfalfa using the CaMV 35S promoter. A similar strategy was used by Tabe and colleagues (1995) in which a cDNA encoding a sunflower seed storage albumin was expressed in alfalfa under the control of constitutive promoters. In plants with the highest expression, protein accumulated to approximately 0.11 percent of total soluble protein; however, the authors estimated that expression needs to be 25- to 50-fold higher to impact wool growth rate in grazing sheep.

Modification of the amount and type of lignin in alfalfa stems has been attempted using several antisense approaches. Baucher and colleagues (1999) decreased cinnamyl alcohol dehydrogenase (CAD) by antisense RNA expression using the CaMV 35S promoter. Activity in plants containing an antisense C-terminal CAD fragment was reduced to 30 percent of the activity in control plants. No significant change in total lignin content was observed, although lignin composition was altered significantly in several lines. In particular, the syringyl units were decreased compared to the control. In situ digestibility was greater in the transgenic lines compared to controls, indicating that modification of lignin composition can impact forage digestibility. Guo and colleagues (Guo, Chen, Inoue, et al., 2001; Guo, Chen, Wheeler, et al., 2001) found that down-regulation of two other genes in the lignin biosynthetic pathway, caffeic acid 3-*O*-methyltransferase (COMT) and caffeoyl CoA-*O*-methyltransferase (CCOMT) can change significantly the lignin composition in alfalfa stems. When the COMT cDNA was expressed in the antisense orientation under the control of the bean PAL2 promoter, syringyl units could be eliminated in xylem tissue while antisense expression of CCOMT reduced guaiacyl units and Klason lignin. Down-regulation of CCOMT led to a 2.8 to 6 percent increase in forage digestibility (Guo, Chen,Wheeler, et al., 2001). Because a 1 percent increase in digestibility can positively impact animal performance, this is a significant achievement in improving forage quality.

Enhancing Tolerance to Biotic and Abiotic Stresses

Transgenic approaches are being applied to enhance tolerance in alfalfa to the biotic and abiotic stresses for which traditional approaches have had limited success. Alfalfa mosaic virus (AMV) infections can cause yield losses of up to 30 percent dry weight in alfalfa and 80 percent in *M. truncatula*. Hill and colleagues (1991) and Jayasena and colleagues (2001) used AMV coat protein expression to confer resistance to AMV in alfalfa and *M. truncatula,* respectively. In alfalfa, transgenic plants were resistant to the virus under greenhouse conditions; however, expression of the coat protein did not confer a yield advantage under field conditions (David Miller, personal communication). *Medicago truncatula* plants expressing the AMV coat protein demonstrated high levels of resistance or immunity to two strains under greenhouse growth conditions (Jayasena et al., 2001).

A number of phenolic compounds have potent antifungal activity. Hipskind and Paiva (2000) demonstrated that a unique antifungal compound, *trans*-resveratrol-3-*O*-D-glucopyranoside (Rgluc) accumulates in transgenic alfalfa expressing the resveratrol synthase gene from peanut. Growth of the fungal pathogen *Phoma medicaginis* was inhibited in agar plate bioassays with purified Rgluc and resveratrol. When Rgluc-containing leaves were inoculated with *P. medicaginis,* disease symptoms were reduced compared to control leaves. Rgluc also reduced the number of spore-producing pycnidia in infected leaves, which could reduce spread and severity of the disease within a field. It is likely that Rgluc will also reduce damage to leaves by other foliar pathogens that directly impact yield and quality of alfalfa.

Winter hardiness is a critical trait in a perennial crop for maintaining stand density and dry matter yield. However, it is a very complex trait, involving tolerance to a multitude of stresses. Because generation of superoxide is a common factor in many plant stresses, expression of genes for superoxide dismutase (SOD) has been investigated as a means of reducing oxidative stress and enhancing winter hardiness in alfalfa (McKersie et al., 1999, 2000). Alfalfa plants were transformed with constructs for Mn-SOD targeted to either mitochondria or chloroplasts (McKersie et al., 1999). In most tissues the SOD activity in the highest expressing transgenic plants was approximately twofold higher than in control plants. In laboratory freezing tolerance assays, the most tolerant transgenic plant had only 1°C more freezing tolerance than the control. Nonetheless, in field trials, most transgenic lines had greater survival after one winter and greater total shoot dry matter yield than nontransgenic control lines. Similarly, expression of Fe-SOD in alfalfa was associated with increased winter survival over two years (McKersie et al., 2000).

Tolerance to salt and acid soils will become important in the coming decades. Many agricultural soils, especially those under irrigation, have plant growth inhibiting concentrations of salt. Salt tolerance is also of key importance in developing crops that can be used for disposal of manure or municipal waste water. Winicov (2000) found that constitutive overexpression of *Alfin1,* a putative zinc-finger transcription factor, increases root growth of alfalfa under normal and saline conditions. Because tolerance to salt is most likely a complex trait, modifying expression of multiple genes by controlling expression of key regulatory factors is a promising strategy for enhancing salt tolerance.

Poor crop growth in mineral acid soils is primarily due to aluminum (Al) toxicity, although toxicity to other minerals and nutrient limitations can occur. In an effort to increase Al tolerance in alfalfa,

Tesfaye and colleagues (2001) generated transgenic alfalfa with gene constructs designed to increase synthesis of organic acids to chelate Al. Several lines overexpressing an alfalfa nodule-enhanced malate dehydrogenase had significantly higher amounts of organic acids in root tissues compared to controls and secreted more malate and citrate from roots. In hydroponic assays and in acid soil, these plants had significantly greater root and shoot growth compared to the controls. Development of varieties using this approach would provide alternative methods for managing soil acidity.

DEVELOPMENT OF ROUNDUP READY ALFALFA: A CASE STUDY

The development of crops that are tolerant to glyphosate started in the 1980s and has resulted in the successful commercial release of Roundup Ready corn, soybean, cotton, and canola. Forage Genetics International (FGI) and Monsanto are jointly developing Roundup Ready (RR) alfalfa. The project was initiated in 1997 when an elite FGI alfalfa clone was transformed with a series of Monsanto gene constructs. RR alfalfa will represent the first application for regulatory approval of a genetically modified trait in alfalfa. After regulatory approval has been obtained, which is anticipated for late 2004 in the United States, FGI intends to release RR cultivars adapted to all major U.S. alfalfa markets. The RR alfalfa trait and trademark purity standard will be 90 percent (i.e., 90 percent of the plants in a RR alfalfa variety will demonstrate the RR phenotype). It is anticipated that RR alfalfa will provide producers with an effective weed management tool while retaining a high-quality harvest. A summary of key steps in the development of RR alfalfa including product development, a unique marker-assisted breeding strategy, and issues of pollen flow are discussed.

Background

Glyphosate (*N*-[phosphonomethyl]glycine) is the active ingredient in the foliar-applied, postemergent herbicide Roundup. Glyphosate provides excellent weed-control capabilities and has very favorable environmental and safety characteristics (Padgette et al., 1996). The target for glyphosate is the chloroplast-localized enzyme, 5-enolpyruvylshikimate-3-phosphate synthase (EPSPS). Although EPSPS enzymes from higher plants are sensitive to glyphosate, enzymes isolated from a number of bacterial sources exhibit tolerance. An EPSPS with high tolerance was identified in the *Agrobacterium* sp. strain CP4 (CP4 EPSPS) (Barry et al., 1992; Padgette et al., 1996). The EPSPS gene was cloned from *Agrobacterium* sp. strain CP4 and its suitability for conferring glyphosate tolerance in plants was examined.

The CP4 EPSPS gene was fused to a chloroplast transit sequence (CTP) and placed under the control of the CaMV 35S promoter (P-E35S) and nopaline synthase transcription terminator (NOS 3') (Padgette et al., 1995). Progeny from a selected soybean line transformed with this vector showed no visual injury after application of up to 1.0 gallon/acre of commercial Roundup formulation (Padgette et al., 1995, 1996). Molecular analysis of this line showed it to contain a single insert of DNA with the glyphosate tolerance trait segregating in a Mendelian fashion. The presence of the transgene resulted in no significant yield reduction under field conditions (Padgette et al., 1995, 1996; Delannay et al., 1995). This work led to the development and commercial release of Roundup Ready soybeans in 1996. During 2002 Roundup Ready soybeans were planted on 54.75 million acres or 75 percent of the soybean acreage (http://usda.mannlib.cornell. edu/reports/nassr/field/pcp-bba/).

Alfalfa Transformation and Event Sorting

A vector containing a version of the CP4 EPSPS gene and CTP targeting sequence under the control of a strong constitutive promoter was used to transform the FGI regenerable alfalfa clone R2336. The line R2336 was selected from an elite high-yielding fall dormant FGI breeding population using a tissue culture screen for callus formation and somatic embryo induction. Transformation and plant regeneration was carried out on SH growth media (Schenk and Hilderbrandt, 1972) using standard *A. tumefaciens*-mediated alfalfa transformation and regeneration conditions employing the two-step procedure described previously. A total of 212 T_0 plants derived from 122 embryogenic lines were recovered from tissue culture. Plants established from stem cuttings from the T_0 lines were evaluated for tolerance to glyphosate, and the best plant from each line was used in an F_1 cross to elite FGI material. In a greenhouse screen the progeny from the F_1 cross were sprayed with Roundup Ultra at a rate equivalent to 1.0 gallon/acre. Eighty-five of the lines exhibited excellent tolerance to glyphosate, of which approximately 75 demonstrated F_1 segregation ratios indicative of single locus T-DNA insertions.

During 1999 an aggressive event-sorting program based on agronomic performance and plant phenotype under glyphosate selection in field and greenhouse conditions was carried out with the remaining lines. Figure 7.3 shows a section of the event-sorting field evaluation. Concurrent with this evaluation a modified back-crossing program was used to introgress the RR transgene into numerous diverse FGI breeding populations. DNA blot analyses combined with F_1 segregation ratios were used to identify lines containing single T-DNA insertions. As a final step in the event sorting process a comprehensive molecular analysis of the transgene region in the remaining elite lines allowed the selection of approximately 12 lead events. From these 12 lead events, four lines were subsequently selected for use in the marker-assisted breeding program and product development program.

FIGURE 7.3. Field evaluation of Roundup Ready alfalfa lines. In May 1999 selected alfalfa lines were transplanted to the field for evaluation. During the initial growing season Roundup Ultra was applied twice at 1 gallon/acre. A row of control nontransgenic alfalfa has been killed. (*Source:* Sharie Fitzpatrick, Forage Genetics International, September 1999.)

Roundup Ready Alfalfa Breeding Strategy

Genetically modified crop cultivars typically carry a transgene from a single transgenic event (A) such that the transgene is present in the hemizygous state at a single location within the genome of a diploid plant species (i.e., A-). After back-crossing the T_0 line to superior agronomic types, high trait purity is easily achieved with the single transgenic event by inbreeding to homozygosity (AA). If these homozygous lines are used as parents in the production of either F_1 hybrids or cultivars, 100 percent of resultant plants will have the transgenic phenotype (A- or AA, respectively). By contrast, alfalfa is an insect pollinated, out-crossing autotetraploid ($2n = 4x$). To obtain very high levels of transgene transmission (> 90 to 95 percent) requires an intercrossing of parents with the transgene in duplex (AA--), triplex (AAA-), and/or quadriplex (AAAA) state at the transgenic locus. Although these genotypes can be generated with multiple cycles of phenotypic recurrent selection (PRS), no high throughput laboratory assay can accurately and precisely distinguish between plants with varying doses of a transgene at a single locus [e.g., simplex (A---) versus triplex (AAA-) or quadriplex (AAAA)]. Therefore, progeny testing is required to identify and discriminate between individuals that share the same transgenic phenotype. The intensive selection/progeny testing activities involved in such a program introduces a significant risk of inbreeding depression and/or genetic drift. Such a program presents a genetic funnel that will likely narrow the germplasm base of RR alfalfa. Consequently, significant financial and product performance risk factors are associated with such a breeding program.

An alternative product development approach based on the use of molecular markers has been developed that overcomes many of the challenges discussed earlier. Implementation of this approach also represents a significant reduction in required resources. The approach uses transgenic plants that contain two independent transgenic events (i.e., A---B---) to achieve high levels of transmission of the transgenic trait in autotetraploid alfalfa cultivars. We propose using the term *multihomogenic* to describe concomitant presence of more than one copy of a single transgene at multiple independent loci within the genome of a single plant. Plants that carry at least one copy of two transgenic events are termed *dihomogenic.* Dihomogenic plants are produced by an F_1 cross between a transgenic plant containing only event A and a transgenic plant containing only event B, each in the simplex condition. Under this convention plants with a single allele at each of two loci (A and B) are classified as 1,1-dihomogenic (i.e., A---B---). A plant that is duplex for the transgene at one locus and triplex at another would be 2,3-dihomogenic (AA--BBB-).

For each of the four elite single-copy RR transgenic alfalfa lines described (designated as events A, B, C, and D), the flanking genomic sequence of the T-DNA insertion position was determined. This was achieved using GenomeWalker technology (Clontech, Palo Alto, California) that utilizes suppressor polymerase chain reaction (PCR) technology to walk from a known point within the transgene cassette into flanking alfalfa sequence. The amplification products were then sequenced and a robust event-specific PCR (ES-PCR) assay was developed for each of the four events where one primer anneals to a sequence within the 5' or 3' region of the transgene, while the other anneals within the corresponding flanking region of the plant genome. The ES-PCR assays were coupled to a high-throughput DNA isolation procedure and have been applied to a breeding program for RR alfalfa variety development using pair-wise combinations of the four transgenic events. This high-throughput technique has been used to rapidly screen thousands of RR plants to identify the desired dihomogenic genotype (A---B---) for use as Syn1 parents. Plants identified using the PCR-aided genotypic recurrent selection procedure (GRS) are referred to as GRS_1Syn0 population. Plants identified as being dihomogenic can be intercrossed and the Roundup tolerance of the progeny (GRS_1Syn1 population) determined. A second cycle of genotypic recurrent selection using ES-PCR can then be used to produce GRS_2Syn0 plants. Cycles of phenotypic recurrent selection prior to intercrossing of the events followed by the identification of dihomogenic progeny ($PRS_n GRS_1$Syn0) can be used to increase trait purity. As with the example described, a second cycle of intercrossing and subsequent use of ES-PCR on the progeny can be used to identify a $PRS_n GRS_2$Syn0 population that will give a further increase in trait purity in the subsequent synthetic generations.

A computer model was developed to predict trait purity and inheritance of a dominant transgene in a two-event autotetraploid system (i.e., two loci). This allows the prediction of both genotypic and phenotypic frequencies in Syn1, Syn2, and Syn3 populations resulting from various crosses and selection programs (Table 7.3). For example, it is predicted that 96.8 percent of the Syn3 (certified class seed) plants will exhibit the RR phenotype if the parental line was selected using two cycles of genotypic recurrent selection (GRS_2). This example assumes phenotypic selection for Roundup tolerance during the Syn2 and Syn3 production cycles.

Validation of the Two-Event RR Breeding Model

Populations of Roundup-tolerant alfalfa plants were derived from F_1 crosses between populations of nondormant plants that had a single copy of an independent RR transgene event (e.g., A--- × B---). In

TABLE 7.3. Comparison of single- and two-event RR product development strategies

		Percent RR progeny (w/selection)		
Selection	**Syn1 parental genotype**	**Syn1**	**Syn2**	**Syn3**
			Single-event strategy	
PRS_1PT_1	Duplex	93.7	94.6	95.4
PRS_3PT_1	Triplex/quadriplex	100.0	99.6	99.4
			Two-event strategy	
GRS_1	Dihomogenic	93.7	92.7	93.8
PRS_1GRS_1	Dihomogenic	95.3	94.5	95.3
PRS_3GRS_1	Dihomogenic	96.8	96.3	96.6
GRS_2	Dihomogenic	97.0	96.4	96.8
PRS_1GRS_2	Dihomogenic	97.8	97.3	97.5

Note: Percent RR phenotype in advanced Syn generations produced by various combinations and number (*n*) of generations of genotypic recurrent selection (GRS_n) and/or phenotypic recurrent selection (PRS_n). PT_1 indicates one generation of progeny testing is required.

this example, three pair-wise event combinations were produced. Approximately two weeks after germination the seedling progeny from the F_1 crosses were sprayed with Roundup Ultra and Roundup tolerance values were obtained (F_1 percent RR). ES-PCR was used to identify 1,1-dihomogenic plants (GRS_1Syn0 plants). The plants identified as being dihomogenic were intercrossed and the Roundup tolerance of the progeny (GRS_1Syn1 populations) was determined (Table 7.4). ES-PCR was used to identify dihomogenic plants in the GRS_1Syn1 populations. These plants (GRS_2Syn0) were intercrossed and Roundup tolerance of the GRS_2Syn1 determined similarly and GRS_3Syn0 plants were generated. Table 7.4 presents the values predicted by the computer model and experimental data for three event combinations for three cycles of genotypic recurrent selection.

The two-event marker-assisted breeding development program may have an additional advantage when compared to a single-event program using triplex and quadriplex plants containing a single event. A slight reduction in the average gene copy number per plant is required to achieve the desired trait purity; this may reduce the occurrence of homology-dependent gene silencing (Vaucheret et al., 1998). By comparing the experimental phenotypic and genotypic values obtained as part of the breeding program with the model predictive values (Table 7.4) we should be able to identify unstable event combinations that are likely to be silenced in subsequent synthetic generations. A careful analysis of the data generated thus far in the alfalfa RR development program shows no evidence for gene silencing with any of the four elite events through multiple back-crossing cycles and through either PRS or GRS cycles.

Using Roundup Ready Alfalfa to Estimate Pollen Flow

Potential pollen flow between adjacent alfalfa seed production fields is a key factor in determining isolation distances required for commercial seed production. The current isolation requirements are based on measurements made using pest resistance genes to monitor pollen flow (Brown et al., 1986). RR alfalfa provides an excellent analytical tool for monitoring pollen-mediated gene flow between transgenic and nontransgenic seed production fields and also potential flow to feral alfalfa populations. In 2000 the FGI in association with

TABLE 7.4. Experimental Roundup tolerance and event-specific data for nondormant alfalfa populations using the two-event breeding scheme for three cycles of genotypic recurrent selection

F_1 cross	F_1 % RR	GRS_1Syn0 % dihomogenic	GRS_1Syn_1 % RR	GRS_2Syn0 % dihomogenic	GRS_2Syn_1 % RR	GRS_3Syn0 % dihomogenic
A × B	75.7	35.6	87.9	58.0	96.5	69.6
B × C	75.6	31.5	91.5	57.2	95.0	71.4
A × C	76.2	31.6	92.3	59.1	95.4	70.0
Predicted value	75.0	33.3	93.7	60.1	97.0	70.1

Note: Values predicted by the computer model are given for comparison. At the time of publication GRS_3Syn_1 data were not available.

the regulatory affairs committee of the North American Alfalfa Improvement Conference (NAAIC) conducted a pollen flow study near Nampa, Idaho (Fitzpatrick et al., 2001). The study measured pollen-mediated gene flow using leafcutter bees as pollinators under seed production conditions typical to the region.

Pollen-mediated gene flow was measured using the CP4 EPSPS gene as a dominant marker in a one-acre pollen source plot to replicated trap plots of conventional alfalfa planted at 500-foot increments (i.e., 500 feet, 1,000 feet, and 1,500 feet) from the source plot. The area between all plots was maintained fallow to provide conditions that would theoretically maximize bee movement. Total pollen-mediated gene flow was 1.39, 0.32, and 0.07 percent at 500, 1,000, and 1,500 feet, respectively. No pollen-mediated gene flow was detected at an additional larger trap plot at 2,000 feet where the intervening land was planted to onions and wheat (Fitzpatrick et al., 2001). A plot of mean gene flow against distance indicates a nonlinear decline (Figure 7.4). However, if the pollen flow values at 500 and 1000 feet are used and the assumption is made that a linear decline in pollen flow occurred between these two points, then it is possible to estimate gene flow at 900 feet (the current isolation distance for foundation class seed). Based on this, pollen flow at 900 feet would be predicted to be 0.53 percent based on a linear decline rate of –0.00214 percent per foot. If pollen-mediated gene flow from a point source decays to zero in a nonlinear manner, then the 0.53 percent predicted value would represent a small overestimation. The calculated value is somewhat lower than the 0.73 percent figure calculated based on a best-fit linear prediction equation (Fitzpatrick et al., 2001). The data from this study support the current foundation class seed field isolation distance of 900 feet as an effective limit for minimal gene flow. As part of the development program for RR alfalfa, FGI will be conducting additional pollen flow studies to determine the required isolation distance for commercial RR seed production.

In another recent alfalfa pollen flow study researchers monitored gene flow from source blocks to alfalfa plants at various distances using a combination of RAPD and gene-specific polymorphic markers (St. Amand et al., 2000). Although it is not possible to compare the results of the two studies directly due to differences in experimental design, several important observations were made in this study. Leaf-

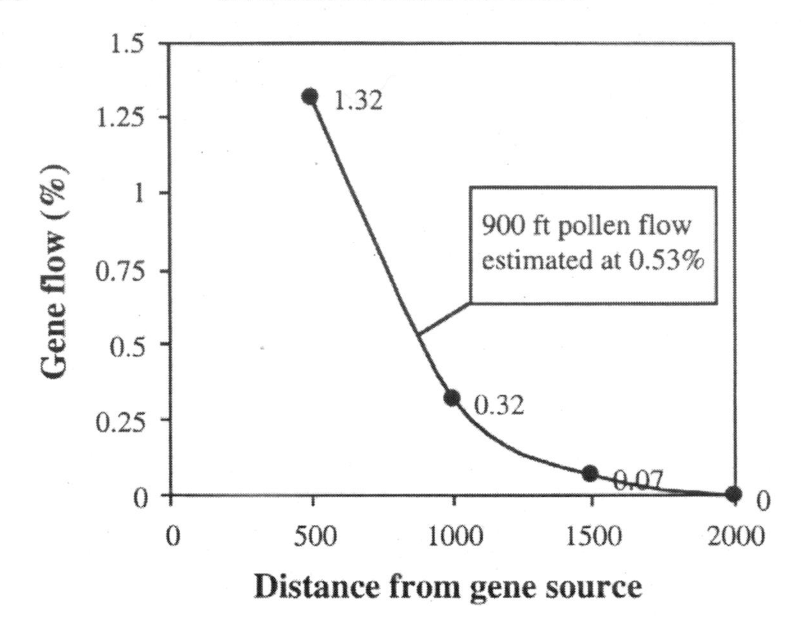

FIGURE 7.4. Plot of gene flow from a Roundup Ready alfalfa source plot to trap alfalfa plots planted at defined distances from the source plot. Pollen flow at 900 feet, the current isolation distance for foundation class seed, is estimated to be 0.53 percent.

cutter bees (*Megachile* spp.) showed a directional, nonrandom bias when pollinating within fields. Within-field movement of pollen was detected only over short distances. Only 0.2 percent gene flow was measured at 4 m and zero at 6 m. Interestingly, 84.6 percent of progeny carrying the marker were located directly north or south of the marker plants transecting the bee domicile, with the majority of pollen movement (61.5 percent) directed toward the domicile. Although the source plot in this study was extremely small, the study does suggest that within-field pollen flow is limited and that as the authors of the study indicated, border areas may provide an effective buffer zone to limit pollen flow (St. Amand et al., 2000). These findings may also affect bee domicile placement. This study also monitored pollen movement from alfalfa fields to small trap plots representing feral alfalfa populations that were planted at regularly spaced distances of up to 1,000 m from the source plot. Under these conditions pollen flow was measured out to 1,000 m from large production fields (St. Amand et al., 2000). The design of this experiment with regularly spaced unidirectional plants may have created a somewhat artificial situation with the trap plots acting as bee bridges between source and more distant trap plots resulting in a worst-case scenario for pollen flow.

CONCLUSIONS

From studies initiated more than 30 years ago to develop tissue culture systems, more recent research has resulted in the evolution of transformation systems for *M. sativa* and *M. truncatula* as tools to express genes that increase the value of crops and improve crop characteristics. The highly efficient transformation methods for *M. truncatula* will accelerate functional genomics approaches to gene discovery that can be applied to improvement of other legume species. Breeding strategies have been developed and validated for production of commercial genetically modified alfalfa varieties. Development of these tools opens up new avenues for variety development and increases the scope of utilization of forages in agricultural systems.

REFERENCES

Austin, S., E.T. Bingham, D.E. Mathews, M.N. Shahan, J. Will, and R.R. Burgess (1995). Production and field performance of transgenic alfalfa (*Medicago sativa* L.) expressing alpha-amylase and manganese-dependent lignin peroxidase. *Euphytica* 85:381-393.

Austin-Phillips, S. and T. Ziegelhoffer (2001). The production of value-added proteins in transgenic alfalfa. In *Molecular Breeding of Forage Crops* (pp. 285-301), Ed. Spangenberg, G. Dordrecht, the Netherlands: Kluwer Academic.

Barnes, D.K., B.P. Goplen, and J.E. Baylor (1988). Highlights in the USA and Canada. In *Alfalfa and Alfalfa Improvement* (pp. 1-24), Eds. Hanson, A.A., D.K. Barnes, and R.R. Hill. Madison, WI: American Society of Agronomy.

Barry, G., G. Kishore, S. Padgette, M. Taylor, K. Kolacz, M. Weldon, D. Re, D. Eichholtz, K. Fincher, and L. Hallas (1992). Inhibitors of amino acid biosynthesis: Strategies for imparting glyphosate tolerance to crop plants. In *Biosynthesis and Molecular Regulation of Amino Acids in Plants* (pp. 139-145), Ed. Singh, B.K. Rockville, MD: American Society of Plant Physiologists.

Baucher, M., M.A. Bernard-Vailhé, B. Chabbert, J.-M. Besle, C. Opsomer, M. Van Montagu, and J. Botterman (1999). Down-regulation of cinnamyl alcohol dehydrogenase in transgenic alfalfa (*Medicago sativa* L.) and the effect on lignin composition and digestibility. *Plant Molecular Biology* 39:437-447.

Bingham, E.T. (1991). Registration of alfalfa hybrid Regen-SY germplasm for tissue culture and transformation research. *Crop Science* 31:1098.

Bingham, E.T., L.V. Hurley, D.M. Kaatz, and J.W. Saunders (1975). Breeding alfalfa which regenerates from callus tissue in culture. *Crop Science* 15:719-721.

Bingham, E.T., T.J. McCoy, and K.A. Walker (1988). Alfalfa tissue culture. In *Alfalfa and Alfalfa Improvement* (pp. 903-929), Eds. Hanson, A.A., D.K. Barnes, and R.R. Hill. Madison, WI: American Society of Agronomy.

Bogorad, L. (2000). Engineering chloroplasts: An alternative site for foreign genes, proteins, reactions, and products. *Trends in Biotechnology* 18:257-263.

Bowley, S.R., G.A. Kielly, K. Anandarajah, B.D. McKersie, and T. Senaratna (1993). Field evaluation following two cycles of backcross transfer of somatic embryogenesis to commercial alfalfa germplasm. *Canadian Journal of Plant Science* 73:131-137.

Brown, D.C.W. and A. Atanassov (1985). Role of genetic background in somatic embryogenesis in *Medicago. Plant Cell, Tissue and Organ Culture* 4:111-122.

Brown, D.E., E.L. Grandstaff, M.R. Hanna, A.A. Hanson, V.L. Marble, and J.B. Moutray (1986). Committee report on alfalfa field isolation. In *Report of the Thirtieth North American Alfalfa Improvement Conference* (p. 123), July 27-31.

Chabaud, M., C. Larsonneau, C. Marmouget, and T. Huguet (1996). Transformation of barrel medic (*Medicago truncatula* Gaertn.) by *Agrobacterium tumefaciens* and regeneration via somatic embryogenesis of transgenic plants with the *MtENOD*12 nodulin promoter fused to the *gus* reporter gene. *Plant Cell Reports* 15:305-310.

Chabaud, M., J. Passiatore, F. Cannon, and V. Buchanan-Wollaston (1988). Parameters affecting the frequency of kanamycin resistant alfalfa obtained by *Agrobacterium tumefaciens* mediated transformation. *Plant Cell Reports* 7:512-516.

Clough, S.J. and A.F. Bent (1998). Floral dip: A simplified method for *Agrobacterium*-mediated transformation of *Arabidopsis thaliana. The Plant Journal* 16:735-743.

Comai, L., P. Moran, and D. Maslyar (1990). Novel and useful properties of a chimeric plant promoter combining CaMV 35S and MAS elements. *Plant Molecular Biology* 15:373-381.

Cook, D.R. (1999). *Medicago truncatula*—A model in the making! *Current Opinions in Plant Biology* 2:301-304.

das Neves, L.O., S.R.L. Duque, J.S. de Almeida, and P.S. Fevereiro (1999). Repetitive somatic embryogenesis in *Medicago truncatula* ssp. Narbonensis and *M. truncatula* Gaertn cv. Jemalong. *Plant Cell Reports* 18:398-405.

Deak, M., G.B. Kiss, C. Koncz, and D. Dudits (1986). Transformation of *Medicago* by *Agrobacterium* mediated gene transfer. *Plant Cell Reports* 5:97-100.

Delannay X., T.T. Bauman, D.H. Beighley, M.J. Buetter, H.D. Coble, M.S. DeFelice, C.W. Derting, T.J. Diedrick, J.L. Griffin, E.S. Hagwood, et al. (1995). Yield evaluation of a glyphosate-tolerant soybean line after treatment with glyphosate. *Crop Science* 35:1461-1467.

Denchev, P., M. Velcheva, and A. Atanassov (1991). A new approach to direct somatic embryogenesis in *Medicago. Plant Cell Reports* 10:338-341.

Desgagnés, R., S. Laberge, G. Allard, H. Khoudi, Y. Castonguay, J. Lapoint, R. Michaud, and L.-P. Vézina (1995). Genetic transformation of commercial breeding lines of alfalfa *(Medicago sativa). Plant Cell, Tissue and Organ Culture* 42:129-140.

D'Halluin, K., J. Botterman, and W. De Greef (1990). Engineering of herbicide-resistant alfalfa and evaluation under field conditions. *Crop Science* 30:866-871.

Du, S., L. Erickson, and S. Bowley (1994). Effect of plant genotype on the transformation of cultivated alfalfa *(Medicago sativa)* by *Agrobacterium tumefaciens. Plant Cell Reports* 13:330-334.

Finstad, K., D.C.W. Brown, and K. Joy (1993). Characterization of competence during induction of somatic embryogenesis in alfalfa tissue culture. *Plant Cell, Tissue and Organ Culture* 34:125-132.

Fitzpatrick S., P. Reisen, and M. McCaslin (2001). Alfalfa pollen-mediated gene flow. Available online at <http://www.foragegenetics.com/News.htm>.

Frugoli, J. and J. Harris (2001). *Medicago truncatula* on the move! *The Plant Cell* 13:458-463.

Fuentes, S.I., R. Suárez, T. Villegas, L.C. Acero, and G. Hernández (1993). Embryogenic response of Mexican alfalfa *(Medicago sativa)* varieties. *Plant Cell, Tissue and Organ Culture* 34:299-302.

Gamborg, O., R. Miller, and K. Ojima (1968). Nutrient requirements of suspension cultures of soybean root cells. *Experimental Cell Research* 50:151-158.

Guilley, H., R.K. Dudley, G. Jonard, E. Balazs, and K.B. Richards (1982). Transcription of cauliflower mosaic virus DNA: Detection of promoter sequences and characterization of transcripts. *Cell* 30:763-773.

Guo, D., F. Chen, K. Inoue, J.W. Blount, and R.A. Dixon (2001). Downregulation of caffeic acid 3-*O*-methyltransferase and caffeoyl CoA 3-*O*-methyltransferase in transgenic alfalfa: Impacts on lignin structure and implications of the biosynthesis of G and S lignin. *The Plant Cell* 13:73-88.

Guo, D., F. Chen, J. Wheeler, J. Winder, S. Selman, M. Peterson, and R.A. Dixon (2001). Improvement of in-rumen digestibility of alfalfa forage by genetic manipulation of lignin *O*-methyltransferases. *Transgenic Research* 10:457-464.

Hernández-Fernández, M.M. and B.R. Christie (1989). Inheritance of somatic embryogenesis in alfalfa (*Medicago sativa* L.). *Genome* 32:318-321.

Hill, K.K., N. Jarvis-Eagan, E.L. Halk, K.J. Krahn, L.W. Liao, R.S. Mathewson, D.J. Merlo, S.E. Nelson, K.E. Rashka, and L.S. Loesch-Fries (1991). The development of virus-resistant alfalfa, *Medicago sativa* L. *Bio/technology* 9:373-377.

Hindson, S., A.R. McElroy, and C. Portelance (1998). Media and genotype effects on the development and conversion of somatic alfalfa (*Medicago sativa* L.) embryos. *In Vitro Cellular and Developmental Biology* 34:181-184.

Hipskind, J.D. and N.L. Paiva (2000). Constitutive accumulation of a resveratrol-glucoside in transgenic alfalfa increases resistance to *Phoma medicaginis. Molecular Plant-Microbe Interactions* 13:551-562.

Hoffmann, B., T.H. Trinh, J. Leung, A. Kondorosi, and E. Kondorosi (1997). A new *Medicago truncatula* line with superior in vitro regeneration, transformation, and symbiotic properties isolated through cell culture selection. *Molecular Plant-Microbe Interactions* 10:307-315.

Jayasena, K.W., M.R. Hajimorad, E.G. Law, A.-U. Rehman, K.E. Nolan, T. Zanker, R.J. Rose, and J.W. Randles (2001). Resistance to *Alfalfa mosaic virus* in transgenic barrel medic lines containing the virus coat protein gene. *Australian Journal of Agricultural Research* 52:67-72.

Kamaté, K., I.D. Rodriguez-Llorent, M. Scholte, P. Durand, P. Ratet, E. Kondorosi, A. Kondorosi, and T.H. Trinh (2000). Transformation of floral organs with GFP in *Medicago truncatula. Plant Cell Reports* 19:647-653.

Keil, M., J.J. Sanchez-Serrano, and L. Wilmitzer (1989). Both wound-inducible and tuber-specific expression are mediated by the promoter of a single member of the potato proteinase inhibitor II gene family. *EMBO Journal* 8:1323-1330.

Khoudi, H., S. Laberge, J.M. Ferullo, R. Bazin, A. Darveau, Y. Castonguay, G. Allard, R. Lemieux, and L.P. Vézina (1999). Production of a diagnostic monoclonal antibody in perennial alfalfa plants. *Biotechnology and Bioengineering* 64:135-143.

Khoudi, H., L.-P. Vézina, J. Mercier, Y. Castonguay, G. Allard, and S. Laberge (1997). An alfalfa rubisco small subunit homologue shares *cis*-acting elements with the regulatory sequences of the RbcS-3A gene from pea. *Gene* 197:343-351.

Kielly, G.A. and S.R. Bowley (1992). Genetic control of somatic embryogenesis in alfalfa. *Genome* 35:474-477.

Li, X.Q. and Y. Demarly (1995). Characterization of factors affecting plant regeneration frequency of *Medicago lupulina* L. *Euphytica* 86:143-148.

Li, X.Q. and Y. Demarly (1996). Somatic embryogenesis and plant regeneration in *Medicago suffruticosa. Plant Cell, Tissue and Organ Culture* 44:79-81.

Mandaci, S. and M.S. Dobres (1997). A promoter directing epidermal expression in transgenic alfalfa. *Plant Molecular Biology* 34:961-965.

Matheson, S.L., J. Nowak, and N.L. MacLean (1990). Selection of regenerative genotypes from highly productive cultivars of alfalfa. *Euphytica* 45:105-112.

McKersie, B.D., S.R. Bowley, and K.S. Jones (1999). Winter survival of transgenic alfalfa overexpressing superoxide dismutase. *Plant Physiology* 119:839-847.

McKersie, B.D., J. Murnagham, K.S. Jones, and S.R. Bowley (2000). Iron-superoxide dismutase expression in transgenic alfalfa increases winter survival without a detectable increase in photosynthetic oxidative stress tolerance. *Plant Physiology* 122:1427-1437.

Micallef, M.C., S Austin, and E.T. Bingham (1995). Improvement of transgenic alfalfa by backcrossing. *In vitro Cellular and Developmental Biology—Plant* 31:187-192.

Mitten, D.H., S.J. Sato, and T.A. Skokut (1984). In vitro regenerative potential of alfalfa germplasm sources. *Crop Science* 24:943-945.

Murashige, T. and F. Skoog (1962). A revised medium for rapid growth and bioassays with tobacco tissue cultures. *Physiologia Plantarum* 15:473-497.

Narváez-Vásquez, J., M.L. Orozco-Cárdenas, and C.A. Ryan (1992). Differential expression of a chimeric CaMV-tomato proteinase inhibitor I gene in leaves of transformed nightshade, tobacco and alfalfa plants. *Plant Molecular Biology* 20:1149-1157.

Ninkovic, S., J. Miljus-Djuki´c, and M. Neskovic (1995). Genetic transformation of alfalfa somatic embryos and their clonal propagation through repetitive somatic embryogenesis. *Plant Cell, Tissue and Organ Culture* 42:255-260.

Nitsch, J.P. and C. Nitsch (1969). Haploid plants from pollen grains. *Science* 163:85-87.

Nolan, K.E. and R.J. Rose (1998). Plant regeneration from cultured *Medicago truncatula* with particular reference to abscisic acid and light treatments. *Australian Journal of Botany* 46:151-160.

Nolan, K.E., R.J. Rose, and J.R. Gorst (1989). Regeneration of *Medicago truncatula* from tissue culture: Increased somatic embryogenesis using explants from regenerated plants. *Plant Cell Reports* 8:278-281.

Oommen, A., R.A. Dixon, and N.L. Paiva (1994). The elicitor-inducible alfalfa isoflavone reductase promoter confers different patterns of developmental expression in homologous and heterologous transgenic plants. *The Plant Cell* 6:1789-1803.

Padgette, S.R., K.H. Kolacz, X. Delannay, D.B. Re, B.J. LaValle, C.N. Tinius, W.R. Rhodes, Y.I. Otero, G.F. Barry, D.A. Eichholtz, et al. (1995). Development, identification, and characterization of a glyphosate-tolerant soybean line. *Crop Science* 35:1451-1461.

Padgette, S.R., D.B. Re, G.F. Barry, D.E. Eichholtz, X. Delannay, R.L. Fuchs, G.M. Kishore, and R.T. Fraley (1996). New weed control opportunities: Development of soybean with a Roundup Ready gene. In *Herbicide Resistant Crops* (pp. 53-84), Ed. Duke, S.O. Boca Raton, FL: CRC Press.

Pathirana, S., D.A. Samac, R. Roeven, C.P. Vance, and S.J. Gantt (1997). Analyses of phosphoenolpyruvate carboxylase gene structure and expression in alfalfa. *The Plant Journal* 12:293-304.

Pereira, L.F. and Erickson, L. (1995). Stable transformation of alfalfa (*Medicago sativa* L.) by particle bombardment. *Plant Cell Reports* 14:290-293.

Ramaiah, S.M. and D.Z. Skinner (1997). Particle bombardment: A simple and efficient method of alfalfa (*Medicago sativa* L.) pollen transformation. *Current Science* 73:674-682.

Reisch, B. and E.T. Bingham (1980). The genetic control of bud formation from callus cultures of diploid alfalfa. *Plant Science Letters* 20:71-77.

Rose, R.J., K.E. Nolan, and L. Bicego (1999). The development of the highly regenerable seed line Jemalong 2HA for transformation of *Medicago truncatula*—Implications for regenerability via somatic embryogenesis. *Journal of Plant Physiology* 155:788-791.

Rosellini, D., M. Pezzoti, and F. Veronesi (2001). Characterization of transgenic male sterility in alfalfa. *Euphytica* 118:313-319.

Samac, D.A. (1995). Strain specificity in transformation of alfalfa by *Agrobacterium tumefaciens*. *Plant Cell, Tissue and Organ Culture* 43:271-277.

Samac, D.A. and D.M. Shah (1991). Developmentally regulated and pathogen inducible expression of the *Arabidopsis* acidic chitinase gene promoter in transgenic plants. *The Plant Cell* 3:1063-1072.

Saruul, P., F. Srienc, D.A. Somers, and D.A. Samac (2002). Production of a biodegradable plastic polymer, poly-ß-hydroxybutyrate, in alfalfa (*Medicago sativa* L.). *Crop Science* 42:919-927.

Saunders, J.W. and E.T. Bingham (1972). Production of alfalfa plants from callus tissue. *Crop Science* 12:804-808.

Scarpa, G.M., F. Pupilli, F. Damiani, and S. Arcioni (1993). Plant regeneration from callus and protoplasts in *Medicago polymorpha. Plant Cell, Tissue and Organ Culture* 35:49-57.

Schenk, B.U. and A.C. Hildebrandt (1972). Medium and techniques for induction and growth of monocotyledonous and dicotyledonous plant cell cultures. *Canadian Journal of Botany* 50:199-204.

Schoenbeck, M.A., D.A. Samac, M. Fedorova, R.G. Gregerson, J.S. Gantt, and C.P. Vance (1999). The alfalfa *(Medicago sativa)* TDY1 gene encodes a mitogen-activated protein kinase homolog. *Molecular Plant-Microbe Interactions* 12:882-893.

Scholte, M., I. d'Erfurth, S. Rippa, S. Mondy, V. Cosson, P. Durand, C. Breda, H. Trinh, I. Rodriguez-Llorente, E. Kondorosi, et al. (2002). T-DNA tagging in the model legume *Medicago truncatula* allows efficient gene discovery. *Molecular Breeding* 10:203-215.

Schroeder, H.E., M. Rafiqul, I. Khan, W.R. Knibb, D. Spencer, and T.J.V. Higgins (1991). Expression of a chicken ovalbumin gene in three lucerne cultivars. *Australian Journal of Plant Physiology* 18:495-505.

Shahin, E.A., A. Spielmann, K. Suhkapinda, R.B. Simpson, and M. Yashar (1986). Transformation of cultivated alfalfa using disarmed *Agrobacterium tumefaciens*. *Crop Science* 26:1235-1239.

Shao, C.Y., E. Russinova, A. Iantcheva, A. Atanassov, A. McCormac, D.F. Chen, M.C. Elliot, and A. Slater (2000). Rapid transformation and regeneration of alfalfa (*Medicago falcata* L.) via direct somatic embryogenesis. *Plant Growth Regulation* 31:155-166.

Shi, L., S.N. Twary, H. Yoshioka, R.G. Gregerson, S. Miller, D.A. Samac, J.S. Gantt, P.T. Unkefer, and C.P. Vance (1997). Nitrogen assimilation in alfalfa: Isolation and characterization of an asparagine synthetase gene showing enhanced expression in root nodules and dark adapted leaves. *The Plant Cell* 9:1339-1356.

Skokut, T.A., J. Manchester, and J. Schaefer (1985). Regeneration in alfalfa tissue culture. *Plant Physiology* 79:579-583.

Smith, S.E., E.T. Bingham, and R.W. Fulton (1986). Transmission of chlorophyll deficiencies in *Medicago sativa:* Evidence for biparental inheritance of plastids. *Journal of Heredity* 77:35-38.

St. Amand, P.C., D.Z. Skinner, and R.N. Peaden (2000). Risk of alfalfa transgene dissemination and scale-dependent effects. *Theoretical and Applied Genetics* 101:107-114.

Tabe, L.M., T. Wardley-Richardson, A. Ceriotti, A. Aryan, W. McNabb, A. Moore, and T.J.V. Higgins (1995). A biotechnological approach to improving the nutritive value of alfalfa. *Journal of Animal Science* 73:2752-2759.

Tesfaye, M., S.J. Temple, C.P. Vance, D.L. Allan, and D.A. Samac (2001). Overexpression of malate dehydrogenase in transgenic alfalfa enhances organic acid synthesis and confers tolerance to aluminum. *Plant Physiology* 127:1836-1844.

Thomas, M.R., R.J. Rose, and K.E. Nolan (1992). Genetic transformation of *Medicago truncatula* using *Agrobacterium* with genetically modified Ri and disarmed Ti plasmids. *Plant Cell Reports* 11:113-117.

Trepp, G.B., M. van de Mortel, H. Yoshioka, S.S. Miller, D.A. Samac, J.S. Gantt, and C.P. Vance (1999). NADH-glutamate synthase (GOGAT) in alfalfa root nodules: Genetic regulation and cellular expression. *Plant Physiology* 119:817-828.

Trieu, A.T., S.H. Burleigh, I.V. Kardailsky, I.E. Maldonado-Mendoza, W.K. Versaw, L.A. Blaylock, H. Shin, T.-J. Chiou, H. Katagi, G.R. Dewbre, et al. (2000). Transformation of *Medicago truncatula* via infiltration of seedlings or flowering plants with *Agrobacterium. The Plant Journal* 22:531-541.

Trieu, A.T. and M.J. Harrison (1996). Rapid transformation of *Medicago truncatula:* Regeneration via shoot organogenesis. *Plant Cell Reports* 16:6-11.

Trinh, T.H., P. Ratet, E. Kondorosi, P. Durand, K. Kamaté, P. Bauer, and A. Kondorosi (1998). Rapid and efficient transformation of diploid *Medicago truncatula* and *Medicago sativa* ssp. *falcata* lines improved in somatic embryogenesis. *Plant Cell Reports* 17:345-355.

Vaucheret, H., C. Beclin, T. Elmayan, F. Feuerbach, C. Goden, J.B. Morel, P. Mourrain, J.C. Palauqui, and S. Vernhettes (1998). Transgene-induced gene silencing in plants. *The Plant Journal* 16:651-659.

Walker, K.A. and S.J. Sato (1981). Morphogenesis in callus tissue of *Medicago sativa:* The role of ammonium ion in somatic embryogenesis. *Plant Cell, Tissue and Organ Culture* 1:109-121.

Walker, K.A., P.C. Yu, S.J. Sato, and E.G. Jaworski (1978). The hormonal control of organ formation in callus of *Medicago sativa* L. cultured in vitro. *American Journal of Botany* 65:654-659.

Wan, Y., E.L. Sorensen, and G.H. Liang (1988). Genetic control of in vitro regeneration in alfalfa (*Medicago sativa* L.). *Euphytica* 39:3-9.

Wang, J.H., R.J. Rose, and B.I. Donaldson (1996). *Agrobacterium*-mediated transformation and expression of foreign genes in *Medicago truncatula. Australian Journal of Plant Physiology* 23:265-270.

Wenzel, C.L. and D.C.W. Brown (1991). Histological events leading to somatic embryo formation in cultured petioles of alfalfa. *In vitro Cellular and Developmental Biology* 27P:190-196.

Wigdorovitz, A., C. Carrillo, M.J. Dus Santos, K. Trono, A. Peralta, M.C. Gómez, R.D. Ríos, P.M. Franzone, A.M. Sadir, J.M. Escribano, and M.V. Borca (1999). Induction of a protective antibody response to foot and mouth disease virus in mice following oral or parenteral immunization with alfalfa transgenic plants expressing the viral structural protein VP1. *Virology* 255:347-353.

Winicov, I. (2000). Alfin1 transcription factor overexpression enhances plant root growth under normal and saline conditions and improves salt tolerance in alfalfa. *Planta* 210:416-422.

Woo, H.-H., M.J. Orbach, A.M. Hirsch, and M.C. Hawes (1999). Meristem-localized inducible epxression of a UDP-glycosyltransferase gene is essential for growth and development in pea and alfalfa. *The Plant Cell* 11:2302-2315.

Yoshioka, H., R.G. Gregerson, D.A. Samac, K.C. Hoevens, G. Trepp, J.S. Gantt, and C.P. Vance (1999). Aspartate aminotransferase in alfalfa nodules: Localization of mRNA during effective and ineffective nodule development and promoter analysis. *Molecular Plant-Microbe Interactions* 12:263-274.

Yu, K. and K.P. Pauls (1993). Identification of a RAPD marker associated with somatic embryogenesis in alfalfa. *Plant Molecular Biology* 22:269-277.

Ziegelhoffer, T., J. Will, and S. Austin-Phillips (1999). Expression of bacterial cellulase genes in transgenic alfalfa (*Medicago sativa* L.), potato (*Solanum tubersosum* L.) and tobacco (*Nicotiana tabacum* L.). *Molecular Breeding* 5:309-318.

Chapter 8

Sorghum Transformation for Resistance to Fungal Pathogens and Drought

S. Muthukrishnan
J. T. Weeks
M. R. Tuinstra
J. M. Jeoung
J. Jayaraj
G. H. Liang

INTRODUCTION

Grain sorghum [*Sorghum bicolor* (L.) Moench] is the fifth most important cereal in the world, with annual planting of 48 million hectares (Food and Agriculture Organization of the United Nations [FAO], 2000). It is important throughout the semiarid tropical and subtropical regions as a staple food source. In fact, grain sorghum is a multiple purpose cereal. Sorghum serves as both food and feed, as well as an ingredient in beverages and for producing ethanol. Food dyes have been extracted from the purple reddish glumes, and stover has been used as fuel, livestock feed, building material, and for charcoal. In the United States, sorghum was planted in approximately 4.6 million hectares with an average yield of 4,387 kg·ha–1 (FAO, 2000). The ability of sorghum to be grown in marginal land, its survival under adverse environmental conditions, its versatility as a food and feed grain, and its high and stable yield ensure the important role of sorghum for millions of people throughout the world.

The genus *Sorghum* is native to tropical and some temperate regions (Morden et al., 1990). Sorghum is a complex genus, comprised of species with various chromosome numbers. The most important species in the genus is *Sorghum bicolor,* a perennial diploid ($2n = 2x = 20$), including *S. bicolor* (cultivated sorghum) and its nearest wild relatives, *S. arundinaceum* (Desv.) de Wet et Harlan (wild sorghums) and *S. drummondii* (Steud.) de Wet (weedy sorghums). The *S. arundinaceum* species also includes sudangrass or *S. sudanense*. In addition to *S. bicolor,* there is a rhizomatous perennial species, *S. halepense* or johnsongrass, at tetraploid ($2n = 4x = 40$) level, and the diploid ($2n = 2x = 20$) rhizomatous perennial species, *S. propinquum* (De Wet, 1978; Guo et al., 1996). Domestication of sorghums likely began in Africa more than 5,000 years ago. Its early domestication and worldwide spread make sorghum one of the most important cereals in the world, and it is cultivated in more than 100 countries (Frederiksen and Odvody, 2000).

The major production limitations for sorghum can be classified into three categories: (1) abiotic stress due to extreme temperatures, drought, acidic soils (which could be associated with aluminum toxicity and deficiency of phosphorus), and high pH soils (which limit the availability of iron leading to iron deficiency); (2) insect pests such as chinch bugs, green aphids (*Schizaphis graminum* Rondani), head midge, shootfly, stem borer, and heliothis *(Helicoverpa armigera);* and (3) diseases caused by pathogenic fungi, bacteria, and viruses, such as stalk rot, downy mildew, maize dwarf mosaic virus, smut, grain mold, leaf blight, leaf rust, and ergot. All these stresses limit yield potential and cause economic loss in excess of $1 billion worldwide (Nwanze et al., 1995).

With recent advances in sorghum tissue culture and transformation, it is feasible to improve sorghum by incorporating agronomically important genes from species that are sexually incompatible with sorghum. Particularly attractive are genes conferring resistance to diseases, herbicides, and insects, or tolerance to drought stress for dryland farming, cold soils for early planting, freezing tolerance for late harvesting, and genes for viral (or other microbial) proteins that can induce mucosal immunity in cattle. Many of these genes are available for transformation, as are the techniques for introducing them into plants, even though the efficiency of transformation is rather low at present.

CONSTRAINTS TO SORGHUM TRANSFORMATION

Sorghum is known to be a recalcitrant material for tissue culture manipulations. Unlike tobacco, rice, wheat, and barley, haploid production via anther and pollen culture has not been successful with sorghum. Unlike wheat and gamagrass, wide-crosses are not feasible with sorghum (Riera-Lizarazu and Mujeeb-Kazi, 1993). Although sorghum tissue culture was reported as early as the 1970s (Masteller and Holden, 1970; Gamborg et al., 1977; Brar et al., 1979), utilization of these techniques for sorghum improvement has been constrained by several factors. Sorghum tissue culture response and regeneration capacity, as with many other monocot cultures, is genotype dependent and restricted to a few inbred lines (Kaeppler and Pedersen, 1996; Godwin and Gray, 2000). Inbred lines that are able to produce calli and regenerate into seedlings include Tx 430, Wheatland, C401 (a Chinese line), CO25 (an Indian line), P898012, P954035, SA281, and M35-1. Several different types of culture media have been used to induce sorghum calli and regenerate plants, including I_6 (Casas et al., 1993), I_3 (Zhao et al., 2000), MS (Murashige and Skoog, 1962), and N6 with low ammonium concentration (Chu et al., 1975). Sorghum cells in culture typically secrete dark purple pgments containing phenolic compounds that can be toxic and inhibit growth of cells and calli. The amount of phenolic compounds produced by cells has been shown to be correlated with seasonal changes (Liang, unpublished). Anthers and immature zygotic embryos collected in the summer perform poorly in tissue culture, while those collected in spring and fall are much better.

Another problem in transformation of sorghum is that cultured cells and tissues may have an inherent tolerance to high concentrations of antibiotics, such as kanamycin and hygromycin. These antibiotics, which often are used as selective agents in dicot species, are not useful as selection agents in sorghum tissue culture. In addition, sorghum transformation has been hampered by the limited availability of promoters with stable, high-level expression. Promoters successfully used in other crop species are not as efficient in sorghum (Hill-Ambroz and Weeks, 2001).

SORGHUM TISSUE CULTURE

A review of sorghum tissue culture was recently presented by Godwin and Gray (2000), and we refer readers to this article for a comprehensive description of the developments in this field. To summarize briefly, most of the success in sorghum regeneration, which is essential for transformation, has come from calli derived from immature embryos and from immature inflorescences. There is one report of sorghum regeneration from mesophyll-derived protoplasts (Sairam et al., 1999). Several reports describe use of sorghum explants (shoot apices, leaf or shoot segments, and anthers) for plant regeneration without the intermediary callus phase, but so far there are no published reports of plant transformation using this organogenic regeneration strategy. However, organogenic regeneration has several advantages, most important, speed and avoidance of tissue-culture-related artifacts, and is being pursued actively in some laboratories. Transformation techniques that do not utilize tissue culture operations may provide a means for improving the efficiency of plant regeneration/transformation. The pollen-tube pathway adopted in rice (Luo and Wu, 1988), pollen-mediated transformation in maize (Wang et al., 2001), and vacuum infiltration approach in *Arabidopsis* (Bechtold et al., 1993), as well as other transformation protocols, should be evaluated for use in sorghum transformation. These transformation techniques do not require tissue culture and could be more effective if transformants can be produced with comparable frequency.

TRANSFORMATION METHODS

Many important crop species have been successfully transformed, but with varying efficiencies. The two different approaches generally used to produce transgenic crop plants are microprojectile bombardment (biolistic transformation) and *Agrobacterium*-mediated transformation. Two others, polyethyleneglycol (PEG) -mediated protoplast transformation and electroporation, have been successfully used in rice and a few other crops. In biolistic transformation, the BioRad PDS 1000/He gun and similar designs or the particle inflow gun have been used to deliver DNA coated on gold or tungsten particles for transformation. The delivery of plasmid DNA into the appropriate layer of the explant (typically the second layer of cells in the explant) results in stable integration of the plasmid DNA into chromosomal or organellar DNA in a small percentage of the cells receiving the DNA. If this initial transformed cell has or acquires embryogenic potential, a transgenic plant results and can be selected in medium containing the appropriate herbicide or antibiotic.

Agrobacterium-mediated transformation utilizes various strains of *Agrobacterium tumefaciens* for introduction of the transgene via the natural infection pathway. In this case, a part of the vector DNA (called the T-DNA) contained between two unique sequences called the left and right borders is transferred as single-stranded DNA coated with DNA binding proteins encoded by *Agrobacterium* and transferred to the nuclear compartment. The T-DNA is converted into double-stranded DNA in the nucleus and is integrated into host DNA at random sites. This transfer process is facilitated by wounding or by the addition of phenolic compounds, such as acetosyringone (Zupan and Zambryski, 1995; Kado, 1998).

Progress in Sorghum Transformation by a Biolistic Protocol

Godwin and Gray (2000) recently summarized the efforts of laboratories worldwide on sorghum transformation and we will briefly review and update recent progress in this area. Battraw and Hall (1991) reported successful transformation of sorghum calli by electroporation, as revealed by selection in kanamycin-containing medium; however, no plants were regenerated from these calli. Similarly Hagio and colleagues (1991) also demonstrated transformation of sorghum cell suspension cultures by selection in hygromycin- or kanamycin-containing medium, but they could not regenerate any plants. Casas and colleagues (1993) were the first to regenerate transgenic sorghum plants after bombardment of immature zygotic embryos of sorghum with a vector containing the *bar* gene. Out of eight genotypes of sorghum tested, three produced embryogenic calli that survived selection in medium containing bialaphos, but plants were regenerated only from genotype P898012. In a later study, they were able to regenerate bialaphos-resistant plants from immature inflorescence of the genotype SRN39 (Casas et al., 1997). The transgenic plants were shown to be resistant to Ignite/Basta and had detectable phosphinothricin acetyltransferase (PAT) activity, indicating a functional *bar* gene. The *bar* gene transmission to the progeny followed Mendelian patterns of inheritance for a single locus. Similarly, Rathus and colleagues (2004) reported the regeneration of plants after selection in medium containing Basta, and leaf extracts from the regenerated plants had PAT activity.

The first successful introduction of an agronomically useful gene into sorghum was reported by Zhu and colleagues (1998). They used a vector containing the *bar* gene as well as a chitinase gene from rice under the control of a CaMV 35S RNA promoter. Six transgenic plants were obtained from a total of about 1,100 immature embryo-derived calli that were bombarded with a plasmid DNA containing the selectable marker, *bar,* and the rice chitinase gene. Stable expression of the *bar* gene and the chitinase gene were demonstrated using enzyme assays and Western blot analyses. There was some evidence of chitinase gene-silencing among the progeny of transformants in subsequent generations (Krishnaveni et al., 2001). Godwin and Gray (2000) also obtained transgenic sorghum plants after bombarding immature embryo explants with a plasmid DNA containing the *bar* gene and *cry1A* gene. They obtained transgenic plants from three different genotypes—SA281, P898012, and M35-1.

Thus, there are now several examples of successful transformation with biolistic transformation protocols even though the efficiency of transformation leaves much to be desired. As indicated earlier, much of the difficulty lies in the excessive production of phenolic pigments that seem to be toxic and inhibitory to regeneration processes, especially under conditions of selection.

Progress in Sorghum Transformation Mediated by **Agrobacterium**

The first suggestion that *Agrobacterium*-mediated transformation is possible with sorghum came from the work of Godwin and Chickwamba (1994). They injected *Agrobacterium* carrying the histochemical marker *gus* into the emerging coleoptile with a hypodermic needle. About 10 percent of the injected seedlings tested GUS positive. Even though the *gus* gene apparently was inherited by the progeny of putative transgenic plants, the segregation ratio did not follow the 3:1 Mendelian ratio and no confirmatory evidence was obtained by Southern blot analyses.

Jeoung (2001) used *Agrobacterium* strains EHA101 and EHA105 containing the plasmid pPZP200(HBT/GFP), which carries the gene encoding green fluorescent protein, to transform calli derived from immature embryos of three sorghum inbreds (Tx 430, Wheatland, and C401). Green fluorescent spots diagnostic of *gfp* gene expression were detected in calli from all three inbreds within 2 h after bombardment. These spots persisted for various lengths of time (up to several days) depending on the particular sample and then disappeared. In some calli, however, the clusters of cells exhibiting GFP fluorescence increased in size, eventually forming a lobelike structure with a bright green fluorescence, from which shoots emerged indicating stable transformation with the *gfp* gene carried in the *Agrobacterium* vector. Southern blot analyses confirmed the presence and integration of the *gfp* gene into sorghum chromosomal DNA. In another series of experiments, a rice *tlp* gene was successfully introduced into the sorghum inbred C401 (Jeoung, 2001). The transgenic plants not only expressed high levels of TLP but also transmitted the transgene to progeny plants (see section titled Transgenic Sorghum Plants with Enhanced Tolerance to Drought). The transformation efficiencies were greater than the values previously reported for biolistic transformation.

Zhao and colleagues (2000) described in detail the transformation of immature embryos of two lines of sorghum under a variety of conditions using the *Agrobacterium* strain LBA4404 carrying a superbinary vector with a *bar* gene or a *gus* gene under the control of maize ubiquitin promoter. Stable transformants were obtained under most of the conditions tested with variable efficiencies (0.17 to 0.25 percent). They reported the isolation of 115 transgenic events (herbicide resistance followed by confirmation by Southern blot analyses in some cases) from 5890 infected embryos for the line P898012 and 16 transgenic events from 285 *Agrobacterium*-infected embryos of line PHI391 as revealed by expression of the *gus* gene. Southern blot analysis revealed that a majority of the transgenic plants had a single insertion site with no apparent rearrangement. Another 25 percent of the transgenic plants had either multiple insertions or rearrangements. Analysis of progeny from five independent transformants indicated typical Mendelian segregation for a single locus of the transgene.

It is clear that *Agrobacterium*-mediated transformation is feasible with at least some sorghum genotypes, even though there may be some genotype preference for some combinations of sorghum inbreds and *Agrobacterium* strains. Further, it appears that the integration patterns of transgenes are less complex with *Agrobacterium* transformation compared to biolistic transformation. Ongoing research should focus on further improving the efficacy of transformation protocols using either the biolistic bombardment or *Agrobacterium*-mediated transformation methods. The experimental details for the two procedures are outlined in the following.

Biolistic Bombardment

Biolistic bombardment involves several steps, beginning with tissue culture and callus development through transformation, selection, plant regeneration, and evaluation. A general schematic listing the steps involved is provided here:

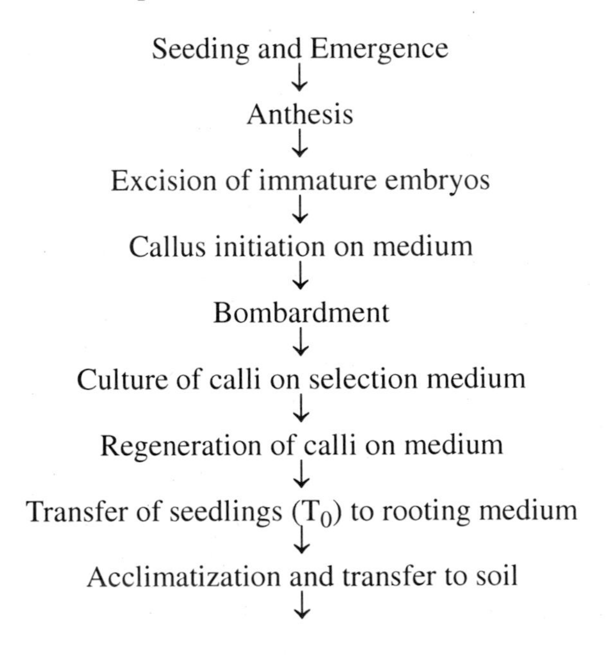

Transfer to greenhouse
↓
Confirmation of putative transgenic plants
↓
Anthesis
↓
Harvest T_1 seeds from mature plants
↓
Progeny analysis of T_1 plants
↓
Selfing of T_1 plants to obtain T_2 seeds
↓
Selection of homozygous T_2 plants for further testing
↓
Selfing of the T_2 plants and selection of homozygous T_3 plants with high expression of the target gene(s)

Various parameters for biolistic transformation, including pressure (ranging from 600 to 1550 psi), particle size (0.6 to 1.1 μm) and type of carrier (gold and tungsten), target distance (7.5 to 10 cm), and target material (cell suspension, callus, meristem, protoplast, immature embryo), vary somewhat depending on the explants involved. The disadvantages of using the biolistic gun include low transformation frequencies, frequent gene silencing resulting from lack of control in transgene copy number, potential intellectual property restrictions, and cost.

Agrobacterium-*Mediated Transformation*

Calli must be induced from the explants, such as immature embryos of sorghum. The procedure used in our laboratory for sorghum is briefly described as follows:

Callus induction from explants and preparation of *Agrobacterium* culture,
resuspension of overnight bacterial culture
in I_6 liquid medium containing 100 to 300 μM acetosyringone
at the desired cell density ($A_{600} = 0.5$)
↓

Coculture/submergence of calli in the bacterial suspension
for 30 min

↓

Blot-drying the calli for a few seconds and placing on semisolid I_6 medium containing 2.5 $g \cdot L^{-1}$ phytagel and 100 μM acetosyringone and culture for three to four days at 25°C in darkness

↓

Washing the calli three times with a cefotaxime (500 $mg \cdot L^{-1}$) solution and culture in I_6 medium without bialaphos for one week in darkness

↓

Transfer of the calli to fresh medium containing 2 $mg \cdot L^{-1}$ bialaphos and culture at 25 to 27°C for six weeks, with subculture to fresh medium once every two weeks

↓

Shoot formation from bialaphos-resistant calli and transfer to rooting medium (½ strength MS medium with 0.2 $mg \cdot L^{-1}$ indole acetic acid [IAA],
0.2 $mg \cdot L^{-1}$ 1= naphthalene acetic
acid [NAA]) containing 3 $mg \cdot L^{-1}$ bialaphos

↓

Transfer of shoots with well-developed root systems to soil and grow in a greenhouse

It is known that DNA delivery efficiency is significantly decreased in the absence of acetosyringone when embryo-derived calli are used in transformation of rice (Hiei et al., 1994) and maize (Ishida et al., 1996). However, when wheat cell suspension was used in transformation with *Agrobacterium,* acetosyringone was not essential (Cheng et al., 1997), suggesting that different tissues may have different competence for *Agrobacterium* infection. However, surfactants, such as Silwet, were reported to improve DNA delivery, due to possible surface-tension-free cells favoring the *Agrobacterium* attachment.

Other Transformation Procedures

Efficiency of tissue-culture-based transformation has been limited by the genotype-specific response in many crops. This is not an easy problem to solve as it involves complex medium composition and culture conditions for each specific genotype. Without adequate amount of callus formation and regenerated plants, production of transgenic plants in large numbers for screening will be difficult. However, transformation techniques that can bypass the tissue culture requirement are known, and some have been evaluated in sorghum transformation experiments.

Vacuum in Filtration

This procedure was proposed by Bechtold and colleagues (1993) for *Arabidopsis*. Sorghum plants (Tx 430) at seven- to nine-leaf stage were submerged for 5 min under vacuum (400 mm Hg) in the *Agrobacterium* solution ($A_{600} = 0.32$) with the construct carrying the *bar* gene. Plants were covered with a plastic film to maintain humidity and kept in a growth chamber at 22°C for 24 h. Then the covers were removed and plants were allowed to grow to maturity and T_1 seeds were harvested. The seeds were germinated in flats and sprayed with 0.2 percent Liberty when they reached four-leaf stage. Among 2,000 seedlings (100 seeds from each treated plant) sprayed, very few maintained green color after four days, and those green seedlings eventually died ten days later. Infiltration time and pressure could be adjusted for positive results, and the appropriate stage of floral development for treatment should also be investigated in future experiments.

Pollen Tube Pathway

It was reported in rice (Luo and Wu, 1998) that the pollen tube could serve as a vector for foreign DNA. We collected fresh sorghum pollen and pollinated ATx 430 (male sterile line). After 45 min, the stigma surface was sprayed with the plasmid DNA carrying *chi11* and *bar* genes without cutting off the stigma surface. Seeds were harvested and planted in a greenhouse. Seedlings were analyzed by Western blotting for *chi11* expression and by polymerase chain reaction (PCR) for detection of the *bar* gene. At present, however, no definitive results are available to confirm whether the technique worked. This approach is natural and simple and it may be desirable to repeat with minor adjustment—timing of DNA application, concentration of DNA solution, and removal of stigma surface before DNA is applied.

Pollen-Mediated Transformation

In maize, Wang and colleagues (2001) reported that plasmid DNA (pGL II-RC-1) containing genes encoding chitinase and hygromycin was mixed with fresh pollen in a sucrose solution: 0.3 g of fresh pollen plus 10 μg of the plasmid DNA in 5 percent sucrose solution for a total volume of 20 ml. Sonication was then applied at an intensity of 300 Watts with a sonicator (Ningbo Xinzi Sci. Instrument Co, China) for 5 s, and the same treatment was repeated eight times at 10 s intervals. The sonicated pollen were used to pollinate the silks of two inbred lines, Tai 9101 and Zong 31, and the progeny from 14 T_1 plants were analyzed using dot blot. Six of them showed a positive band, whereas the controls were all negative. PCR analysis of the T_2 progeny showed positive dot blots; 19 of 24 T_2 plants examined were positive. PCR-positive T_2 plants also were analyzed by Southern blotting, and 11 of the 15 plants assayed were positive while the control was negative. The positive T_2 plants were able to germinate and grow normally in hygromycin solution (25 $mg \cdot L^{-1}$) whereas all controls grew slowly, with inhibited root development, and eventually died. We are currently experimenting with a similar approach for sorghum transformation with Tx 430.

TRANSGENIC SORGHUM PLANTS WITH INCREASED STRESS TOLERANCE

Transgenic Sorghum Plants with Enhanced Resistance to **Fusarium** *Stalk Rot*

Sorghum plants are susceptible to many diseases caused by fungal, bacterial, and viral pathogens (Frederiksen and Odvody, 2000). In the Great Plains region of the United States where sorghum is grown year after year, it is not uncommon that sorghum plants are under attack by many pathogens. In Kansas, the most severe disease is induced by *Fusarium thapsinum* and *F. verticillioides,* which cause stalk rots, seed rots, and root rots in grain and forage sorghum. On average,

Fusarium stalk rot contributes to a 5 to 10 percent reduction in yield; however, these losses can reach 100 percent in localized areas. Loss of yield potential is attributed to poor grain filling and lodging of infected stalks. The severity of the disease is dependent on sorghum genotype, soil type, fertility, drainage, temperature, moisture, and the prevalence of other diseases. In Kansas alone, the annual loss due to *Fusarium* stalk rot can reach $80 million (Jardine, 1996, personal communication). Many inbred lines that are currently used in producing hybrids are susceptible to *Fusarium* stalk rot.

Lesions of infected plants vary in size from small circular spots to elongated streaks and also vary in color from light red to dark purple. Lesions may appear on the exterior and interior or the cortex region of the stalks. Typically, the lower two to three internodes will show large areas of discoloration, showing the reddish pith. Premature plant death, usually during grain development, is diagnostic for *Fusarium* stalk rot.

Rotation, fallow, and good management practices can reduce losses from stalk rot. However, the most important component in disease control is hybrid selection. Some hybrids express superior stalk quality and better lodging resistance than others. More aggressive disease management strategies could include use of transgenic sorghum hybrids that incorporate resistance genes, such as those encoding chitinase, β-glucanase, and thaumatin-like protein (TLP) .

Plant chitinases are ubiquitous plant protection enzymes against fungal pathogens (Neuhaus, 1999). These enzymes degrade chitin, the structural polysaccharide of the cell wall of many fungi, through the hydrolysis of the β-1,4-linkages of chitin. A rice chitinase gene, *chi11*, has been successfully incorporated into the sorghum inbred Tx 430 via biolistic bombardment (Zhu et al., 1998). The homozygous T_2 plants grown in the greenhouses were inoculated with spores of *Fusarium thapsinum* and examined for symptoms 20 days after inoculation. The symptoms of stalk rot in transgenic sorghums were significantly lower than the control (Figure 8.1), indicating a positive effect of the introduced chitinase gene (Krishnaveni et al., 2001). However, there was variation in the level of resistance among the transgenic lines and the disease resistance appeared to correlate positively with the level of chitinase in these plants as indicated by Western blot analysis. The rice chitinase was also shown to protect roots of

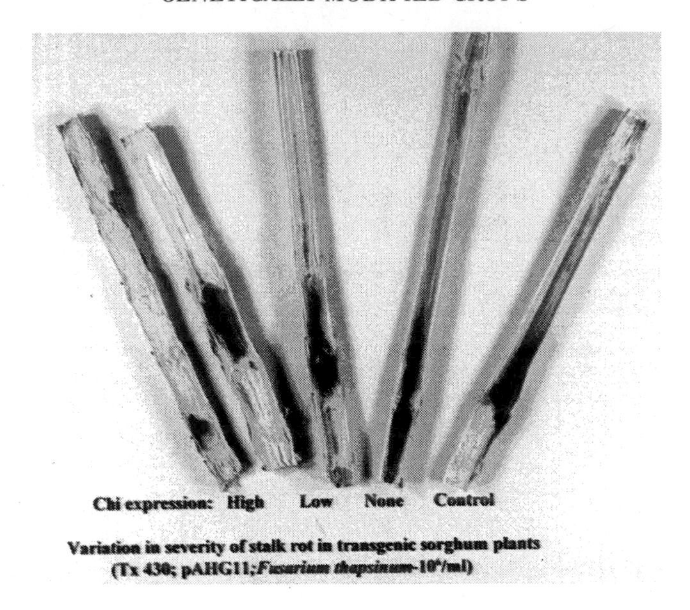

FIGURE 8.1. Reaction of Tx 430, a susceptible sorghum inbred to stalk rot, and the chitinase transgenic plants after inoculation with *F. thapsinum* (10^6 spores/mL)

transgenic seedlings from *Fusarium* infection. Transgenic sorghums had much less root discoloration when the seedlings were inoculated with a conidial suspension of the pathogen, whereas control seedlings exhibited reddish purple color on all roots (Krishnaveni et al., 2001).

Likewise, the gene encoding a rice thaumatin-like protein, *tlp,* also exhibited a positive effect on disease resistance (Jeoung, 2001). Thaumatin-like protein is a PR-5 protein that is known to influence the permeability of fungal plasma membrane (Roberts and Selitrennikoff, 1990). The TLPs are normally not detected in leaves of young healthy plants, but they are rapidly induced to high levels in response to biotic or abiotic stress (Velazhahan et al., 1999). The TLPs are generally highly soluble proteins with molecular weights in the range of 22 to 26 kDa (201 to 229 amino acids) or 16 kDa (148 to 151 amino acids). They are resistant to proteases and heat-induced denaturation.

High concentrations of TLPs can lyse fungal membranes and inhibit hyphal extension, as demonstrated with osmatin isolated from tobacco and zeamatin from maize. Transgenic sorghum lines containing the *tlp* gene were developed and tested to determine the antifungal effects of TLP from rice on stalk rot of sorghum. These results indicated that transgenic plants containing the *tlp* gene had better disease resistance than the control following inoculation with the pathogen (Jeoung, 2001).

Transgenic Sorghum Plants with Enhanced Tolerance to Drought

Due to the diversity of environments in which the crop is grown, sorghum is constantly challenged not only by pathogens but also by abiotic stresses, especially drought. TLPs such as osmatin are proteins that can be induced by drought stress and accumulate in plant cells. In our efforts to determine whether overexpression of a TLP in sorghum would help protect against drought stress, transgenic sorghum plants containing a rice *tlp* gene have been developed in our laboratory. Three-week-old transgenic sorghum plants withstood drought stress much better than control plants 12 to 16 days after water supply ceased. Other parameters, such as photosynthetic efficiency or the ratio of F_v (variable fluorescence) to F_m (maximum fluorescence), osmotic potential, and relative water content, also indicated that some transgenic lines with high levels of rice TLP were more drought tolerant than the control group. In this regard, TLP, similar to the osmatin-like proteins (OLP) from tobacco, may have multiple beneficial effects such as enhancing drought tolerance as well as disease resistance. Some TLPs also are known to protect against freezing (Zhu et al., 1996).

Transgenic Plants with Other Agronomically Useful Genes

In addition to *Fusarium* stalk rot and drought tolerance, sorghum plants suffer from many other pests and abiotic stresses. Numerous insect pests attack and feed on sorghum. These include greenbugs and chinch bugs, fall armyworm, and sorghum head midge. Important diseases including ergot and potyvirus also contribute to crop damage and yield loss. Stalk strength may be increased using genes of the biosynthetic pathway of cellulose and lignins. Sorghum is of tropical origin and does not tolerate cold temperature. If genes conferring cold or freezing tolerance could be transferred, it would lengthen the growing season by allowing earlier planting and late harvest without concern over frost damage. Because grain and forage sorghums are primarily used for livestock feed, transgenic plants expressing genes for viral or microbial antigenes can serve as vaccines capable of inducing mucosal immunity. This strategy would be an economical and convenient way to prevent epidemics. Further, since sorghums are typically grown on poor soil, it would be very desirable to introduce nitrogen-fixing genes from *Rhizobium* species and other microorganisms into sorghum. However, this may be a technically demanding and complicated goal because several operons comprising at least 20 genes are subject to NodD-flavonoid or other early symbiotic regulation (Buchanan et al., 2000).

As more genes become available for transgenic plants and methods of delivering genes into sorghum become refined and improved, the potential to utilize the genetic engineering approach to improve sorghum appears to be unlimited.

UTILIZING TRANSGENIC SORGHUMS FOR CROP IMPROVEMENT

Although basic genetic research to develop methods for producing transgenic sorghum plants has been successful, numerous regulatory and social concerns must be addressed before these products can be brought to the field. Since sorghum plants can cross with several noxious weeds, including shattercane and johnsongrass, the likelihood that transgenes introduced into sorghum will escape into wild populations of these weeds remains high if genetically modified sorghum hybrids are grown commercially (Arriola and Ellstrand, 1996). Although transgene escape is not a problem unique to sorghum, the economic impact and ubiquitous nature of sorghum weeds in crop production environments around the world will require considerable attention to risk assessment before transgenic sorghums are commercialized.

Nearly all plant transformation systems require the use of a selectable marker gene in combination with the transgene of interest in order to easily identify and select plant cells that have been transformed. The most commonly used selectable markers include genes for resistance to antibiotics or herbicides. The use of these selectable marker genes allows for the identification of putative transformants that may contain the transgene of interest by exposure to appropriate antibiotic or herbicide. Although these systems were very useful for developing transformation procedures and methodology, transfer of these genes to wild populations of sorghum where they could confer a selective advantage is highly undesirable. For example, *bar* gene is one of the most commonly used selectable marker genes and confers resistance to the herbicide Liberty. This herbicide is commonly used to control sorghum weeds in corn and soybean production fields. If this transgene were to escape into weedy sorghum populations, plants containing these genes would have a strong selective advantage through resistance to herbicide application. The development of sorghum transformation systems that do not utilize antibiotic or herbicide selectable markers are being developed to address this problem. One promising alternative is a new selectable marker gene, cyanamide hydratase *(Cah),* derived from the soil fungus *Myrothecium verrucaria*. This gene has been isolated and produces an enzyme that converts cyanamide into urea by catalyzing the addition of a water molecule. It has been demonstrated that the *Cah* gene can be used as a selectable marker for monocot transformation (Weeks et al., 2000). Because the *Cah* selectable marker gene does not confer resistance to a herbicide or antibiotic, there will likely be decreased public concern about its presence in the environment as compared to herbicide or antibiotic selectable marker genes. In addition, the presence of the *Cah* gene at the whole-plant level could offer a number of benefits and contribute to plant improvement.

In contrast to developments in other crops, genetically modified sorghum will not focus on herbicide resistance genes for improved management. Greenhouse experiments evaluating the effects of transgenic plants with host-plant resistance and environmental adaptation genes have indicated promise in enhancing resistance to biotic and abiotic stress. However, in view of the regulatory concerns about transgenic sorghum, field testing of genetically modified sorghum hybrids has not yet been realized. Extensive testing of these hybrids will be required to ensure that no adverse pleiotropic effects or yield penalties are associated with expression of the transgene.

If regulatory concerns about transgenic sorghum are addressed, the likelihood is good that genetically modified sorghums with proven performance will be adopted in the commercial sorghum seed industry. Production of transgenic sorghum hybrids with expression of chitinases, TLP, β-1,3 glucanases, and other transgenes will provide a powerful tool to augment crop improvement strategies of plant breeders in the public and private sectors, by providing broad-spectrum resistance to several diseases.

Given the current focus of researchers on transformation of elite parent lines, agronomically useful transgenes can be rapidly incorporated into the male and female germplasm pools. Experiments have shown that the elite U.S. pollinator and seed parent lines Tx 430 and Wheatland can be readily transformed and can be used as donors of transgenes for the male parent and female parent germplasm pools. Given the use of these lines as transgene donors, the lag time between identification of agronomically useful genes and production of agronomically acceptable transgenic hybrids should be decreased.

REFERENCES

Arriola, P.E. and N.C. Ellstrand (1996). Crop-to-weed flow in the genus *Sorghum* (Poaceae): Spontaneous interspecific hybridization between johnsongrass, *Sorghum halepense,* and crop sorghum, *S. bicolor. American Journal of Botany* 83:1153-1160.

Battraw, M. and T.C. Hall (1991). Stable transformation of *Sorghum bicolor* protoplasts with chimeric neomycin phosphotransferase II and β-glucuronidase genes. *Theoretical and Applied Genetics* 82:161-168.

Bechtold, N., J. Ellis, and G. Pelletier (1993). *Agrobacterium* mediates gene transfer by infiltration of adult *Arabidopsis thalina. Plant Journal* 5:421-427.

Brar, D.S., S. Rambold, O. Gamborg, and F. Constabel (1979). Tissue culture of corn and sorghum. *Zeitschrift fur Pflanzenphysiologie* 95:377-388.

Buchanan, B.B., W. Gruissem, and R.L. Jones (2000). *Biochemistry and Molecular Biology of Plants.* Rockville, MD: American Society of Plant Physiology.

Casas, A.M., A.K. Kononowicz, T.G. Haan, L. Zhang, D.T. Tomes, R.A. Bressan, and P.M. Hasegawa (1997). Transgenic sorghum plants obtained after microprojectile bombardment of immature inflorescences. *In Vitro Cellular Developmental Biology—Plant* 33:92-100.

Casas, A.M., A.K. Kononowicz, U.B. Zehr, D.T. Tomes, J.D. Axtell, L.G. Butler, R.A. Bressan, and P.M. Hasegawa (1993). Transgenic sorghum plants via microprojectile bombardment. *Proceedings of National Academy of Sciences USA* 90:11212-11216.

Cheng, M., J.E. Fry, S. Peng, H. Zhu, C.M. Hironaka, D.R. Duncan, T.W. Conner, and Y. Wan (1997). Genetic transformation of wheat mediated by *Agrobacterium tumefaciens*. *Plant Physiology* 115:971-980.

Chu, C.C., C.C. Wang, C.S. Sun, C. Hsu, K.C. Yin, C.Y. Chu, and F.Y. Bi (1975). Establishment of an efficient medium for anther culture of rice through comparative experiments on the nitrogen sources. *Science Sinica* 18:659-668.

De Wet, J.M.J. (1978). Systematics and evolution of *Sorghum* Sect. *Sorghum* (Gramineae). *American Journal of Botany* 65:477-484.

Food and Agriculture Organization of the United Nations (FAO) (2000) FAOSTAT Agriculture Database Web site, <http://apps.fao.org/cgi-bin/nph-db.pl?subset=agriculture>.

Frederiksen, R.A. and G.N. Odvody (2000). *Compendium of Sorghum Disease,* Second Edition. St. Paul, MN: APS Press, The American Phytopathology Society.

Gamborg, O., J.P. Shyluk, D.S. Brar, and F. Constabel (1977). Morphogenesis and plant regeneration from callus of immature embryos of sorghum. *Plant Science Letters* 10:67-74.

Godwin, I.D. and R. Chickwamba (1994). Transgenic grain sorghum *(Sorghum bicolor)* plants via *Agrobacterium*. In *Improvement of Cereal Quality by Genetic Engineering* (pp. 47-93), Eds. Henry, R.J. and J.A. Ronalds. Sydney, Australia: Cereal Chemistry Division Symposium on Improvement of Cereal Quality by Genetic Engineering.

Godwin, I.D. and S.J. Gray (2000). Overcoming productivity and quality constraints in sorghum: The role for genetic engineering. In *Transgenic Cereals* (pp. 153-177), Eds. O'Brien, L. and R.J. Henry. St. Paul, MN: American Association of Cereal Chemists.

Guo, J.H., D.Z. Skinner, and G.H. Liang (1996). Phylogenetic relationships of sorghum taxa inferred from mitochondrial DNA restriction fragment analysis. *Genome* 39:1027-1034.

Hagio, T., A.D. Blowers, and E.D. Earle (1991). Stable transformation of sorghum cell cultures after bombardment with DNA-coated microprojectiles. *Plant Cell Reports* 10:260-264.

Hiei, Y., S. Ohta, T. Komari, and T. Kumashiro (1994) . Efficient transformation of rice (*Oryza sativa* L.) mediated by *Agrobacterium* and sequence analysis of the boundaries of the T-DNA. *The Plant Journal* 6:271-282.

Hill-Ambroz, K.L. and J.T. Weeks (2001). Comparison of constitutive promoters for sorghum [*Sorghum bicolor* (L.) Moench] transformation. *Cereal Research Communication* 29:17-24.

Ishida, Y., H. Saito, S. Ohta, Y. Hieie, T. Komari, and T. Kumashiro (1996). High efficiency of maize (*Zea mays* L.) mediated by *Agrobacterium tumefaciens*. *Nature Biotechnology* 14:745-750.

Jeoung, J.M. (2001). *Agrobacterium*-mediated gene transformation in grain sorghum. PhD dissertation. Genetics Program, Department of Agronomy, Kansas State University, Manhattan, Kansas.

Kado, C.I. (1998). *Agrobacterium*-mediated horizontal gene transfer. In *Genetic Engineering,* Volume 20 (pp. 1-24), Ed. Setlow, J.K. New York: Plenum Press.

Kaeppler, H.F. and J.F. Pedersen (1996). Media effects on phenotype of callus cultures initiated from photoperiod-insensitive, elite inbred sorghum lines. *Maydica* 41:83-89.

Krishnaveni, S., S. Muthukrishnan, and G.H. Liang (2001). Transgenic sorghum plants constitutively expressing a rice chitinase gene show improved resistance to stalk rot. *Journal of Genetics and Breeding* 55:151-158.

Luo, Z.X. and R. Wu (1988). A simple method for the transformation of rice via the pollen tube pathway. *Plant Molecular Biology Reporter* 6:165-174.

Masteller, V.J. and D.J. Holden (1970). The growth of and organ formation from callus tissue of *Sorghum. Plant Physiology* 45:362-364.

Morden, C.W., Doebley, J.F., and K.F. Schertz (1990). Allozyme variation among the sontaneous species of *Sorghum* section *Sorghum* (Poaceae). *Theoretical and Applied Genetics* 80:296-304.

Murashige, T. and F. Skoog (1962). A revised medium for rapid growth and bioassays with tobacco tissue cultures. *Physiology Plantarum* 15:473-497.

Neuhaus, J.-M. (1999). Plant chitinases (PR-3, PR-4, PR-8, PR-11). In *Pathogenesis-Related Proteins in Plants* (pp. 77-105), Eds. Datta, S.K. and S. Muthukrishnan. New York: CRS Press.

Nwanze, K.F., N. Seetharama, H.C. Sharma, and J.W. Stenhouse (1995). Biotechnology in pest management improving resistance in sorghum to insect pests. *African Crop Science Journal* 3:209-215.

Rathus, C., T.V. Nguyen, J.A. Able, S.J. Gray, and I.D. Godwin (2004). Optimizing sorghum transformation technology via somatic embryogenesis. In *Sorghum Tissue Culture Transformation and Genetic Engineering,* Eds. Seetharama, N. and I.D. Godwin. Patancheru, India: ICRISAT and Oxford Publishers.

Riera-Lizarazu, O. and A. Mujeeb-Kazi (1993). Polyhaploid production in the Triticeae: Wheat × *Tripsacum* crosses. *Crop Science* 33:973-976.

Roberts, W.K. and C.P. Selitrennikoff (1990). Zeamatin, an antifungal protein from maize with membrane permeability activity. *Journal of General Microbiology* 136:1771-1778.

Sairam, A.V., N. Seetharama, P.S. Devi, A. Verma, U.R. Murthy, and I. Potrykus (1999). Culture and regeneration of mesophyll-derived protoplasts of sorghum [*Sorghum bicolor* (L.) Moench]. *Plant Cell Reports* 18:972-977.

Velazhahan, R., S.K. Datta, and S. Muthukrishnan (1999). The PR-5 family: Thaumatin-like proteins. In *Pathogenesis-Related Proteins in Plants* (pp. 107-129), Eds. Datta, S.K. and S. Muthukrishnan. New York: CRS Press.

Wang, J.X., Y. Sun, G.M. Cui, and J.J. Hu (2001). Transgenic maize plants obtained by pollen-mediated transformation. *Acta Botanica Sinica* 43:275-279.

Weeks, J.T., K.Y. Koshiyama, U. Maier-Greiner, T. Schäeffner, and O.D. Anderson (2000). Wheat transformation using cyanamide as a new selective agent. *Crop Science* 40:1749-1754.

Zhao, Z., T. Cai, L. Tagliana, M. Miller, N. Wang, H. Pang, M. Rudert, S. Schroeder, D. Hondred, J. Seltzer, and D. Pierce (2000). *Agrobacterium*-mediated sorghum transformation. *Plant Molecular Biology* 44:789-798.

Zhu, B., T.H. Chen, and P.H. Li (1996). Analysis of late-blight disease resistance and freezing tolerance in transgenic potato plants expressing sense and antisense genes for an osmotin-like protein. *Planta* 198:70-77.

Zhu, H., S. Muthukrishnan, S. Krishnaveni, G. Wilde, J.M. Jeoung, and G.H. Liang (1998). Biolistic transformation of sorghum using a rice chitinase gene. *Journal of Genetics and Breeding* 52:243-252.

Zupan, J.R. and P. Zambryski (1995). Transfer of T-DNA from *Agrobacterium* to the plant cell. *Plant Physiology* 107:1041-1047.

Chapter 9

Rice Transformation: Current Progress and Future Prospects

James Oard
Junda Jiang

INTRODUCTION

Dramatic improvements in rice varietal development have been achieved by traditional breeding and genetic approaches in the past forty years that include hybrid seed production, induced mutations, and anther culture. Nevertheless, a plateau in grain yield has been observed in the past ten years for many rice-growing regions of the world. The "new plant type" (Peng et al., 1999) has been proposed as a solution to move beyond this stagnation in grain yield, but this approach has yet to be adequately demonstrated even under field-plot conditions. Moreover, various biotic and abiotic constraints often limit the genetic potential of new breeding material. For example, no suitable sources for resistance have been developed by traditional breeding methods to certain pests such as the rice water weevil *(Lissorhoptrus oryzophilus)* or the sheath blight fungus *(Rhizoctonia solani),* and the rice blast fungus *(Magnaporthe grisea)* is constantly changing its genetic makeup to thwart advances made by conventional breeding efforts. Gene transfer technology and recent advances in sequencing of the rice genome (Eckardt, 2000) offer new avenues to overcome certain problems facing future germplasm enhancement.

Several recent reports (Giri and Laxmi, 2000; Ignacimuthu et al., 2000; Jain and Jain, 2000; Rasco-Gaunt et al., 2000; Repellin et al., 2001; Roy et al., 2000; Tyagi and Mohanty, 2000; Upadhyaya et al., 2000) have provided excellent summaries of rice transformation technology that cover various aspects of gene delivery methods, promoters and selectable markers, basic and applied transgenic studies, gene silencing, functional genomics, etc. The review presented here will focus primarily on new information not covered or emphasized in previous reports.

PRODUCTION OF U.S. TRANSGENIC RICE LINES AND FIELD EVALUATION OF HERBICIDE RESISTANCE

Various laboratories have reported the production of transgenic rice for different agronomic and basic research purposes (Upadhyaya et al., 2000), but very few have actually evaluated transgenic material under field conditions, most probably due to government or legal restrictions that block or impair such investigations. For this reason a summary of research conducted to produce and evaluate U.S. transgenic rice lines under field conditions (Jiang et al., 2000) is presented in Tables 9.1 through 9.3.

The transformation method previously described (Jiang et al., 2000) using particle bombardment and the CaMV 35S promoter resulted in relatively high transformation efficiency (24 percent; Table 9.1) as compared to other published procedures (Sivamani et al., 1996; Chen et al., 1998). Key components in obtaining this level of efficiency included the use of CC culture medium (Potrykus et al., 1979), in both the callus initiation and regeneration phases, and proper levels of the hygromycin and bialophos selection agents. Successful application of the herbicide bialophos to produce transgenic plants will be beneficial in future studies when incorporation of hygromycin or other antibiotic resistance genes is not desirable.

The use of two vectors combined during bombardment experiments did result in the lowest transformation rate observed (3 percent), but this was not always the case as shown with the 'Cocodrie' line (Table 9.1). The long-grain material ('Cocodrie', LA9502002, 96URN085) was in general more efficient in transformation rates (33 percent) than the medium-grain lines LA9502065 and 96URN131 (10 percent), but additional research will be required to determine if these results apply in general to other genotypes.

Segregation of glufosinate resistance fit either a one- or two-gene model for each of the five independent long-grain transgenic lines in the R_1 generation (Table 9.2). The same genetic pattern was observed in the subsequent R_2 generation with the exception of two cases that displayed either unknown or unexpected genetic ratios. This aberrant behavior, whose origins may be traced to phenomena such as gene silencing (Matzke et al., 2000), has been detected in other unrelated transformed rice lines (Oard, unpublished results). Multiple integration sites may help explain the two-gene inheritance pattern found in all R_1 lines and in ~50 percent of the R_2 generation.

As was observed for 'Cocodrie', glufosinate resistance did segregate in the R_1 for the medium-grain line in a one- or two-gene ratio (Table 9.3). However, ~40 percent of the succeeding R_2 generation exhibited strange or unexplainable genetic behavior. For example, the R_2 of line $L0_1$ produced a majority of plants (78 percent) that were glufosinate susceptible, an unexpected result which nonetheless satisfactorily fit a 1(resistant) : 3 (susceptible) ratio. In a similar case, 64 percent of the plants in line $L0_2$ were sensitive to the herbicide, but these results were not consistent with any known genetic pattern. It may be currently fashionable to attribute the unexpected reversal in herbicide resistance to gene silencing, but other events such as gene conversion previously observed in rice (Xie et al., 1995) could also explain these results.

NEW RICE POLYUBIQUITIN PROMOTER AND POTENTIAL FOR ENHANCED TRANSGENE EXPRESSION

Success in transgenic plant development depends on several factors such as optimization of cell culture conditions and choice of suitable transfer methods. In addition, use of an appropriate promoter is crucial to ensure the successful production of transgenic plants.

With this is mind, we have recently isolated strong, constitutive promoters from the rice polyubiquitin family of genes (Wang et al., 2000). These regulatory elements harbor, in addition to TATA box sequences, enhancer core and heat shock motifs that appear to modulate transcriptional activity under varying temperature and other conditions. One promoter, designated RUBQ2, was fused to the *gus* reporter gene

TABLE 9.1. Mean transformation efficiency, number of calli bombarded, hygromycin- or bialaphos-resistant calli, hygromycin- or bialaphos- and glufosinate-resistant plants for five transgenic elite U.S. rice lines

Rice line	Pedigree	Plasmid[a]	No. of calli bombarded (A)	No. of Hyg-R or Bl-R calli (B)	No. of calli producing Hyg-R or Bl-R plants (C)	No. of glufosinate-resistant plants (D)	Callus transformation frequency (B/A)	Transformation efficiency	
								Method 1[b] (C/A)	Method 2[c] (D/A)
Cocodrie	Cypress//82CAY21/Tebonnet	pAHC25+ pTRA151	1056	44	29	146	0.037	0.024	0.131
		pPAT63	230	43	35	288	0.128	0.105	0.886
		pB2/35SAcK	210	20	13	44	0.064	0.041	0.141
LA9502065	Bengal/Rico1	pAHC25+ pTRA151	1415	96	49	392	0.079	0.035	0.027
		pPAT63	1257	65	45	157	0.051	0.038	0.141

LA9502002	Maybelle//Lemont/2001-5	pPAT63	204	12	10	48	0.057	0.048	0.225
96URN085	Cypress//L201/7402003	pPAT63	214	35	24	53	0.171	0.125	0.282
96URN131	Mercury/Rico 1	pPAT63	180	33	21	34	0.183	0.117	0.139
Total			4766	348	226	1162	0.073	0.047	0.244

[a] Plasmids described in Jiang et al., 2000
[b] Refers to ratio of bombarded calli that produced Hyg-R or Bl-R plants
[c] According to Sivamani et al., 1996

TABLE 9.2. Segregation of glufosinate resistance under field conditions in R_1 and R_2 transgenic lines derived from the long-grain cultivar cocodrie

Generation	R_1 line and no. of R_2 plants	No. of plants resistant	No. of plants susceptible	P^2 ratios[a]	P^2 values[b]
R_1	617-1	11	3	3:1 (9:7)	0.09 (2.83)
R_2	93	42	20	3:1 (9:7)	1.73 (2.45)
	127	65	62	9:7 (1:1)	1.32 (0.07)
R_1	617-2	9	4	3:1 (9:7)	0.23 (0.89)
R_2	193	137	56	3:1	1.65
R_1	617-6	13	5	3:1 (9:7)	0.07 (1.87)
R_2	204	160	44	3:1	1.28
	239	152	87	–	–
R_1	617-8	25	10	3:1 (9:7)	0.238 (3.27)
R_2	445	328	117	3:1 (9:7)	0.11
R_1	A01	12	3	3:1 (9:7)	0.20 (3.43)
R_2	124	82	42	3:1	2.95

[a] Dash indicates segregation did not fit any known ratio.
[b] df = 1, P^2 = 3.841 at the 0.05 probability level.

and shown to drive high levels of transient gene expression in cultured suspension cells. Under these conditions RUBQ2 produced expression levels two- to tenfold greater than other DNA constructs containing either the CaMV 35S or maize ubiquitin promoter. Similarly, *gus* activity in stably transformed calli was 45-fold greater with RUBQ2 compared to CaMV 35S (Wang and Oard, unpublished results). In stably transformed plants, GUS expression varied from 8- to 35-fold higher at the tillering and heading stages with RUBQ2 than corresponding levels produced from CaMV 35S. A significantly higher percentage of putative *gus* gene silencing events were observed in transgenic plants containing CaMV 35S (52 percent) as compared to

TABLE 9.3. Segregation of glufosinate resistance under field conditions in R_1 and R_2 transgenic lines derived from the U.S. medium-grain breeding line LA9502065

Generation	R_1 line and no. of R_2 plants	No. of plants resistant	No. of plants susceptible	P^2 ratios[a]	P^2 values[b]
R_1	L01	16	10	3:1 (9:7)	2.51 (0.29)
R_2	644	141	503	1:3	3.31
R_1	L02	20	4	3:1 (9:7)	0.44 (3.57)
R_2	282	171	111	9:7	2.20
	270	192	78	3:1	2.14
	528	188	340	–	–
R_1	L03	21	4	3:1	1.08
R_2	281	213	68	3:1	0.31
	242	155	87	–	–
R_1	L04	11	1	3:1 (15:1)	1.78 (0.09)
R_2	136	49	87	7:9	3.29
	72	14	58	1:3	1.17

[a] Dash indicates segregation did not fit any known ratio.
[b] df = 1, P^2 = 3.841 at the 0.05 probability level.

RUBQ2 (8 percent). Southern blot analysis showed that three to seven copies of the *gus* gene driven by RUBQ2 were stably inserted after particle bombardment and segregated in a monogenic fashion in the R_1 generation. High levels of *gus* activity were detected in all tissues examined including seed, roots, leaves, shoot apex, and pollen. Recent studies have also demonstrated that RUBQ2 can drive strong *gus* expression in stably transformed sugarcane callus (Liu and Oard, unpublished results).

Sequence analysis of RUBQ2 from two distinct rice lines has revealed a novel 270 bp element located 210 bp upstream of the TATA box (Yang et al., 2001). This element displayed standard features of miniature inverted transposable elements (MITEs) (Bureau et al., 1996). The MITE did exhibit unique characteristics and was shown by database mining to represent a new family of MITEs that occur within ~530 bp of coding sequences in the rice genome. Presence of the MITE element, which contains a G-box factor binding site, in RUBQ2 was postulated to augment transcriptional activity of this promoter. Taken together these results demonstrate the potential of RUBQ2 as a useful new promoter for rice transformation systems.

It will be interesting to further evaluate temporal and spacial expression of RUBQ2 under field conditions since recent studies have reported tissue specificity or reduced transgene expression in maturing plants when driven by the maize ubiquitin promoter (Rasco-Gaunt et al., 2000; Rooke et al., 2000). The maize ubiquitin element was also found to produce high levels of the soybean ferritin protein in rice vegetative tissues, but not in seeds (Drakakai et al., 2000), which suggests that seed-specific promoters, such as the rice GluB-1 seed storage promoter (Goto et al., 1999) may be needed to augment levels of iron and other nutrients in cereal grains. Inadequate expression levels in this case may also be influenced by differences in guanine-cytosine (GC) content and codon usage between a dicot transgene and monocot host plant (Rasco-Gaunt et al., 2000).

Other new promoters evaluated include the 5' region from rice NADH-glutamate synthase that was shown to drive high levels of *gus* expression in leaf blades and vascular bundles of developing grains (Kojima et al., 2000). A ferredoxin-NADP(+)-oxidoreductase promoter from spinach has produced high expression levels specifically in chlorophyll-containing tissue (Mohanty et al., 2000). The rice seed storage RP5 prolamin promoter was used to drive high levels of tissue-specific expression in the aleurone layer of transgenic endosperm (Su et al., 2001). Very few studies have explored basic issues of monocotyledonous plant regeneration from callus at both the biochemical and molecular level. In one such study, expression levels driven by the strong maize ubiquitin and the relatively weak CaMV 35S promoters produced contrasting results in production and metabolism of the polymines spermidine and putrescine which are believed to play a key role in morphogenesis of rice callus cells (Bassie et al., 2000).

TRANSGENE DELIVERY AND EXPRESSION STUDIES

Indica and *japonica* rice lines were used to investigate integration, expression, and inheritance of transgenes mediated by *Agrobacterium tumefaciens* (Azhakanandam et al., 2000). Phenotypically normal transgenic plants were observed in only a small proportion (10 to

31 percent) of the regenerated lines. Although the hygromycin resistance trait was inherited in a Mendelian fashion at a single locus in the T_1 and T_2 generations, the hygromycin resistance gene routinely inserted at multiple sites in the genome. The effect of different *Agrobacterium* virulence genes and T-DNA insert size on transformation efficiency has been evaluated (Park et al., 2000). A 50 percent reduction in size of the T-DNA insert and the *virE/virG* virulence gene combination of the transformation vector were found to dramatically enhance efficiency in gene transfer and expression. In another study the transfer of cells of engineered *A. tumefaciens* into cultured plant cells via particle bombardment has resulted in high transformation rates, single copy insertion, and Mendelian inheritance of the *gus* reporter gene (Ming et al., 2001).

Transgenic rice plants obtained via *Agrobacterium* or particle bombardment methods were compared in the greenhouse for copy number, integration pattern, and stability of transgene expression in the T_0-T_2 generations (Dai et al., 2001). Transgene copy number was slightly higher in transformed lines produced by particle bombardment (2.7 copies) compared to the *Agrobacterium* method (1.8 copies). Integration of intact copies was higher and *gus* expression was more stable via *Agrobacterium* than the physical delivery method. However, a higher percentage of cosegregation of marker genes in the T_1 generation was observed using the gene gun versus the *Agrobacterium* method. A greater number of seeds was produced by plants via *Agrobacterium* than by the gene gun technique. The authors have developed transgenic plants via *Agrobacterium* and particle bombardment methods to compare transgene expression and phenotypic stability under field conditions (Jiang and Oard, unpublished results). Results from this field trial were consistent with those of Dai and colleagues (2000) in that we observed significantly less variability for plant height and maturity for lines produced by *Agrobacterium* versus the particle bombardment method. In addition, copy number for the *pat* herbicide resistance gene was fourfold greater in transgenic lines derived from particle bombardment than the *Agrobacterium* method. No differences between the two techniques were detected for transgene expression levels, but in contrast to Dai and colleagues (2000), seed fertility in *Agrobacterium*-derived lines was only 55 percent of that in lines produced by particle bombardment. Callus derived from embryos has been a favorite target for *Agrobacterium*-mediated transformation, and now the young inflorescence has been recruited to develop an efficient transformation protocol with Mendelian inheritance patterns of the transgene (Dong et al., 2001). Behavior and expression of multiple genes transferred to rice by particle bombardment were investigated in different genetic backgrounds (Gahakwa et al., 2000). Extensive variation in expression for six different genes was detected. Expression levels nevertheless were faithfully transmitted over four generations. In other experiments, a total of 14 different vectors were introduced via the gene gun (Chen et al., 1998). No transformant contained all introduced genes, but 17 percent of the lines expressed nine of the transgenes and integration of the genes occurred at one or two sites with inheritance conforming to a single locus model.

TRANSGENIC ADVANCES IN PEST AND STRESS TOLERANCE

Disease and Insect Pests

Sheath blight caused by *Rhizoctonia solani* is a major fungal disease in the southern United States and Asia. No germplasm sources showing high levels of resistance have been identified. The PR-3 rice chitinase gene was recently transferred to five *indica* cultivars susceptible to *R. solani* (Datta et al., 2001). Highly variable levels of the chitinase protein were detected in primary transgenic lines which was reflected in different levels of resistance when challenged with the sheath blight fungus. In a separate study the antimicrobial peptide cecropin B from the moth *Bombyx mori* was transferred to rice and shown to confer high levels of resistance to the bacterial pathogen *Xanthomonas oryzae* pv. *oryzae* (Sharma et al., 2000). Maximum resistance levels were obtained when the rice chitinase signal peptide sequence was incorporated into the transformation vector. Resistance to the rice dwarf virus has been developed by transfer of the gene for the Rz hammerhead ribozyme that targets the mRNA of this virus (Han et al., 2000). Resistance or attenuated viral symptoms were clearly produced with Rz, although loss of resistance detected in several progeny was postulated to occur by silencing mechanisms. Rice blast disease caused by fungus *Magnaporthe grisea* is the principal foliar disease on a worldwide basis. The *Rir1b* rice gene encodes extracellular proteins that accumulate in response to challenge with a nonhost pathogen such as *Pseudomonas syringae* pv. *syringae*. The *Rir1b* gene was transferred to rice and expressed in the T_0-T_2 generations (Schaffrath et al., 2000). Inoculation of T_1 and homozygous T_2 plants displayed increased resistance over controls to the rice blast fungus which demonstrated potential of this strategy to help combat this important rice pathogen.

Bacterial blight caused by *Xanthonmonas oryzae* pv. *oryzae* is the principal bacterial disease in southeast Asia and other humid rice-growing regions. The *Xa21* gene confers resistance to many races of this pathogen and has been previously isolated by map-based cloning procedures (Song et al., 1995). *Xa21* was transferred to the susceptible cultivar IR72 and evaluated for resistance for two years under field conditions (Tu, Datta, et al., 2000). Transgenic lines carrying the engineered *Xa21* genes showed resistance to six isolates of *X. oryzae* and displayed smaller lesions compared to the original untransformed donor line of *Xa21*. However, homozygous transgenic material produced a yield drag when grown in the same field with the untransformed 'IR72'. The leaffolder *(Cnaphalocrocis medinalis)* and yellow stem borer *(Scirpophaga incertulas)* insects can cause substantial reductions in grain yield and seed quality under heavy infestations. To combat these insect pests, a fusion of the *cryIA(b)* and *cryI(c)* genes from *Bacillus thuringiensis* was inserted into rice and shown to confer high resistance levels when tested under manual and natural infestations of the insects (Tu, Zhang, et al., 2000). Fortunately, no yield drag was observed in the transgenic lines compared to the untransformed controls.

Agronomic Traits

Grain and milling quality are of paramount importance for development of new cultivars in the U.S. rice market and are gaining increased attention in China and other Asian countries where grain yield has traditionally received primary consideration. The wheat puroindoline genes *pinA* and *pinB* are believed to influence grain texture. Both *pinA* and *pinB* were introduced into rice under control of the maize ubiquitin promoter (Krishnamurthy and Giroux, 2001). Transgenic lines expressing *pinA* and/or *pinB* produced substantially softer grain than nontransgenic controls and reduced starch damage after milling. Plant characteristics such as height and maturity are critical considerations for breeding commercial rice cultivars. To modify these traits, regulatory rice genes (referred to as MADS-box) were engineered and transferred to rice by *Agrobacterium*-mediated transformation (Jeon et al., 2000). Expression of the MADS-box genes resulted in reduced plant height and heading date by varying degrees that were correlated with different expression levels of the introduced genes. An alternative approach to modulate heading date in rice was demonstrated by transfer and expression of the *Arabidopsis LEAFY* gene which was known to accelerate maturity in dicotyledonous plants (He et al., 2000). Constitutive expression of *LEAFY* resulted in transgenic rice exhibiting heading dates 24 to 36 days earlier than untransformed controls.

Abiotic Stress

Increased stress caused by high salinity levels has become a major constraint to maximize production for certain rice-growing regions of the world. The oat arginine decarboxylase gene was introduced into rice under control of an abscisic acid (ABA)-inducible promoter and was shown to increase biomass production under salt stress conditions (Roy and Wu, 2001). Availability of certain micronutrients in certain soils can be an important limiting factor to obtain optimum grain yields. Under high soil pH environment, iron is often not accessible in a form for efficient uptake by the rice plant. To increase tolerance to low iron availability, two barley nicotianamine aminotransferase genes were introduced into rice that exhibited high transgene activity and increased secretion of phytosiderophores to solubilize iron in the soil compared to nontransformed lines (Takahashi et al., 2001). Importantly, the transgenic lines produced over four times the number of seeds compared to normal lines when grown in an alkaline, iron-poor soil.

Output Traits

Rice grains lack sufficient levels of the amino acid lysine and must be supplemented by other food sources. To increase lysine content, the lysine-rich protein gene from *Psophocarpus tetragonolobus* was transferred to rice via particle bombardment under control of the maize ubiquitin promoter (Gao et al., 2001). Lysine content of the transgenic lines was reported to be enhanced by up to 16 percent when compared to corresponding untransformed lines. Uptake of sufficient amounts of iron in the diet of small children is a serious health issue in certain poor regions of the world. To enhance iron levels in rice, the soybean ferritin gene under control of the rice *GluB-1* promoter was transferred to rice by *Agrobacterium* transformation (Goto et al., 1999). Iron levels in the T_1 generation were detected in seeds at three times that of the untreated cultivar Kita-ake.

Gene Silencing

Native or heterologous genes may be stably incorporated into the rice genome by various delivery techniques, but stable expression of transgenes over different generations is not guaranteed with the current level of technology. Mechanisms of gene silencing may be ascribed to different sources such as methylation, structure of transgene and transformation vector, nonhomologous recombination, etc. Gene suppression or other like phenomena that in essence nullify the intended results of transformation pose a serious impediment to routine and efficient gene delivery and expression. To study silencing events in transgenic rice, various sense, antisense, and inverted-repeat transformation vectors with the *gus* gene were engineered and transferred to callus (Wang and Waterhouse, 2000). Inverted-repeat constructs were substantially more efficient at GUS suppression than all other configurations. Methylation of coding sequences and low RNA levels were detected in all silenced lines, but in contrast to other research, methylation by itself was not considered essential for gene silencing at the posttranscriptional level.

Gene silencing can take on different forms as shown by recent studies (Fu, Duc, et al., 2000). Three distinct classes of gene suppression were identified in R_1-R_3 transgenic plants based on different patterns of methylation with no apparent interaction operating among them. Another study of silencing in rice showed that high transgene expression levels, rather than high insert copy number, may promote silencing events (Chareonpornwattana et al., 1999). The degree of suppression was substantially higher at eight weeks versus three-week-old seedlings, but silencing was not restricted to specific organs or tissues.

The form of the transformation vector may also influence insertion and expression of the transgene. To examine this possibility, circular and linear DNA fragments, containing only the promoter, gene, and terminator elements, were inserted into rice by particle bombardment (Fu, Kholi, et al., 2000). Insertion of the linear fragments over four generations produced simple and stable integration events, low copy number, and no gene silencing, whereas the reverse was observed for transgenic lines developed via supercoiled or linearized plasmids containing backbone vector sequences.

The authors have produced herbicide-resistant transgenic lines containing only promoter-gene-terminator fragments (Jiang and Oard, unpublished results). Evaluation of T_1 plants in the field revealed stable expression of herbicide resistance, but variation for plant height, plant architecture, and presence of pubescent leaves was observed across different independent transformed lines. Specific DNA elements that bind to the nuclear matrix are known as matrix attachment regions (MARs) and are considered by some investigators to mitigate the effects of gene silencing. The subject of MARs and their effect of gene silencing has been extensively reviewed in a recent report (Allen et al., 2000).

Selectable and Screenable Gene Markers

The hygromycin gene is currently the most common and efficient selectable marker for rice transformation, but due to potential environmental concerns, its use will be reduced or eliminated in development of future commercial transgenic lines. An alternative to the hygromycin selectable marker, such as the green fluorescent protein *(gfp)* gene, has been evaluated and reported to act as an efficient nondestructive marker in callus and various transformed tissues (Vain et al., 1998; Chung et al., 2000), although a separate study concluded that *gfp* was susceptible to high levels of gene silencing (Upadhyaya et al., 2000). A new selectable marker gene, *pmi,* produces phosphomannose isomerase from *Escherichia coli* and allows growth of transformed calli on solid medium containing the sugar mannose. (Lucca et al., 2001) The *pmi* gene segregated in Mendelian fashion in the T_1 generation. The use of double, right-border *Agrobacterium* binary vectors has been demonstrated as a viable strategy to eliminate selectable markers from transgenic rice (Lu et al., 2001).

NOVEL METHODS OF GENE TRANSFER FOR RICE

As stated in this review, transfer and expression of heterologous genes in certain rice genotypes has become a routine laboratory procedure. Except in certain instances where protoplasts can be used, particle bombardment and *Agrobacterium*-mediated transformation are the current methods of choice. Both techniques generally employ a callus phase that is lengthy, labor intensive, and often results in a fairly high proportion of transformed lines that vary under field conditions in one or more traits when compared to their untransformed counterparts. To overcome these undesirable features, alternative gene delivery methods for rice are needed. Gene transfer to the model plant *Arabidopsis thaliana* is routinely accomplished without a callus phase by vacuum infiltration of engineered *Agrobacterium* strains into young floral buds (Bechtold and Pelletier, 1998) which results in ~1 percent transformation rate. Further modifications that produce reduced, but acceptable, transfer rates have eliminated the need for the vacuum treatment (Chung et al., 2000; Desfeux et al., 2000). The infiltration technique has been extended recently to the legume *Medicago truncatula* and *Petunia hybrida* (Tjokrokusumo et al., 2000; Trieu et al., 2000). Another noncallus method uses *Agrobacterium* to infect and transfer genes to the young vegetative shoot apex of rice (Park et al., 1996). Promoters such as CAMV 35S are not expressed in meristematic cells of the apex, but strong, nonviral promoters, such as RUBQ2 discussed previously in this review, function very efficiently based on transient GUS assays (Samuels and Oard, unpublished results). Gene transfer to young tissues and organs via electroporation is another non-tissue culture approach, but it has been criticized as being technically demanding at both pre- and post-transformation handling stages (Rasco-Gaunt et al., 2000). In spite of these drawbacks, transgenic plants via electroporation of wheat and a barley-wheat hybrid have been produced (Sorokin et al., 2000; He et al., 2001), so clearly this technique merits further evaluation. Given that certain drawbacks limit the efficiency and flexibility of current rice transformation techniques, research in these and other novel approaches should be given high priority.

CONCLUSIONS

Traditional rice breeding and genetics in the past decade has brought about remarkable advances in grain yield and quality to a rapidly growing world population, and this technology will continue to do the same in the foreseeable future. However, yield plateaus have been reached in certain rice-growing regions, and several chronic problems in pest resistance and stress tolerance have yet to be overcome. For these challenges transgenic technology will play a crucial role in germplasm improvement. Future success in the new areas of functional genomics, metabolic engineering, and production of biomedical compounds will rely heavily on continued improvements in gene delivery systems to realize the desired outcome or product. Several limitations in the current rice transformation methodology, as described in this review, could be reduced or avoided if a callus-free approach were developed similar to the infiltration method successfully used with *A. thaliana*. Clearly, enhanced research efforts directed to this research area should be promoted by federal, state, and private research funding agencies.

Rapid advances in rice transgenic research will occur if researchers are allowed to evaluate transgenic material under field conditions. Current government regulations and policies in certain countries restrict proper field analysis of genetically modified lines. Relaxation of these policies should occur with time, but for now only certain laboratories are permitted to examine transformed material in small, field-plot settings.

The issues of patents and intellectual property have a direct impact on the practical application of rice transformation technology. Virtually all genes and delivery systems currently used by rice researchers have been patented by private or public entities. This means that only private companies have the financial wherewithal to purchase "freedom to operate" from various patent holders and bring the benefits of transgenic material directly to the public and the marketplace. An additional situation occurs in the United States where a single transgenic "event" must undergo extensive and very expensive evaluation before it is to be registered with the federal government. Due to the high cost, only private companies can afford registration of events. As a consequence, public rice researchers will participate in future development of transgenic material for benefit of farmers and the general public only if they enter into cooperative research efforts with private companies.

REFERENCES

Allen, G.C., S. Spiker, and W.F. Thompson (2000). Use of matrix regions (MARs) to minimize transgene silencing. *Plant Molecular Biology* 43:361-376.

Azhakanandam, K., M.S. McCabe, J.B. Power, K.C. Lowe, E.C. Cocking, and M.R. Davey (2000). T-DNA transfer, integration, expression and inheritance in rice: Effects of plant genotype and *Agrobacterium* supervirulence. *Journal Plant Physiology* 157:429-439.

Bassie, L., M. Noury, O. Lepri, T. Lahaye, P. Christou, and T. Capell (2000). Promoter strength influences polyamine metabolism and morphogenic capacity in transgenic rice tissue expressing the oat adc cDNA constitutively. *Transgenic Research* 9:33-42.

Bechtold, N. and G. Pelletier (1998). *In planta Agrobacterium*-mediated transformation of adult *Arabidopsis thaliana* plants by vacuum infiltration. *Methods Molecular Biology* 82:259-266.

Bureau, T.E., P.C. Ronald, and S.R. Wessler (1996). A computer-based systematic survey reveals the predominance of small inverted-repeat elements in wild-type rice genes. *Proceedings of the National Academy Sciences USA* 93:8524-8529.

Chareonpornwattana, S., K.V. Thara, L. Wang, S.K. Datta, W. Panbangred, and S. Muthukrishnan (1999). Inheritance, expression and silencing of a chitinase transgene in rice. *Theoretical and Applied Genetics* 98:371-378.

Chen, L., P. Marmey, N.J. Taylor, J.-P. Brizard, C. Espinoza, P. D'Cruz, H. Huet, S. Zhang, A. de Kochko, R.N. Beachy, and C.M. Fauquet (1998). Expression and inheritance of multiple transgenes in rice plants. *Nature Biotechnology* 16:1060-1064.

Chung, B.C., J.K. Kim, B.H. Nahm, and C.H. Lee (2000). *In planta* visual monitoring of green fluorescent protein in transgenic rice plants. *Molecular Cell* 10:411-414.

Dai, S.H., P. Zheng, P. Marmey, S.P. Zhang, W.Z. Tian, S.Y. Chen, R.N. Beachy, and C. Fauquet (2001). Comparative analysis of transgenic rice plants obtained by *Agrobacterium*-mediated transformation and particle bombardment. *Molecular Breeding* 7:25-33.

Datta, K., J. Tu, N. Oliva, I. Ona, R. Velazhahan, T.W. Mew, S. Muthukrishnan, and S.K. Datta (2001). Enhanced resistance to sheath blight by constitutive expression of infection-related rice chitinase in transgenic elite indica rice cultivars. *Plant Science* 160:405-414.

Deong, J., P. Kharb, W. Teng, and T.C. Hall (2001). Characterization of rice transformed via an *Agrobacterium*-mediated inflorescense approach. *Molecular Breeding* 7:187-194.

Desfeux, C., S.J. Clough, and A.F. Bent (2000). Female reproductive tissues are the primary target of *Agrobacterium*-mediated transformation by the *Arabidopsis* floral-dip method. *Plant Physiology* 123:895-904.

Drakakai, G., P. Christou, and E. Stoger (2000). Constitutive expression of soybean ferritin cDNA in transgenic wheat and rice results in increased iron levels in vegetative tissues but not in seeds. *Transgenic Research* 9:445-452.

Eckardt, N.A. (2000). Sequencing the rice genome. *Plant Cell* 12:2011-2016.

Fu, X., L.T. Duc, S. Fontana, B.B. Bong, P. Tinjuangjun, D. Sudhakar, R.M. Twyman, P. Christou, and J. A. Kohli (2000). Linear transgene constructs lacking vector backbone sequences generate low-copy-number transgenic plant with simple inheritance patterns. *Transgenic Research* 9:11-19.

Fu, X., A. Kohli, R.M. Twyman, and P. Christou (2000). Alternative silencing effects involve distinct types of non-spreading cytosine methylation at a three-gene, single-copy transgenic locus in rice. *Molecular General Genetics* 263:106-118.

Gahakwa, D., S.B. Maqbool, X. Fu, D. Sudhakar, P. Christou, and A. Kohli (2000). Transgenic rice as a system to study the stability of transgene expression: Multiple heterologous transgenes show similar behavior in diverse genetic backgrounds. *Theoretical and Applied Genetics* 101:388-399.

Gao, Y.F., Y.X. Jing, S.H. Shen, S.P. Tian, T.Y. Kuang, and S.S.M. Sun (2001). Transfer of lysine-rich protein gene into rice and production of fertile transgenic rice. *Acta Botanica Sinica* 43:506-511.

Giri, C.C. and G.V. Laxmi (2000). Production of transgenic rice with agronomically useful genes: An assessment. *Biotechnology Advances* 18:653-683.

Goto, F., T. Yoshihara, N. Shigemoto, S. Toki, and F. Takaiwa (1999). Iron fortification of rice seed by the soybean ferritin gene. *Nature Biotechnology* 17:282-286.

Han, S., Z. Wu, H. Yang, R. Wang, Y. Yie, L. Xie, and P. Tien (2000). Ribozyme-mediated resistance to the rice dwarf virus and the transgene silencing in the progeny of transgenic plants. *Transgenic Research* 9:195-203.

He, G.Y., P.A. Lazzere, and M.E. Cannell (2001). Fertile transgenic plants obtained from tritordeum inflorescences by tissue electroporation. *Plant Cell Reports* 20:67-72.

He, Z., Q. Zhu, T. Dabi, D. Li, D. Weigel, and C. Lamb (2000). Transformation of rice with the *Arabidopsis* floral regulator LEAFY causes early heading. *Transgenic Research* 9:223-227.

Ignacimuthu, S., S. Arockiasamy, and R. Terada (2000). Genetic transformation of rice: Current status and future prospects. *Current Science* 79:186-195.

Jain, R.K. and S. Jain (2000). Transgenic strategies for genetic improvement of Basmati rice. *Indian Journal of Experimental Biology* 38:6-17.

Jeon, J.S., S. Lee, K.H. Jung, W.S. Yang, G.H. Yi, B.G. Oh, and G.H. An (2000). Production of transgenic rice plants showing reduced heading date and plant height by ectopic expression of rice MADS box genes. *Molecular Breeding* 6:581-592.

Jiang, J., S.D. Linscombe, J. Wang, and J.H. Oard (2000). High efficiency transformation of U.S. rice lines from mature seed-derived calli and segregation of glufosinate resistance under field conditions. *Crop Science* 40:1729-1741.

Kojima, S., M. Kimura, Y. Nozaki, and T. Yamaya (2000). Analysis of a promoter for the NADH-glutamate synthase gene in rice *(Oryza sativa):* Cell type-specific expression in developing organs of transgenic rice plants. *Australian Journal Plant Physiology* 27:787-793.

Krishnamurthy, K. and M.J. Giroux (2001). Expression of wheat puroindoline genes in transgenic rice enhances grain softness. *Nature Biotechnology* 19:162-166.

Lu, H.J., X.R. Shou, Z.X. Gong, and N.M. Upadhyaya (2001). Generation of selectable marker-free transgenic rice using double right-border (DRB) binary vectors. *Australian Journal of Plant Physiology* 28:241-248.

Lucca, P., X.D. Ye, and I. Potrykus (2001). Effective selection and regeneration of transgenic plants with mannose as selective agent. *Molecular Breeding* 7:43-49.

Matzke, M.A., M.F. Mette, and A.J. Matzke (2000). Transgene silencing by the host genome defense: Implications for the evolution of epigenetic control mechanisms in plants and vertebrates. *Plant Molecular Biology* 43:401-415.

Ming, X.T., H.Y. Yuan, L.J. Wang, and Z.L. Chen (2001). *Agrobacterium*-mediated transformation of rice with help of bombardment. *Acta Botanica Sinica* 43:72-76.

Mohanty, A., M. Grover, A. Chaudhury, Q. Rizwan-ul Haq, A.K. Sharma, S.C. Maheshwari, and A.K. Tyagi (2000). Analysis of the activity of promoters from two photosynthesis-related genes psaF and petH of spinach in a monocot plant, rice. *Indian Journal of Biochemistry Biophysics* 37:447-452.

Park, S.H., B.M. Lee, M.G. Salas, M. Srivatanakul, and R.H. Smith (2000). Shorter T-DNA or additional virulence genes improve *Agrobacterium* transformation. *Theoretical and Applied Genetics* 101:1015-1020.

Park, S.H., S.R.M. Pinson, and R.H. Smith (1996). T-DNA integration into genomic DNA of rice following *Agrobacterium* inoculation of isolated shoot apices. *Plant Molecular Biology* 32:1135-1148.

Peng, S., K.G. Cassman, S.S. Virmani, J. Sheehy, and G.S. Kush (1999). Yield potential trends of tropical rice since the release of IR8 and the challenge of increasing rice yield potential. *Crop Science* 39:1552-1559.

Potrykus, I., C.T. Harms, and H. Lorz (1979). Callus formation from cell culture protoplast of corn (*Zea mays* L.). *Theoretical and Applied Genetics* 54:209-214.

Rasco-Gaunt, S., C. Thorpe, P.A. Lazzeri, and P. Barcele (2000). Advances in cereal transformation technologies: Transgenic rice (pp. 178-251). In *Transgenic*

Cereals (pp. 178-251), Eds. O'Brien, L. and R.J. Henry. St. Paul, MN: American Association of Cereal Chemists.

Repellin, A., M. Baga, P.P. Jahar, and R.N. Chibbar (2001). Genetic enrichment of cereal crops via alien gene transfer: New challenges. *Plant Cell Tissue Organ Culture* 64:159-183.

Rooke, L., D. Byrne, and S. Salgueiro (2000). Marker gene expression driven by the maize ubiquitin promoter in transgenic wheat. *Annals of Applied Biology* 136:167-172.

Roy, M., R.K. Jain, J.S. Rohila, and R. Wu (2000). Production of agronomically superior transgenic rice plants using *Agrobacterium* transformation methods: Present status and future perspectives. *Current Science* 79:954-960.

Roy, M. and R. Wu (2001). Argenine decarboxylase transgene expression and analysis of environmental stress tolerance in transgenic rice. *Plant Science* 160:869-875.

Schaffrath, U., F. Mauch, E. Freydl, P. Schwizer, and R. Didler (2000). Constitutive expression of the defense-related Rir1b gene in transgenic rice plants confers enhanced resistance to the rice blast fungus *Magnaporthe grisea. Plant Molecular Biology* 43:59-66.

Sharma, A., R. Sharma, M. Imamura, M. Yamakawa, and H. Machii (2000). Transgenic expression of cecropin B, an antibacterial peptide from *Bombyx mori,* confers enhanced resistance to bacterial leaf blight in rice. *FEBS Letters* 484:7-11.

Sivamani, E., P. Chen, N. Opalka, R.N. Beachy, and C.M. Fauquet (1996). Selection of large quantities of embryogenic calli from indica rice seeds for production of fertile transgenic plants using the biolistic method. *Plant Cell Reports* 15:322-277.

Song, W.Y., G.L. Wang, L.L. Chen, H.S. Kim, L.Y. Pi, T. Holsten, J. Gardner, B. Wang, W.X. Zhai, L.H. Zhu, et al. (1995). A receptor kinase-like protein encoded by the rice disease resistance gene, Xa21. *Science* 270:1804-1806.

Sorokin, A.P., X.Y. Ke, D.F. Chen, and M.C. Elliot (2000). Production of fertile transgenic wheat plants via tissue electroporation. *Plant Science* 156:227-233.

Su, P.H., S.M. Yu, and C.S. Chen (2001). Spatial and temporal expression of a rice prolamin gene RP5 promoter in transgenic tobacco and rice. *Journal of Plant Physiology* 158:247-254.

Takahashi, M., H. Nakanishi, S. Kawasaki, N.K. Nishizawa, and S. Mori (2001). Enhanced tolerance of rice to low iron availability in alkalaine soils using barley nicotianamine aminotransferase genes. *Nature Biotechnology* 19:466-469.

Tjokrokusumo, D., T. Heinrich, S. Wylie, R. Potter, and J. McComb (2000). Vacuum infiltration of *Petunia hybrida* pollen with *Agrobacterium tumefaciens* to achieve plant transformation. *Plant Cell Reports* 19:792-797.

Trieu, A.T., S.H. Burleigh, I.V. Kardailsky, I.E. Maldonado-Mendoza, W.K. Versaw, L.A. Blaylock, H. Shin, T.J. Chiou, H. Katagi, G.R. Dewbre, et al. (2000). Transformation of *Medicago truncatula* via infiltration of seedlings or flowering plants with *Agrobacterium. Plant Journal* 22:531-541.

Tu, J., K. Datta, G.S. Kush, Q. Zhang, and S.K. Datta (2000). Field performance of Xa21 transgenic indica rice (*Oryza sativa* L.), IR72. *Theoretical Applied Genetics* 101:15-20.

Tu, J.M., G.A. Zhang, K. Datta, K.C.G. Xu, Y.Q. He, Q.F. Zhang, G.S. Kush, S.K. Datta (2000). Field performance of transgenic elite commercial hybrid rice expressing *Bacillus thuringiensis* delta-endotoxin. *Nature Biotechnology* 18:1101-1104.

Tyagi, A.K. and A. Mohanty (2000). Rice transformation for crop improvement and functional genomics. *Plant Science* 158:1-18.

Upadhyaya, N.M., X.-R. Zhou, Q.-H. Zhu, A. Eamens, M.B. Wang, P.M. Waterhouse, and E.S. Dennis (2000). Transgenic rice. In *Transgenic Cereals* (pp. 28-87), Eds. O'Brien, L. and R.J. Henry. St. Paul, MN: American Association of Cereal Chemists.

Vain, P., B. Worland, A. Kohli, J.W. Snape, and P. Christou (1998). The green fluorescent protein (GFP) as a vital screenable marker in rice transformation. *Theoretical Applied Genetics* 96:164-169.

Wang, J., J. Jiang, and J.H. Oard (2000). Structure, expression and promoter activity of two polyubiquitin genes from rice (*Oryza sativa* L.). *Plant Science* 156:201-211.

Wang, M.B. and P.M. Waterhouse (2000). High-efficiency silencing of a beta-glucuronidase gene in rice is correlated with repetitive transgene structure but is independent of DNA methylation. *Plant Molecular Biology* 43:67-82.

Xie, Q.J., J.H. Oard, and M.C. Rush (1995). Genetic analysis of an unstable, purple-red hull rice mutation derived from tissue culture. *Journal of Heredity* 86:154-165.

Yang, G., J. Dong, M.B. Chandrasekharan, and T.C. Hall (2001). *Kiddo,* a new transposable element family closely associated with rice genes. *Molecular Genetics and Genomics* 266:417-424. [DOI 10.1007/s004380100530]

Chapter 10

Cotton Transformation: Successes and Challenges

Roberta H. Smith
James W. Smith
Sung Hun Park

DEVELOPMENT OF COTTON TRANSFORMATION TECHNOLOGY

Agrobacterium-*Mediated Transformation*

Agracetus and Monsanto first reported cotton, *Gossypium hirsutum* L., transformation in 1987 (Umbeck et al., 1987; Firoozabady et al., 1987). The development of a cotton transformation system had been challenging due to the difficulty in developing an efficient tissue culture regeneration system and an efficient selectable marker system for cotton. The report by Umbeck and colleagues (1987) preceded that of Firoozabady and colleagues (1987). Umbeck and colleagues (1987) described a system involving cocultivation of seedling hypocotyls of *G. hirsutum* cultivar Coker 312 with *Agrobacterium tumefaciens* LBA4404. Kanamycin at 5 to 50 mg/m·L^{-1} was the selective agent along with cefotaxime and carbenicillin to control bacterial growth. Later, Umbeck and colleagues (1989) reported inheritance and gene expression in progeny from the three primary transgenic plants derived in the 1987 study. The inheritance of the neomycin phosphotransferase II *(nptII)* and chloramphenicol acetyltransferase *(cat)* genes was reported in progeny from selfed or backcrossed material. Seeds were germinated on medium containing kanamycin, and surviving seedlings were tested for NPT-II and CAT enzyme activity. Selfed material gave a 3:1 ratio, and backcrossed progeny had a 1:1

ratio which would be expected for a single gene trait. However, only the *nptII* gene was present in the progeny from two primary plants; the third primary plant progeny had both genes present, but the *cat* gene was not expressed.

The report by Firoozabady and colleagues (1987) used cotyledon explants from *G. hirsutum* cultivar Coker 201, and *A. tumefaciens* 15955 and LBA4404. Numerous transgenic plants were obtained. Kanamycin at 15 to 35 $mg \cdot L^{-1}$ was used as the selective agent along with carbenicillin to control bacterial growth. The authors compared cotyledon and hypocotyl explants and found better contact with the selection medium and a reduced level of chimeric callus using cotyledon explants. In addition, glucose was the preferred carbohydrate because it eliminated or reduced tissue phenolic production from cotton explants. Inheritance and expression of the *nptII* gene in the progeny was not reported.

A very useful report (Cousins et al., 1991) on the transformation of an elite Australian cotton cultivar, Siokra 1-3, provided detailed information on development of the transformation protocol using *Agrobacterium.* A clear description of the regeneration protocol including the visual selection and sieving of the embryogenic suspension, culture density, type of wrap on the culture container, hypocotyl comparison to cotyledon explants, and cocultivation procedure was presented. The authors did obtain cotton plants expressing the neomycin phosphotransferase or the β-glucuronidase gene within nine to 12 months. They also noted some fertility problems.

Up to this point, cotton transformation had not progressed very fast compared to other plant species. Major problems with the genotype specificity in plant regeneration limited regeneration feasibility mainly to the Coker cultivars. Trolinder and Goodin (1987) reported the first somatic embryogenesis and plant regeneration from 'Coker 312', and this procedure with modifications was used by Umbeck and colleagues (1987). The medium used for callus induction from hypocotyls was Murashige and Skoog (1962) inorganic salts with 100 $mg \cdot L^{-1}$ inositol, B_5 vitamins (Gamborg et al., 1968), 30 $g \cdot L^{-1}$ glucose, 0.1 $mg \cdot L^{-1}$ 2,4-dichlorophenoxyacetic acid (2,4-D), 0.1 $mg \cdot L^{-1}$ 6-furfurylaminopurine, solidified using 1.6 gA Gelrite and 0.75 $g \cdot L^{-1}$ $MgCl_2$. The procedure also involved a suspension culture phase without plant growth regulators and sieving suspension cultures for subculture followed by plating the embryos on solid medium for maturity. Mature embryos were then placed on vermiculite saturated with Stewart and Hsu's medium (1977) with 0.1 mg·L^{-1} indole acetic acid. The authors reported that over a one-year period 139 plants were obtained, 60 to 80 percent matured, and 78 flowered, but only 15.4 percent set seed due to pollen infertility problems. Firoozabady and colleagues (1987) developed a similar medium for 'Coker 201' containing the Murashige and Skoog (1962) inorganic salts, 100 mg·L^{-1} myo-inositol, 0.4 mg·L^{-1} thiamine HCl, 5 mg·L^{-1} 6-(γ,γ,-dimethylallylamino)-purine, 0.1 mg·L^{-1} α-naphthaleneacetic acid, and 3 percent glucose solidified with 0.2 percent Gelrite. Callus was subcultured to embryogenic medium, the same as initiation medium, without plant growth regulators.

Many laboratories found that even within the regenerable 'Coker lines', individual seedlings could vary dramatically in their ability to form embryos. Close observation of the cultures was necessary to identify explants that would form embryos. In addition, cotton was sensitive to kanamycin selection and did not tolerate the higher levels routinely used in other plant transformation systems. Cotton transformation and regeneration was certainly not as easy to repeat as the tobacco, potato, or petunia transformation and regeneration systems. The Coker cultivars, moreover, were not the cultivars being commercially used; therefore, transgenic 'Coker' plants had to be put into a breeding program consisting of a series of backcrossing and selection to move the transgenes into commercial cottons. This could take from three to six years depending on how many generations could be matured in a year. Laboratories focused on making improvements in cotton transformation by identifying genotypes that would regenerate in culture and decreasing the time spent in cell culture to reduce somaclonal variation. Stelly and colleagues (1989) established that cotton had a high level of genotype and cytogenetic variation in plants regenerated from callus culture. It became important to reduce the time in vitro to reduce this somaclonal variation.

Zapata and colleagues (1999) used the seedling shoot apex of a Texas cultivar, CUBQHRPIS, as the target tissue for *Agrobacterium*-mediated transformation. The seedling shoot apex as an explant circumvented the problem of the genotype limitation in cotton regeneration, as it had been established that many cotton cultivars could be cultured from the shoot apex explant. Whole plants could be obtained in three months. Progeny from two generations showed expression of the β-glucuronidase gene by Southern blot analysis. The shoot meristem transformation system was not limited to genotype and was rapid in going from explant to plant without somaclonal variation; however, the efficiency of transformation was low.

Biolistic Transformation in Cotton

Other cotton transformation systems were also being developed. Finer and McMullen (1990) reported particle bombardment and transformation of a 'Coker 310' embryogenic suspension culture. Hygromycin at 100 $\mu g \cdot mL^{-1}$ was used as the selection agent, and picloram was used at 0.5 $mg \cdot L^{-1}$ to initiate an embryogenic suspension with subsequent culture on 2 $mg \cdot L^{-1}$ 2,4-D to maintain the callus. Primary plants were obtained within five months, reducing transformation time from the 6 to 12 months reported (Umbeck et al., 1987; Firoozabady et al., 1987) using *Agrobacterium* and somatic embryogenesis as the regeneration system. The incorporation of the hygromycin-resistance gene was confirmed by Southern analysis.

Another report of biolistic transformation of embryonic cell suspension cultures of 'Acala B1654' and 'Coker 315' resulted in plants expressing the foreign gene (Rajasekaran et al., 2000). The authors used a liquid medium containing 2 $mg \cdot L^{-1}$ α-naphthaleneacetic acid and geneticin as the selectable antibiotic.

McCabe and Martinell (1993) targeted meristem tissue in the embryonic axes of germinated seeds to develop a genotype-independent system to transform cotton. This approach eliminated the callus regeneration requirement that is so genotype specific in cotton. 'Sea Island', 'Pima S6', 'Delta Pine 90', and 'Delta Pine 50' were all transformed using this approach, and one 'Pima' transformant was carried out to the R_3 generation to verify stable integration and transmission to progeny. Chlan and colleagues (1995) also reported a method for biolistic transformation of cotton shoot meristem tissue.

INSECT- AND HERBICIDE-RESISTANT COTTON

Insect-Resistant Cotton

The main gene for control of lepidopteran larvae is from a soil bacterium, *Bacillus thuringiensis.* The Bt gene derived from this bacterium inserted into the cotton cells produces the Bt protein toxin that kills the insects. Perlak and colleagues (1990) from Monsanto reported 'Coker 312' transformation by *A. tumefaciens* A208 containing the foreign gene for the insect toxin, cryIA(b) or cryIA(c). Progeny from transgenic parent plants were verified using Western blot analysis to determine protein expression and insect feeding bioassays using the beet armyworm, cabbage looper larvae, and cotton bollworm larvae. This demonstration had a strong impact on rapidly moving cotton transformation technology to commercial application. Bollgard cotton was developed by Monsanto, was commercialized in 1996, and was effective against larvae of tobacco budworm, cotton bollworm, and pink bollworm.

Herbicide-Resistant Cotton

The first herbicide-resistant cotton was reported by Bayley and colleagues (1992) using 'Coker 312' and a modified bacterial gene encoding 2,4-D monooxygenase for 2,4-dichlorophenoxyacetic acid resistance. This *Agrobacterium*-mediated transformation followed the procedure of Umbeck and colleagues (1987) with some modification. Primary plants that resisted three times the recommended rate of 2,4-D were obtained, and progeny showed a 3:1 segregation pattern indicating a singe gene insertion.

A second report of herbicide-resistant 'Acala' and 'Coker' cottons was reported (Rajasekaran et al., 1996). The gene was a mutant form of a native acetohydroxyacid synthase that gave resistance to imidazolinone and sulfonylurea herbicides depending on the mutation site. This report compared biolistic and *Agrobacterium* methods of transformation and found the biolistic method was more efficient. Progeny resistant to five times the field application rate of imidazolinone herbicide, imazaquin, were obtained from Acala transformed using *Agrobacterium.* The efficiency in transforming 'Acala' using *Agrobacterium* was lower as compared to the 'Coker' varieties.

Herbicide-tolerant cotton was released for commercial sales by Calgene and Monsanto. The Calgene cottons, 'Stoneville BXN' varieties, were tolerant to the herbicide bromoxynil (Buctril). Monsanto Roundup Ready cottons can be used with glyphosate (Roundup) during the first four leaves of growth. After this stage, direct exposure to glyphosate results in a delay of boll production.

COMMERCIAL USE OF GENETICALLY ENGINEERED COTTON

Benefits of Using Genetically Engineered Cotton

Farmers in the United States started growing insect- and herbicide-resistant cottons in 1996. Their acceptance, with some minor problems, was good, and the technology was rapidly adopted. In the 1999-2000 crop, 2.63 million hectares of genetically modified cotton worldwide were planted in Argentina, Australia, China, Mexico, and South Africa. The United States accounted for 12 percent of the cotton grown (International Cotton Advisory Committee, 2000).

The genes for insect and herbicide resistance had a positive impact on reducing the number of herbicide or pesticide applications needed. For example, it has been estimated that an average of only 3.6 sprays per crop season is necessary using Bt cotton varieties, which translates to a reduction of 1.0 to 7.7 sprays per growing season (International Cotton Advisory Committee, 2000). This in turn has had positive economic, environmental, and social benefits. According to the International Cotton Advisory Committee (2000), the direct and indirect benefits of using genetically modified cotton were as follows:

- Reduced pesticide use
- Improved crop management effectiveness
- Reduced production costs
- Improved yield and profitability
- Reduced farming risk
- Improved opportunity to grow cotton in areas of severe pest infestation
- Reduced pesticide spraying improved populations of beneficial insects and wildlife in cotton fields
- Reduced pesticide runoff, air pollution, and waste from the use of insecticides
- Improved farm worker and neighbor safety
- Reduced labor costs and time

- Reduced fossil fuel use
- Improved soil quality due to less soil compaction and erosion caused by tractors moving across the field

Concerns About Using Genetically Engineered Cotton

In the process of developing genetically modified cotton, concerns have been raised regarding loss of genetic diversity. The problem arises in part due to the limited number of cotton cultivars that will regenerate in culture. Because some Coker cultivars have been most regenerative, they have been used as the recipient variety for the foreign genes. Following this initial transformation event is a series of backcrossing and selection to move the gene into a useable commercial cultivar. This backcrossing is thought to limit or decrease genetic diversity in commercial lines (International Cotton Advisory Committee, 2000). Further improvement in cotton transformation systems is needed to enable direct transformation of elite commercial lines.

A question regarding the escape of herbicide resistance into weedy relatives or other plants has been raised. However, since cotton is normally self-pollinated, wild populations of cotton are rare or grow in nonagricultural areas. Therefore, escape of herbicide resistance into weedy relatives has been evaluated to be low to nonexistent (International Cotton Advisory Committee, 2000).

A major concern regarding insect resistance in cotton is that of insects developing resistance to the Bt protein toxin and development of a strain of super insects (for a detailed review see Hardee et al., 2001). The Bt toxins are insect specific and are considered innocuous to other life forms. The Bt toxin must be orally ingested by an insect with a midgut that has an alkaline pH for mortality to be induced. Thus, only specific taxonomic groups of insects with chewing mouthparts and alkaline midguts are susceptible to the Bt toxin. In general, the insect toxic plants express the toxin during plant growth and reproduction, and thus the pest is continuously exposed to the toxin. The rare preadapted individuals in the extant population that carry the gene to metabolize the toxin are relentlessly selected. Only those few individuals that detoxify the toxin survive. Under conditions where constant selection pressure exists and the surviving resistant individuals mate with similar genotypes, populations resistant to the toxin are rapidly selected. Once the gene that confers resistance becomes prevalent in the population, the transformed plants no longer effectively suppress the target pests. Thus, the rapidity and prevalence of pest resistance to the toxin is constrained by the magnitude of the selection pressure and mating frequency among individuals carrying the resistant gene (Georghiou and Saito, 1983).

In an attempt to circumvent the selection of resistant populations of insects in the transgenic crop, population ecologists devised the refuge-resistant management plan. This plan is part of an agreement between the cotton producer and the seed company. The seed must be used in accordance with the agreement. The agreement specifically outlines the refuge-resistant management plan requirements that the grower must follow. The company that sells the genetically engineered cotton seed is required to monitor grower compliance with the agreement. The 2002 Bollgard Refuge Requirements (http://www.plantprotection.org/IRMtraining/cotton/cottonindex.cfm) have been modified from the original requirement of growers to plant 20 acres of non-Bollgard cotton for every 80 acres of Bollgard cotton.

The refuge-resistance management plan is based in population genetics theory that the few surviving individuals from the genetically transformed cotton which carry the resistant gene(s), e.g., RR alleles, have a high probability of mating with the predominant susceptible genotype (SS alleles) from the refuge planting (Hardee et al., 2001). In general, the susceptible genotype (SS) predominates because these individuals express greater fitness under natural conditions (no exposure to Bt toxin gene in engineered plants) than do SR or RR (predominant in the transformed crop because they possess greater fitness under selection pressure for resistance to the Bt toxin).

Several assumptions are associated with the refuge-resistant management plan. The first assumption is that the Bt toxin is expressed in the insect-resistant plants at a concentration high enough to kill all individuals possessing the SS and SR genotype and that only a very few of RR genotype survive; the RR genotype must be a rare occurrence. The second assumption, random mating, presumes that the few survivors arising from the genetically transformed field, RR genotype, will mate with the vastly numerically predominate SS genotype from the refuge planting. The extent that the model will predict the outcome is based on how well the behavioral ecology and toxicology of the specific target pests satisfies all the model assumptions.

Currently, most commercially marketed genetically transformed crops may not satisfy all of the assumptions. The first assumption, expression of the Bt toxin at a high enough concentration, is seldom met. When the mortality is not sufficient to kill 99+ percent of the pests, the ratios of SS, SR, and RR alleles in the population will be different than predicted. A higher prevalence of SR and RR in the population will generate resistant pest progeny at an accelerated rate. However, a saving caveat may be that the proportion of transformed crops expressing Bt toxins planted across the agricultural landscape is presently low (Gould et al., 1992). A greater presence of SS individuals arising from noncrop habitats and conventional crop varieties could contribute a preponderance of SS genotypes and delay of selection of resistant populations. The second assumption, the random mating model, requires that sexually mature adults emerging from both the transformed and conventional crop be temporally and spatially coincident for mating. Survivors (RR, and often SR genotypes if the toxin level is not high) that have fed on the transgenic cotton often express increased generation times compared to the generation time for those having fed on nontransgenic cotton and native plants. Differences in the developmental periods of pests on the different plant hosts make mating between individuals from the genetically transformed and the refuge crops temporally asynchronous. Without temporal coincidence, the survivors of the toxin (RR) may have a greater probability of mating among themselves because the alternative mates (SS) are not as available. Spatial coincidence for random mating requires that mating adults from the conventional and transformed cotton adequately mix prior to copulation. Prenuptial movement of adult insects and distance traveled is often a species-specific or environmentally induced trait. For cotton, unless unmated individuals of the targeted lepidopteran pests move at least half a mile prior to mating, then the lack of coincidence in space is violated. The use of a refuge management plan, the potential ability to pyramid Bt toxin genes, and the many different forms of the Bt gene are some strategies that can be used to delay or avoid the problem of insects becoming resistant.

In addition, the concern arises that nontarget insect species will be harmed, and tests have been conducted to assess this impact. The sprayable form of Bt toxin has been used extensively and has demonstrated safety on many nontarget organisms (International Cotton Advisory Committee, 2000). In fact, due to the reduction of pesticide use on insect-resistant cotton, observations indicate increases in populations of other insects. Many other studies have also been conducted including evaluation of the pollen from insect-resistant cotton on monarch butterflies. No detrimental effects were noted (International Cotton Advisory Committee, 2000).

REFERENCES

Bayley, C., N. Trolinder, C. Ray, M. Morgan, J.E. Quisenberry, and D.W. Ow (1992). Engineering 2,4-D resistance into cotton. *Theoretical and Applied Genetics* 83:645-649.

Chlan, C.A., J. Lin, J.W. Cary, and T.E. Cleveland (1995). A procedure for biolistic transformation and regeneration of transgenic cotton from meristematic tissue. *Plant Molecular Biology Reporter* 13(1):31-37.

Cousins, Y.L., B.R. Lyon, and D.J. Llewellyn (1991). Transformation of an Australian cotton cultivar: Prospects for cotton improvement through genetic engineering. *Australian Journal of Plant Physiology* 18:481-494.

Finer, J.J. and M.D. McMullen (1990). Transformation of cotton (*Gossypium hirsutum* L.) via particle bombardment. *Plant Cell Reports* 8:586-589.

Firoozabady, E., D.L. De Boer, D.J. Merlo, E.L. Halk, L.N. Amerson, K.E. Rashka, and E.E. Murray (1987). Transformation of cotton (*Gossypium hirsutum* L.) by *Agrobacterium tumefaciens* and regeneration of transgenic plants. *Plant Molecular Biology* 10:105-116.

Gamborg, O.L., R.A. Miller, and K. Ojima (1968). Nutrient requirements of suspension cultures of soybean root cells. *Experimental Cellular Research* 50:151-158.

Georghiou, G.P. and T. Saito, Eds. (1983). *Pest Resistance to Pesticides.* New York: Plenum Press.

Gould, F., A. Martinez-Ramirez, and A. Anderson (1992). Broad-spectrum resistance to *Bacillus thuringiensis* toxins in *Heliothis virescens. Proceedings of the National Academy of Sciences USA* 89:7986-7990.

Hardee, D.D., J.W. Van Duyn, M.B. Layton, and R.D. Bagwell (2001). *Bt Cotton and Management of the Tobacco Budworm-Bollworm Complex.* ARS-154. Stoneville, MD: United States Department of Agriculture, Agricultural Research Service.

International Cotton Advisory Committee (2000). Report of an Expert Panel on Biotechnology in Cotton. Washington, DC: Author.

McCabe, D. and B.J. Martinell (1993). Transformation of elite cotton cultivars via particle bombardment of meristems. *Bio/Technology* 11:596-598.

Murashige, T. and F. Skoog (1962). A revised medium for rapid growth and bioassays with tobacco tissue cultures. *Physiologia Plantarum* 15:473-497.

Perlak, F.J., R.W. Deaton, T.A. Armstrong, R.L. Ruchs, S.R. Sims, J.T. Greenplate, and D.A. Fischhoff (1990). Insect resistant cotton plants. *Bio/Technology* 8:939-943.

Rajasekaran, K., J.W. Grula, R.L. Hudspeth, S. Pofelis, and D.M. Anderson (1996). Herbicide-resistant Acala and Coker cottons transformed with a native gene encoding mutant forms of acetohydroxyacid synthase. *Molecular Breeding* 2:307-319.

Rajasekaran, K., R.L. Hudspeth, J.W. Cary, D.M. Anderson, and T.E. Cleveland (2000). High-frequency stable transformation of cotton (*Gossypium hirsutum* L.) by particle bombardment of embryogenic cell suspension cultures. *Plant Cell Reports* 19:539-545.

Stelly, D.M., D.W. Altman, R. Kohel Jr., T.S. Rangan, and E. Commiskey (1989). Cytogenetic abnormalities of cotton somaclones from callus cultures. *Genome* 32:762-770.

Stewart, J.McD. and C.L. Hsu (1977). In ovulo embryo culture and seedling development of cotton (*Gossypium hirsutum* L.). *Planta* 137:113-117.

Trolinder, N.L. and J.R. Goodin (1987). Somatic embryogenesis and plant regeneration in cotton (*Gossypium hirsutum* L.). *Plant Cell Reports* 6:231-234.

Umbeck, P., G. Johnson, K. Barton, and W. Swain (1987). Genetically transformed cotton (*Gossypium hirsutum* L.) plants. *Bio/Technology* 5:263-265.

Umbeck, P., W. Swain, and N.-S. Yang (1989). Inheritance and expression of genes for kanamycin and chloramphenicol resistance in transgenic cotton plants, *Crop Science* 29:196-201.

Zapata, C., S.H. Park, K.M. El-Zik, and R.H. Smith (1999). Transformation of a Texas cotton cultivar by using *Agrobacterium* and the shoot apex. *Theoretical and Applied Genetics* 98:252-256.

Chapter 11

Progress in Transforming the Recalcitrant Soybean

Jack M. Widholm

INTRODUCTION

Since soybean [*Glycine max* (L.) Merrill] is one of the most important crops worldwide and the second most important in the United States, a great amount of effort has been expended in developing transformation methods. Soybean, similar to other large-seeded legumes, was initially a difficult material for plant regeneration and to date is still a difficult plant to transform, at least in comparison to most other species.

Soybean is a desirable crop since the seeds contain about 40 percent protein and about 20 percent oil. As a legume, soybean can fix most of the nitrogen needed and leave a significant amount for next year's crop. Soybean oil is the most abundant vegetable oil on earth, and the high-protein meal left after oil extraction is a very important livestock protein supplement. Soybeans are also an important human food that is increasing in popularity due to the clearly beneficial functional food characteristics which include anticancer and general circulatory health properties.

Several reviews of soybean transformation progress have been conducted since the first transformed plants were produced in 1988, but new improvements and innovations continue to be made. Since

---

Funds for the unpublished results reported here were provided by the United Soybean Board, the Illinois Soybean Program Operating Board, U.S. Department of Agriculture, and the Illinois Agricultural Experiment Station. I thank Shiyun Chen for reading the manuscript.

no technique is very efficient, a new summary of these is worthwhile. The recent reviews include Finer and colleagues (1996), Trick and colleagues (1997), Widholm (1995, 1999), Chee and Hu (2000), and Dinkins and colleagues (2002). This review will emphasize recent advances, problems, and possible solutions and will not cover some of the older literature that has been discussed in the past reviews.

The first successful soybean transformations that produced fertile plants were accomplished by Hinchee and colleagues (1988), who used *Agrobacterium tumefaciens* with the cotyledonary node plant regeneration system, and by McCabe and colleagues (1988) using particle bombardment of meristems of immature seeds. This work was followed by success using embryogenic suspension cultures and particle bombardment (Finer and McMullen, 1991). These methods or modifications thereof have produced most of the transformed soybean plants to date.

ORGANOGENIC SYSTEMS

Cotyledonary Node

As mentioned earlier, one of the first successful soybean transformation systems was the *A. tumefaciens*-mediated DNA transfer to wounded cotyledonary nodes of germinating seeds from which shoots can regenerate by organogenesis on cytokinin-containing medium (Hinchee et al., 1988). Early work showed a range of susceptibility for crown gall formation on different soybean genotypes (Owens and Cress, 1985), and this trait was heritable in progeny of crosses of high by low lines (Bailey et al., 1994; Mauro et al., 1995). Stable transformation is also soybean genotype dependent, so the genotype used can be very important (Meurer et al., 1998).

There is also a range of transformation effectiveness with different *A. tumefaciens* strains or mutants for crown gall formation (Owens and Cress, 1985; Donaldson and Simmonds, 2000), transient expression (Ke et al., 2001), and for stable transformation to produce soybean plants (Meurer et al., 1998).

The selectable marker gene is also an important factor, and *nptII* which encodes neomycin phosphotransferase for kanamycin resistance has been used successfully (Hinchee et al., 1988; Di et al., 1996; Donaldson and Simmonds, 2000). In recent work, the *bar* gene that encodes phosphinothricin acetyltransferase which detoxifies the herbicide glufosinate has been used (Zhang et al., 1999; Xing et al., 2000). The CP4, enolpyruvylshikimate-3-phosphate synthase (EPSPS), gene which encodes an enzyme that is resistant to the herbicide glyphosate has also been used successfully (Clemente et al., 2000).

Recently it has been reported that the addition of thiol-containing compounds such as L-cysteine during the cocultivation of cotyledonary nodes with *A. tumefaciens* increases transformation efficiency markedly (Olhoft and Somers, 2001; Olhoft et al., 2001). The positive response is apparently due to decreased wound- and pathogen-defense responses by the explant (less browning is observed).

The *A. tumefaciens*-cotyledonary node system was used to produce soybean plants containing an antisense soybean vegetative storage protein gene construct which markedly suppressed the vegetative storage protein exprcssion (Staswick et al., 2001). No significant changes in productivity were reported when these plants were grown under various conditions, indicating that these proteins were not essential for normal plant growth and reproduction.

The cotyledonary node system was also used to obtain transgenic soybean plants with markedly higher than normal levels of oleic acid in the seed (Buhr et al., 2002). The plants were transformed with antisense fatty acid desaturase (FAD-2) terminated by a self-cleaving ribozyme in the 3' untranslated region that should cause nuclear retention of the truncated transcript. A decrease in the FAD-2 enzyme would decrease the conversion of the monounsaturated fatty acid oleic acid to linoleic acid. An even larger increase in oleic acid was obtained when sense constructs of both FAD-2 and palmitoyl-thioesterase genes terminated by the ribozyme were used to down-regulate both enzymes. Palmitic acid levels were decreased as expected due to the down-regulation of the thioesterase. Evidence was presented indicating that gene silencing involving small interfering RNA was causing the down-regulation.

Apical Meristem

The particle bombardment of immature seed meristems using the *gusA* gene as a reporter without selection for a resistance marker was used by McCabe and colleagues (1988) to produce shoots on a cytokinin-containing medium. The transformed tissues were followed by assessing GUS expression in regenerating shoots and plants. Some transformed plants were produced that produced transformed progeny. Although this method is very labor and cost intensive and is not very efficient as far as the number of transformants recovered, this method was used to produce the glyphosate (Roundup) herbicide-resistant line carrying the CP4 EPSPS gene that has been used by breeders to produce the commercial Roundup Ready varieties grown on more than half the total U.S. soybean acres (Padgette et al., 1995).

A similar system, whereby meristems from imbibed mature seeds were bombarded and then placed on the shoot formation medium containing the herbicide imazapyr (Arsenal) as the selection agent, has successfully produced transformed soybean plants (Aragão et al., 2000). The tissues were bombarded with an *Arabidopsis* mutant acetolactate synthase gene that should impart resistance to Arsenal, and indeed many recovered plants produced progeny that were transformed.

We have been able to produce some transgenic soybean plants using the same selectable marker and plant regeneration system as described by Aragão and colleagues (2000) with *A. tumefaciens* as the DNA transfer agent (S. Chen and J.M. Widholm, unpublished). We have also attempted to use a new selectable marker gene, *ASA2* (feedback-resistant anthranilate synthase from tobacco, Song et al., 1998) which should impart resistance to toxic tryptophan analogs. Numerous experiments were carried out using *A. tumefaciens*-mediated DNA transfer, and about 25 shoots came through the selections with 5-methyltryptophan or α-methyltryptophan as selection agents. However, only one *ASA2*-transformed plant was recovered as confirmed by Southern analysis. However, *Astragalus sinicus* hairy roots that express *ASA2* were clearly tryptophan analog resistant (Cho, Brotherton, et al., 2000).

Seedling Hypocotyls

A shoot regeneration system using soybean seedling hypocotyl explants (Dan and Reichert, 1998) has been transformed via particle bombardment followed by selection on glufosinate (*bar* gene) to produce shoots that contain the *bar* gene as shown by polymerase chain reaction (PCR) (Chen et al., 1999). Further evidence for the reproducibility of this system has not been reported.

Protoplasts

Soybean plants have been regenerated from immature cotyledon protoplasts via organogenesis (Wei and Xu, 1988), and protoplasts, which lack cell walls, can readily be used for direct DNA uptake by electroporation or treatment with an agent such as polyethylene glycol (PEG) (Dhir et al., 1991). This system has been used to insert an insecticidal protein gene from *Bacillus thuringiensis* using polyethylene glycol to produce transformed protoplasts following hygromycin or kanamycin selection that were regenerated into plants (Wei et al., 1997). Southern blot analysis showed that the plants contained the *hph* or *nptII* genes.

The difficulty in regenerating plants from the protoplasts makes this system problematic and irreproducible in many labs, so widespread use is unlikely unless improvements are made.

Embryogenic Systems

The embryogenic suspension culture system which consists of clumps of replicating globular-stage embryos initiated from immature cotyledons on a very high 2,4-dichlorophenoxyacetic acid (2,4-D)-containing medium (40 $mg \cdot L^{-1}$) was first used by Finer and McMullen (1991) to produce transformed plants. The liquid medium contains 5 $mg \cdot L^{-1}$ 2,4-D and asparagine or glutamine. This work and most other research reported to date used particle bombardment to deliver the DNA with the hygromycin phosphotransferase gene *(hph)* as the selectable marker gene. Once hygromycin-resistant clumps have been selected, plants are regenerated when the 2,4-D is removed from the medium and the embryos that form are desiccated and germinated.

A series of reports have described improvements in the culturing aspects of the embryogenic system which have increased the percent germination of the embryos, increased the culture growth rate, and decreased the time needed to obtain transformed plants. The optimized liquid medium for embryogenic suspension cultures, summarized by Samoylov, Tucker, and Parrott (1998), has lower nitrogen and sucrose levels was therefore named FN Lite. The embryogenic cultures can also be grown continuously on a semisolid medium called MSD20 (20 mg·L^{-1} 2,4-D) as described by Santarem and Finer (1999). The liquid-based-medium protocol for regenerating plants from these embryogenic cultures (Samoylov, Tucker, Thibaud-Nissen, and Parrott, 1998) has recently been improved by adding sorbitol (Walker and Parrott, 2001). Embryogenic cultures grown either in liquid medium or on semisolid medium have been successfully used to obtain transformed plants.

The embryogenic culture response, as measured by induction percent, growth rate and maintenance of embryogenic morphology, and yield of embryos that germinate and convert into plants has been shown to vary with genotype, although most genotypes can respond well enough to form cultures and plants (Bailey et al., 1993; Simmonds and Donaldson, 2000; Meurer et al., 2001).

Embryogenic suspension cultures of 'Jack' and 'Asgrow' A2872 were reported to not be very transformable by particle bombardment until about six months after initiation, and the transient expression that was measured correlated with mitotic activity which peaked about four days after transfer into fresh medium (Hazel et al., 1998). The most transformable cultures consisted of embryogenic clumps with cytoplasmic-rich outer cell layers. Many of us have routinely used much younger cultures to obtain reasonable transformation frequencies, however. The use of newly initiated cultures might be important since plant regeneration frequency declines and plant sterility increases when cultures become too old.

'Jack' suspension cultures were used to show that the *gus* gene driven by the soybean seed lectin gene *(Le1)* promoter was not expressed in the embryogenic clumps but was highly expressed in the developing embryos (Cho et al., 1995). The seed-specific expression controlled by the *Le1* promoter was further demonstrated when transformed soybean plants were produced following particle bombardment with a *Le1* promoter-bovine β-casein gene construct of 'Jack' embryogenic cultures grown on solid culture medium (Maughan et al., 1999). Northern analysis showed very high bovine β-casein mRNA in seed, a small amount in leaves, and none in the embryogenic cultures. Western analysis showed high bovine β-casein in seed and none in leaves. The bovine β-casein is processed and localized in protein bodies in the transformed soybean seed using the lectin signal sequence (Philip et al., 2001). Subsequent studies showed that four copies of the β-casein gene were inserted at a single locus no larger than 40 kb in size with some plant genomic DNA interspersed between each copy (Choffnes et al., 2001). A high frequency of recombination (16 percent) occurred in T_1 and T_2 progeny as shown by novel fragments produced by restriction endonuclease digestion.

Soybean embryogenic suspension cultures have also been transformed with the *B. thuringiensis cryIA(b)* (cultivar F376, Parrott et al., 1994) and *cryIAc* genes (cultivar Jack, Stewart et al., 1996) to produce plants that expressed the genes and had decreased defoliation caused by several different insects.

The bean pod mottle virus (BPMV) coat protein precursor gene was inserted into soybean plants via the *A. tumefaciens*-cotyledonary node transformation system, and some progeny plants were found to be completely resistant to infection by BPMV (Di et al., 1996). The transgene was not stable in advanced generations, so this transformation was repeated using particle bombardment of embryogenic 'Jack' cultures on solid medium (Reddy et al., 2001). Progeny of regenerated transgenic plants did exhibit systemic resistance to BPMV.

A number of soybean lines have been produced that lack the β-conglycinin seed storage protein α and/or α' subunits via cosuppression following transformation with either the coding regions or 5' untranslated leader sequences (Kinney et al., 2001). The one homozygous line studied lacked both the α and α' subunits and also had increased oleic acid since the soybean fatty acid desaturase gene *Fad2* had also been inserted which caused cosuppression of this gene. The total amounts of seed protein and oil were not changed. The β subunit of β-conglycinin was increased some, but overall increases in the glycinin storage protein compensated for the decrease in β-conglycinin. The storage proteins accumulated in a new endoplasmic reticulum-derived vesicle not seen in normal soybean seeds. This work clearly demonstrates that cosuppression can occur and can be used to modify soybean seed composition.

Embryogenic cultures of the cultivar Jack and the line F173 obtained from a cross of 'Jack' × PI417138 were bombarded with constructs containing the maize 15 kd zein protein gene driven by the β-phaseolin promoter for seed-specific expression (Dinkins et al., 2001). Two transgenic lines did produce zein in seeds that led to increases of from 12 to 20 percent in methionine and 15 to 35 percent in cysteine concentrations in mature seeds.

Because embryogenic cultures form embryos that are generally similar to developing seeds, the effects of certain genes such as those involved in seed fatty acid biosynthesis can be studied without regenerating plants from the transgenic cultures. Full-length cDNAs for unusual fatty acid desaturases were cloned from two plant species that accumulate 18 carbon fatty acids with three and four conjugated double bonds, and these cDNAs, driven by the seed-specific β-conglycinin α' subunit promoter, were bombarded into embryogenic suspension cultures of the cultivar A2872 and transformed clones selected with hygromycin (Cahoon et al., 1999). Up to 17 percent of the total fatty acids of the transformed embryos could be the two unusual conjugated double-bond fatty acids. Genes encoding a Δ^5 fatty acid desaturase and a fatty acid elongase isolated from *Limnanthes douglasii,* a plant that accumulates high levels of a 20 carbon fatty acid with one Δ^5 double bond, were both expressed using the β-conglycinin α' subunit promoter in 'A2872' embryogenic cultures (Cahoon et al., 2000). Up to 12 percent of the total fatty acids were $20{:}1\Delta^5$ and $22{:}1\Delta^5$ in the transgenic embryos compared with none in the untransformed control.

In most cases only one plasmid at a time has been used in the bombardment of embryogenic cultures. However, Cahoon and colleagues (2000) mixed two plasmids to obtain the line expressing the two fatty acid biosynthesis genes and the hygromycin selectable marker. Hadi and colleagues (1996) bombarded embryogenic cultures with 12 different plasmids mixed together and selected hygromycin-resistant transformed clones. Most of the clones contained all the plasmids including the *gus* gene at varying copy numbers, although less than half showed GUS expression. Simmonds and Donaldson (2000) did report the selection of two transgenic embryogenic suspension-cultured lines following bombardment with two mixed plasmids using phosphinothricin as the selection agent. One of the plasmids contained the *pat* gene encoding phosphinothricin acetyltransferase and the other contained *gus*. Both clones expressed GUS, and the one tested was Southern hybridization–positive for *pat*.

We have bombarded embryogenic soybean cultures with gene cassettes with the promoter and terminator that were cut out of the plasmid and then purified on gels (O. Zernova, V. Lozovaya, and J.M. Widholm, unpublished). Transgenic lines have been recovered that contain all four genes or other combinations when mixtures were used, including the selectable marker gene *bar*. These results indicate that multiple genes can be inserted simultaneously and that plasmid DNA sequences can be excluded using the particle bombardment procedure.

Yan and colleagues (2000) used *A. tumefaciens* to transfer DNA into embryos that were induced directly on immature zygotic cotyledon explants after wounding and cocultivation and selection using hygromycin. Three GUS-positive fertile plants were obtained that passed the genes to progeny.

A new protocol for wounding tissue before cocultivation with *A. tumefaciens* called sonication-assisted *Agrobacterium*-mediated transformation (SAAT) was described by Trick and Finer (1997). Very short sonication treatments of soybean embryogenic suspension cultures and immature cotyledons caused wounding that allowed very high transient GUS expression following cocultivation with A. tumefaciens (Trick and Finer, 1997; Santerem et al., 1998). When SAAT was applied to embryogenic suspension cultures, hygromycin-resistant clones were selected that showed GUS expression and were Southern hybridization positive for both the *hph* and *gus* genes (Trick and Finer, 1998).

NEW TRANSFORMATION METHOD POSSIBILITIES AND IMPROVEMENTS

In Planta *via Infiltration*

A transformation method that does not require the use of tissue culture is *A. tumefaciens* infiltration, which was first used with *Arabidopsis* seeds by Feldmann and Marks (1987) and later with flowering tissues by Bechtold and colleagues (1993). Recent improvements which have eliminated the vacuum infiltration step can produce up to 3 percent transformed seeds from the plants that were dipped in the *A. tumefaciens* suspension containing 5 percent sucrose and a surfactant (reviewed by Bent, 2000). The vacuum infiltration method has also been successful with another crucifer, *Brassica rapa* L. ssp. *chinensis* (Qing et al., 2000), and with legume *Medicago truncatula* (Trieu et al., 2000). Although no details were given (unpublished), a similar technique was apparently used by de Ronde and colleagues (2000) to insert an antisense construct of a proline biosynthesis gene into soybean to decrease proline biosynthesis and also in turn to decrease drought tolerance.

In Planta *via Pollen-Tube Pathway*

Another *in planta* method for inserting foreign genes into plants is called the pollen-tube pathway whereby the style is severed following normal pollination and DNA applied to the cut end. In theory, the DNA then flows down the pollen tube into the fertilized zygotic cells to be integrated into the host genome. In the case of soybean, Liu and colleagues (1992) found that about 3 percent of the plants grown from the seed produced following application of DNA from wild and semiwild soybean and pea showed phenotypic variation. This variation included changes in growth habit, pubescence, leaf and seed shape, flower and hilum color, and maturity. In reports summarized by Hu and Wang (1999), when Lei and colleagues (1989) applied DNA from *Glycine gracilis* with high seed protein and early maturity to soybean cut styles, progeny were produced that were earlier in maturity and had higher seed protein levels. Some plants also had new peroxidase isoenzyme patterns in comparison with the recipient cultivar and new random amplified polymorphic DNA (RAPD) bands, some of which matched those of the donor. Although these and other reports with other species show phenotypic changes that appear to be due to the applied DNA, in most cases no molecular evidence is presented to actually show that the foreign DNA has been integrated. However, Luo and Wu (1988) showed by Southern hybridization with progeny plant DNA of rice that some plants did contain copies of the *nptII* marker gene that was used. Some of the plants also produced measurable NPTII enzyme activity.

We have carried out pollen-tube pathway experiments with soybean using plasmids that carried the *gus* or *bar* genes so that a clear phenotype could be assayed: GUS expression or glufosinate resistance, respectively (Li et al., 2002). A total of 1,942 seeds were produced from 1,295 flowers treated with the *bar* gene and 3,167 seeds from 2,375 flowers treated with the *gus* gene. None of the plants grown from the seed produced by the *bar* gene-treated flowers were found to be resistant to glufosinate (Liberty herbicide); 14 showed intermediate resistance, but this was not seen in the next generation after self-pollination. Some of the seeds produced by the *gus* gene-treated flowers did show light blue GUS staining on chips cut from the cotyledons, but the plants grown from these seeds were not GUS positive. PCR analysis using two different pairs of *gus* gene primers that did amplify positive control DNA did not detect the *gus* gene in any of a number of plants tested. Some plants grown from seed from treated flowers had abnormal leaf numbers or shape, but these phenotypes were not seen in progeny produced by these plants. Negative results were also reported by Moore and colleagues (1996) with soybean using the *bar* gene as the selectable marker with the pollen-tube pathway method.

Although the pollen-tube pathway method would appear in some cases to be promising since some phenotypic changes occur and the technique does not involve specialized laboratory techniques such as tissue culture, the overall lack of unequivocal proof of DNA integration in controlled experiments and irreproducibility in many labs leaves doubt about the validity and utility.

Plastid DNA Transformation

Plastid DNA transformation has been possible with tobacco for many years (Maliga et al., 1993) but has not been successful with any other species to produce fertile plants, with the exception of the recent successful transformation of tomato (Ruf et al., 2001). Plastid DNA transformation relies on homologous recombination to integrate the foreign genes into the plastid genome so the flanking sequences need to be homologous to a plastid DNA region. We bombarded 984 'Jack' embryogenic suspension culture samples (each with about 100, 1 mm diameter clumps) with a construct carrying the *aadA* gene that confers spectinomycin resistance and the *Chlamydomonas reinhardtii* Rubisco large and small subunit genes. Following selection by using 100-300 mg/ml spectinomycin, three growing green clumps were recovered; PCR analysis of the only one that grew sufficiently confirmed the integration of the transgenes (Zhang, Portis, and Widholm, 2001). No plants were recovered, however.

The plastid transformation project prompted us to characterize the photosynthetic properties of the 'Jack' embryogenic suspension cultures to compare with the soybean photoautotrophic suspension culture (SB-P) (cultivar Corsoy) and soybean leaves (Zhang, Widholm, and Portis, 2001). The embryogenic culture had very low chlorophyll, Rubisco protein, and photosynthesis levels. The plastids were smaller with less well developed thylakoid membranes. Although the nuclear DNA content was near normal, the plastid DNA amount was much less than that found in the SB-P cells. The smaller chloroplasts of the embryogenic culture caused us to use 0.6 μm diameter gold particles for the plastid transformation attempts instead of larger ones.

Hairy Roots

Hairy roots can be readily induced by inoculating wounded soybean tissues, such as mature seed cotyledons, with *Agrobacterium rhizogenes* (Cho, Farrand, et al., 2000). Rapidly growing roots of clonal origin that are *A. rhizogenes*-free can be obtained within about two months. Expression of genes carried on a binary vector can be easily studied in soybean hairy roots (Mazarei et al., 1998), and the roots can be used to study *Bradyrhizobium japonicum* nodulation (Cheon et al., 1993) and to propagate the soybean cyst nematode (Savka et al., 1990; Cho, Farrand, et al., 2000). Hairy roots of *M. truncatula* have been used by Boisson-Dernier and colleagues (2001) to study symbiotic nitrogen fixing and endomycorrhizal associations.

Attempts to induce shoot formation from ten different soybean genotypes in media containing different concentrations of 2,4-D, benzyladenine, and bialaphos were unsuccessful (Cho, Farrand, et al., 2000).

Soybean hairy roots provide a convenient system to express genes and to determine if these genes affect certain root-associated phenomena such as pathogenesis, N fixation, and nematode attack.

Selectable Marker Genes

A key element in most transformation methods is the selectable marker system that is used to select out the rare transformed cell from the untransformed ones. As summarized earlier, most selectable marker genes in use encode enzymes that detoxify the toxic selection agent (kanamycin, hygromycin, glufosinate) or are resistant to the toxic agent (glyphosate, imazapyr). Due to consumer concerns, antibiotic resistance cannot now be used for commercial applications, and other selectable marker genes are patented and therefore must be licensed. The presently used selectable marker genes also are not very efficient with soybean, so searches for new unobjectionable genes that might also increase the selection efficiency are in progress.

Some selectable marker gene systems such as phosphomannose isomerase do not detoxify a compound but allow the cells that express the encoded gene to use a compound as a carbon source, in this case mannose, that plant cells normally cannot use. This gene has been found to be an effective selectable marker for several species including sugar beet, potato, wheat, oil seed rape, and maize (reviewed by Joersbo, 2001). A similar system, xylose isomerase that converts xylose, which also cannot be used as a sole carbon source, to xylulose which can, was used by Haldrup and colleagues (1998) as a selectable marker gene to produce transformed potato plants at high frequency.

The *gus* gene that is usually used as a reporter gene has also been used as a direct selectable marker with tobacco by using benzyladenine *N*-3-glucuronide as the cytokinin source which will be active only if hydrolyzed by the GUS enzyme to release benzyladenine (Joersbo, 2001). Transformed cells that express GUS grow and regenerate into plants.

Other enzymes also detoxify compounds that might be useful with soybean; these include cyanamide dehydratase which was effective with wheat (Weeks et al., 2000) and betaine aldehyde dehydrogenase which was used by Rathinasabapathy and colleagues (1994) for tobacco nuclear transformation and by Daniell and colleagues (2001) for tobacco plastid transformation. In the latter case the plastid transformation efficiency was much higher than that found with the conventional selectable marker gene, aminoglycoside adenyltransferase *(aadA)*.

We have been able to use the tobacco feedback insensitive form of the tryptophan biosynthetic control enzyme anthranilate synthase (ASA2) as a selectable marker gene since expression can impart resistance to toxic tryptophan analogs such as 5-methyltryptophan. Hairy roots that express ASA2 can be selected with inhibitory concentrations of 5-methyltryptophan of both *Astragalus sinicus* (Cho, Brotherton, et al., 2000) and soybean (H.-J. Cho, J.E. Brotherton, and J.M. Widholm, unpublished). Attempts are being made to use ASA2 as a selectable marker with soybean transformation systems, but as described, limited success has been attained.

The use of the *gus* reporter gene to follow the transformed tissues without the use of a selectable marker has been successful for producing transgenic soybean plants (McCabe et al., 1988). Recently a new reporter gene *(gfp)* from a jellyfish that encodes a green fluorescent protein has been used successfully with several cereals to identify transformed callus sectors that are then selectively transferred to obtain transformed plants without the use of a selection agent (for example, Kaeppler et al., 2000, with oats). The *gfp* gene is more convenient than *gus* since only blue light needs to be shined on the tissue to excite the fluorescence in a nondestructive manner. There are, however, often problems in clearly differentiating GFP fluorescence from chlorophyll in green tissues. Kamaté and colleagues (2000) reported that they were able to detect *gfp* expression in *M. truncatula* callus under kanamycin selection.

Xing and colleagues (2000) used a two T-DNA binary *A. tumefaciens* system with the cotyledonary node to produce soybean plants without objectionable marker genes. Two separate T-DNA regions are integrated into the host genomic DNA and if they are on different chromosomes then the two could separate in subsequent generations. Xing and colleagues (2000) found that the *bar* selectable marker gene from one T-region did indeed separate from the *gus* gene found in the other T-DNA in four of the ten individual soybean transgenic events studied. Thus this system which can produce plants without any marker genes appears to be feasible using *A. tumefaciens*-mediated soybean transformation.

CONCLUSIONS

A number of laboratories are producing transformed soybean plants using the organogenic systems, especially the cotyledonary node, using several different selectable marker genes. The cotyledonary node system uses easily obtained materials, germinating seedlings, but seems to require skill for wounding and selection. In many reports, it is possible to obtain transformed plants from about 1 percent of the cotyledon explants within a few months after initial treatment, while Olhoft and Somers (2001) report that addition of L-cysteine to the cocultivation medium can double the transformation frequency to 2 percent. An advantage of this system is that *A. tumefaciens* is the DNA transfer agent so usually one or a low number of gene copies are inserted without rearrangement or breakage. The two T-DNA binary system described by Xing and colleagues (2000) that did produce some selectable marker-free soybean plants also seems to be very useful for producing commercially acceptable plants and for having lines that can be transformed again since they carry no resistance genes.

The imbibed mature seed apical meristem system transformed by particle bombardment using a herbicide-resistant acetolactate synthase as selectable marker described by Aragão and colleagues (2000) may be an ideal system if reproducible in other laboratories.

As mentioned previously, the protoplast system suffers from difficulty in plant regeneration and will not be used widely unless the techniques are improved substantially. Many laboratories are successfully using the embryogenic culture system with particle bombardment to obtain transformed plants. The cultures, however, are initiated from immature seed cotyledons, so flowering plants must be grown and the initiation and culturing process requires experience and effort. An antibiotic resistance gene, *hph*, has been used as the selectable marker in all cases, which will prevent commercialization of the plants produced due to consumer concerns. In addition, particle bombardment often results in multiple and rearranged gene copies but can also allow the insertion of several genes at one time. The time from bombardment until plants are obtained is on the order of six months. It seems important that new, acceptable selectable marker genes be developed and that *A. tumefaciens* be used as the DNA transfer agent if possible.

Breakthroughs could be made to bypass tissue culture altogether using methods such as *in planta* infiltration of flowering plants, which may already have been accomplished with soybean as indicated by de Ronde and colleagues (2000). Any promising new method, including several listed in this review, must pass the test of being valid and reproducible in other laboratories. This is especially clear in the case of the pollen-tube pathway method.

Plastid DNA transformation is one method to obtain plants with defined specific gene insertions that lead to very high expression levels. However, the gene of interest needs to encode a protein that is localized in the plastid and the transformation methods need to be improved greatly for this to be a viable choice. Plastid transformation is likely to be more acceptable by society since the plastid genome of most crop plants is not passed through the pollen to progeny, thus the transgene is contained in the original plant.

One way to circumvent the soybean transformation and plant regeneration problems is to produce hairy roots that express the gene of interest. These roots can be used only to study gene expression and cannot be used directly for plant improvement since plants cannot be regenerated from the roots at present.

It is clear that many laboratories are transforming soybean so that plant improvement and basic research can be accomplished. However, because the present methods are not very efficient and require experienced personnel, they are not going to be used in all laboratories that would like to transform soybean. It is possible that some new method similar to *in planta* infiltration could be developed to allow widespread insertion of genes into soybean. It is also possible that some new selectable marker could be found which would make the selection process much more efficient with tissue cultures. It is clear that new nonantibiotic selectable marker genes are needed if the embryogenic suspension culture system is to be used to obtain commercially acceptable products. Thus although steady and real progress has been made since the first reports of soybean transformation, there is still a need for improvement in the techniques so that soybean is no longer a recalcitrant species.

REFERENCES

Aragão, F.J.L., L. Sarokin, G.R. Vianna, and E.L. Rech (2000). Selection of transgenic meristematic cells utilizing a herbicidal molecule results in the recovery of fertile transgenic soybean [*Glycine max* (L.) Merrill] plants at a high frequency. *Theoretical and Applied Genetics* 101:1-6.

Bailey, M.A., H.R. Boerma, and W.A. Parrott (1993). Genotype effects on proliferative embryogenesis and plant regeneration of soybean. *In Vitro Cellular Developmental Biology—Plant* 29:102-108.

Bailey, M.A., H.R. Boerma, and W.A. Parrott (1994). Inheritance of *Agrobacterium tumefaciens*-induced tumorigenesis of soybean. *Crop Science* 34:514-519.

Bechtold, N., J. Ellis, and G. Pelletier (1993). *In planta Agrobacterium*-mediated gene transfer by infiltration of adult *Arabidopsis thaliana* plants. *Comptes Rendue Academy Science Paris Life Sciences* 316:1194-1199.

Bent, A.F. (2000). *Arabidopsis in planta* transformation: Uses, mechanisms, and prospects for transformation of other species. *Plant Physiology* 124:1540-1547.

Boisson-Dernier, A., M. Chabaud, F. Garcia, G. Becard, C. Rosenberg, and D.G. Barker (2001). *Agrobacterium rhizogenes*-transformed roots of *Medicago truncatula* for the study of nitrogen-fixing and endomycorrhizal symbiotic associations. *Molecular Plant-Microbe Interactions* 14:695-700.

Buhr, T., S. Sato, F. Ebrahim, A. Xing, Y. Zhou, M. Mathiesen, B. Schweiger, A. Kinney, P. Staswick, and T. Clemente (2002). Ribozyme termination of RNA transcripts down-regulate seed fatty acid genes in transgenic soybean. *Plant Journal* 30:155-163.

Cahoon, E.B., T.J. Carlson, K.G. Ripp, B.J. Schweiger, G.A. Cook, S.E. Hall, and A.J. Kinney (1999). Biosynthetic origin of conjugated double bonds: Production of fatty acid components of high-value drying oils in transgenic soybean embryos. *Proceedings of the National Academy of Sciences USA* 96:12935-12940.

Cahoon, E.B., E.-F. Marillia, K.L. Stecca, S.E. Hall, D.C. Taylor, and A.J. Kinney (2000). Production of fatty acid components of meadowfoam oil in somatic soybean embryos. *Plant Physiology* 124:243-251.

Chee, P.P. and C.-Y. Hu (2000). Transgenic soybean *(Glycine max)*. In *Transgenic Crops I, Biotechnology in Agriculture and Forestry,* Volume 46 (pp. 268-282), Ed. Bajaj, Y.P.S. New York: Springer Verlag.

Chen, L., Y. Dan, J.M. Tyler, and N.A. Reichert (1999). Transgenic soybean generated from hypocotyl explants. *In Vitro Cellular Developmental Biology* 35:61A.

Cheon, C.-I., N.-G. Lee, A.B.M. Siddique, A.K. Bal, and D.P.S. Verma (1993). Roles of plant homologs of Rab1p and Rab7p in the biogenesis of the peribacteroid membrane, a subcellular compartment formed de novo during root nodule symbiosis. *EMBO Journal* 12:4125-4135.

Cho, H.-J., J.E. Brotherton, H.S. Song, and J.M. Widholm (2000). Increasing tryptophan synthesis in a forage legume *Astragalus sinicus* by expressing the tobacco feedback insensitive anthranilate synthase (ASA2) gene. *Plant Physiology* 123:1069-1076.

Cho, H.-J., S.K. Farrand, G.R. Noel, and J.M. Widholm (2000). High-efficiency induction of soybean hairy roots and propagation of the soybean cyst nematode. *Planta* 210:195-204.

Cho, M.-J., J.M. Widholm, and L.O. Vodkin (1995). Cassettes for seed-specific expression tested in transformed embryogenic cultures of soybean. *Plant Molecular Biology Reporter* 13:255-269.

Choffnes, D.S., R. Philip, and L.O. Vodkin (2001). A transgenic locus in soybean exhibits a high level of recombination. *In Vitro Cellular Developmental Biology—Plant* 37:756-762.

Clemente, T.E., B.J. LaValle, A.R. Howe, D. Conner-Ward, R.J. Rozman, P.E. Hunter, D.L. Broyles, D.S. Kasten, and M.A. Hinchee (2000). Progeny analysis of glyphosate selected transgenic soybeans derived from *Agrobacterium*-mediated transformation. *Crop Science* 40:797-803.

Dan, Y. and N.A. Reichert (1998). Organogenic regeneration of soybean from hypocotyl explants. *In Vitro Cellular Development Biology—Plant* 34:14-21.

Daniell, H., B. Muthukumar, and S.B. Lee (2001). Marker free transgenic plants: Engineering the chloroplast genome without the use of antibiotic selection. *Current Genetics* 39:109-116.

de Ronde, J.A., M.H. Spreeth, and W.A. Cress (2000). Effect of antisense L-Δ'-pyrroline-5 carboxylate reductase transgenic soybean plants subjected to osmotic and drought stress. *Plant Growth Regulation* 32:13-26.

Dhir, S.K., S. Dhir, A. Hepburn, and J.M. Widholm (1991). Factors affecting transient gene expression in electroporated *Glycine max* protoplasts. *Plant Cell Reports* 10:106-110.

Di, R., V. Purcell, G.B. Collins, and S.A. Ghabrial (1996). Production of transgenic soybean lines expressing the bean pod mottle virus coat protein precursor gene. *Plant Cell Reports* 15:746-750.

Dinkins, R.D., M.S.S. Reddy, C.A. Meurer, C.T. Redmond, and G.B. Collins (2002). Recent advances in soybean transformation. In *Focus on Biotechnology,* Volume 10B (pp. 1-22), Eds. Jaiwal, P.K. and R.P. Singh. Dordrecht, the Netherlands: Kluwer Academic Publishers.

Dinkins, R.D., M.S.S. Reddy, C.A. Meurer, B. Yan, H. Trick, F. Thibaud-Nissen, J.J. Finer, W.A. Parrott and G.B. Collins (2001). Increased sulfur amino acids in soybean plants overexpressing the maize 15 kDa zein protein. *In Vitro Cellular Developmental Biology—Plant* 37:742-747.

Donaldson, P.A. and D.H. Simmonds (2000). Susceptibility to *Agrobacterium tumefaciens* and cotyledonary node transformation in short-season soybean. *Plant Cell Reports* 19:478-484.

Feldmann, K.A. and M.D. Marks (1987). *Agrobacterium*-mediated transformation of germinating seeds of *Arabidopsis thaliana:* A non-tissue culture approach. *Molecular and General Genetics* 208:1-9.

Finer, J.J., T.-S. Cheng, and D.P.S. Verma (1996). Soybean transformation: Technologies and progress. In *Soybean: Genetics, Molecular Biology and Biotechnology* (pp. 249-262), Eds. Verma, D.P.S. and R.C. Shoemaker. Wallingford, UK: CAB International.

Finer, J.J. and M.D. McMullen (1991). Transformation of soybean via particle bombardment of embryogenic suspension culture tissue. *In Vitro Cellular Developmental Biology—Plant* 27:175-182.

Hadi, M.Z., M.D. McMullen, and J.J. Finer (1996). Transformation of 12 different plasmids into soybean via particle bombardment. *Plant Cell Reports* 15:500-505.

Haldrup, A., S.G. Petersen, and F.T. Okkels (1998). Positive selection: A plant selection principle based on xylose isomerase, an enzyme used in the food industry. *Plant Cell Reports* 18:76-81.

Hazel, C.B., T.M. Klein, M. Anis, H.D. Wilde, and W.A. Parrott (1998). Growth characteristics and transformability of soybean embryogenic cultures. *Plant Cell Reports* 17:765-772.

Hinchee, M.A.W., D.V. Conner-Ward, C.A. Newell, R.E. McDonnell, S.J. Sato, C.S. Gasser, D.A. Fischhoff, D.B. Re, R.T. Fraley, and R.B. Horsch (1988). Production of transgenic soybean plants using *Agrobacterium*-mediated DNA transfer. *Bio/Technology* 6:915-922.

Hu, C.Y. and L.Z. Wang (1999). *In planta* soybean transformation technologies developed in China: Procedure, confirmation, and field performance. *In Vitro Cellular and Developmental Biology—Plant* 35:417-420.

Joersbo, M. (2001). Advances in the selection of transgenic plants using non-antibiotic marker genes. *Physiologia Plantarum* 111:269-272.

Kaeppler, H.F., G.R. Menon, R.W. Skadsen, A.M. Nuutila, and A.R. Carlson (2000). Transgenic oat plants via visual selection of cells expressing green fluorescent protein. *Plant Cell Reports* 19:661-666.

Kamaté, K., I.D. Rodriguez-Llorente, M. Scholte, P. Durand, P. Ratet, E. Kondorosi, A. Kondorosi, and T.H. Trinh (2000). Transformation of floral organs with GFP in *Medicago truncatula*. *Plant Cell Reports* 19:647-653.

Ke, J., R. Khan, T. Johnson, D.A. Somers, and A. Das (2001). High-efficiency gene transfer to recalcitrant plants by *Agrobacterium tumefaciens*. *Plant Cell Reports* 20:150-156.

Kinney, A.J., R. Jung, and E.M. Herman (2001). Cosuppression of the α subunits of β-conglycinin in transgenic soybean seeds induces the formation of endoplasmic reticulum derived protein bodies. *Plant Cell* 13:1165-1178.

Lei, B.J., G.C. Yin, S.L. Wang, C.H. Lu, H. Qian, S.J. Zhou, K.W. Zhang, D.Z. Sui, and Z.P. Lin (1989). Variation occurring by introduction of wild soybean DNA into cultivar soybean. *Oil Crops of China* 3:11-14.

Li, Z., R.L. Nelson, J.M. Widholm, and A. Bent (2002). Soybean transformation via the pollen tube pathway. *Soybean Genetics Newsletter* 29:1-11 (www. soygenetics.org).

Liu, D., Y. Yuan, and H. Sun (1992). A study on exogenous DNA introduction into cultivated soybean. In *Advances in Molecular Breeding Research of Agriculture* (pp. 134-140), Eds. Zhou, G., J. Huang, and S. Chen. Beijing: Chinese Agriculture Science Press.

Luo, Z. and R. Wu (1988). A simple method for the transformation of rice via the pollen-tube pathway. *Plant Molecular Biology Reporter* 6:165-174.

Maliga, P., H. Carrer, I. Kanevski, J. Staub, and Z. Svab (1993). Plastid engineering in land plants: A conservative genome is open to change. *Philosophical Transactions of the Royal Society of London* B 342:203-208.

Maughan, P.J., R. Philip, M.-J. Cho, J.M. Widholm, L.O. Vodkin (1999). Biolistic transformation, expression, and inheritance of bovine β-casein in soybean *(Glycine max). In Vitro Cellular Developmental Biology—Plant* 35:344-349.

Mauro, A.O., T.W. Pfeiffer, and G.B. Collins (1995). Inheritance of soybean susceptibility to *Agrobacterium tumefaciens* and its relationship to transformation. *Crop Science* 35:1152-1156.

Mazarei, M., Z. Ying, and R.L. Houtz (1998). Functional analysis of the Rubisco large subunit N-methyltransferase promoter from tobacco and its regulation by light in soybean hairy roots. *Plant Cell Reports* 17:907-912.

McCabe, D.E., W.F. Swain, B.J. Martinell, and P. Christou (1988). Stable transformation of soybean *(Glycine max)* by particle acceleration. *Bio/Technology* 6:923-926.

Meurer, C.A., R.D. Dinkins, and G.B. Collins (1998). Factors affecting soybean cotyledonary node transformation. *Plant Cell Reports* 18:180-186.

Meurer, C.A., R.D. Dinkins, C.T. Redmond, K.P. McAllister, D.T. Tucker, D.R. Walker, W.A. Parrott, H.N. Trick, J.S. Essig, H.M. Frantz, et al. (2001). Embryogenic response of multiple soybean [*Glycine max* (L.) Merr.] cultivars across three locations. *In Vitro Cellular Developmental Biology—Plant* 37:62-67.

Moore, S., T. Croughan, G. Myers, and R. Vidrine (1996). Investigation of transferring the *bar* gene into soybean via the pollen-tube pathway. *Soybean Genetics Newsletter* 23:167-168.

Olhoft, P.M., K. Lin, J. Galbraith, N.C. Nielsen, and D.A. Somers (2001). The role of thiol compounds in increasing *Agrobacterium*-mediated transformation of soybean cotyledonary-node cells. *Plant Cell Reports* 20:731-737.

Olhoft, P.M. and D.A. Somers (2001). L-Cysteine increases *Agrobacterium*-mediated T-DNA delivery into soybean cotyledonary-node cells. *Plant Cell Reports* 20:706-711.

Owens, L.D. and D.E. Cress (1985). Phenotypic variability of soybean response to *Agrobacterium* strains harbouring the Ti or Ri plasmids. *Plant Physiology* 77:87-94.

Padgette, S.R., K.H. Kolacz, X. Delannay, D.B. Re, B.J. LaVallee, C.N. Tinius, W.K. Rhodes, Y.I. Otero, G.F. Barry, D.A. Eichholtz, et al. (1995). Development, identification, and characterization of a glyphosate-tolerant soybean line. *Crop Science* 35:1451-1461.

Parrott, W.A., J.N. All, M.J. Adang, M.A. Bailey, H.R. Boerma, and C.N. Stewart (1994). Recovery and evaluation of soybean plants transgenic for a *Bacillus thuringiensis* var. *kurstaki* insecticidal gene. *In Vitro Cellular Developmental Biology—Plant* 30:144-149.

Philip, R., D.W. Darnowski, P.J. Maughan, and L.O. Vodkin (2001). Processing and localization of bovine, β-casein expressed in transgenic soybean seeds under control of a soybean lectin expression cassette. *Plant Science* 161:323-335.

Qing, C.M., L. Fan, Y. Lei, D. Bouchez, C. Tourneur, L. Yan, and C. Robaglia (2000). Transformation of pakchoi (*Brassica rapa* L. ssp. *chinensis*) by *Agrobacterium* infiltration. *Molecular Breeding* 6:67-72.

Rathinasabapathy, B., K.F. McCue, D.A. Gage, and A.D. Hanson (1994). Metabolic engineering of glycine betaine synthesis: Plant betaine aldehyde dehydrogenases lacking typical transit peptides are targeted to tobacco chloroplasts where they confer aldehyde resistance. *Planta* 193:155-162.

Reddy, M.S.S., S.A. Ghabrial, C.T. Redmond, R.D. Dinkins, and G.B. Collins (2001). Resistance to bean pod mottle virus in transgenic soybean lines expressing the capsid polyprotein. *Phytopathology* 91:831-838.

Ruf, S., M. Hermann, I.J. Berger, H. Carrer, and R. Bock (2001). Stable genetic transformation of tomato plastids—High-level foreign protein expression in fruits. *Nature Biotechnology* 19:870-875.

Samoylov, V.M., D.M. Tucker, and W.A. Parrott (1998). Soybean embryogenic cultures: The role of sucrose and nitrogen content on proliferation. *In Vitro Cellular Developmental Biology—Plant* 34:8-13.

Samoylov, V.M., D.M. Tucker, F. Thibaud-Nissen, and W.A. Parrott (1998). A liquid medium-based protocol for rapid regeneration from embryogenic soybean cultures. *Plant Cell Reports* 18:49-54.

Santarem, E.R. and J.J. Finer (1999). Transformation of soybean [*Glycine max* (L.) Merrill] using proliferative embryogenic tissue maintained on semi-solid medium. *In Vitro Cellular Developmental Biology—Plant* 35:451-455.

Santarem, E.R., H.N. Trick, J.S. Essig, and J.J. Finer (1998). Sonication-assisted *Agrobacterium*-mediated transformation of soybean immature cotyledons: Optimization of transient expression. *Plant Cell Reports* 17:752-759.

Savka, M.A., B. Ravillion, G.R. Noel, and S.K. Farrand (1990). Induction of hairy roots on cultivated soybean genotypes and their use to propagate the soybean cyst nematode. *Phytopathology* 80:503-508.

Simmonds, D.H. and P.A. Donaldson (2000). Genotype screening for proliferative embryogenesis and biolistic transformation of short-season soybean genotypes. *Plant Cell Reports* 19:485-490.

Song, H.-S., J.E. Brotherton, R.A. Gonzales, and J.M. Widholm (1998). Tissue culture specific expression of a naturally occurring tobacco feedback-insensitive anthranilate synthase. *Plant Physiology* 117:533-543.

Staswick, P.E., Z. Zhang, T.E. Clemente, and J.E. Specht (2001). Efficient down-regulation of the major vegetative storage protein genes in transgenic soybean does not compromise plant productivity. *Plant Physiology* 127:1819-1826.

Stewart, C.N., M.J. Adang, J.N. All, H.R. Boerma, G. Cardineau, D. Tucker, and W.A. Parrott (1996). Genetic transformation, recovery and characterization of fertile soybean transgenic for a synthetic *Bacillus thuringiensis cryIAc* gene. *Plant Physiology* 112:121-129.

Trick, H.N., R.D. Dinkins, E.R. Santarem, R. Di, V. Samoylov, C.A. Meurer, D.R. Walker, W.A. Parrott, J.J. Finer, and G.B. Collins (1997). Recent advances in soybean transformation. *Plant Tissue Culture Biotechnology* 3:9-26.

Trick, H.N. and J.J. Finer (1997). SAAT: Sonication-assisted *Agrobacterium*-mediated transformation. *Transgenic Research* 6:329-336.

Trick, H.N. and J.J. Finer (1998). Sonication-assisted *Agrobacterium*-mediated transformation of soybean [*Glycine max* (L.) Merrill] embryogenic suspension culture tissue. *Plant Cell Reports* 17:482-488.

Trieu, A.T., S.H. Burleigh, I.V. Kardailsky, I.E. Maldonado-Mendoza, W.K. Versaw, L.A. Blaylock, H. Shin, T.-J. Chiou, H. Katagi, G.R. Dewbre, et al. (2000). Transformation of *Medicago truncatula* via infiltration of seedlings or flowering plants with *Agrobacterium*. *Plant Journal* 22:531-541.

Walker, D.R. and W.A. Parrott (2001). Normalizing soybean somatic embryo development in liquid medium. *Plant Cell Tissue Organ Culture* 64:55-62.

Weeks, J.T., K.Y. Koshiyama, U. Maier-Greiner, T. Schaeffner, and O.D. Anderson (2000). Wheat transformation using cyanamide as a new selective agent. *Crop Science* 40:1749-1754.

Wei, Z. and Z. Xu (1988). Plant regeneration from protoplasts of soybean (*Glycine max* L.). *Plant Cell Reports* 7:348-351.

Wei, Z.-M., J.-Q. Huang, S.-P. Xu, and Z.H. Xu (1997). The transformation of protoplasts of *Glycine max* with B.T. gene and regeneration of transgenic plants. Abstracts, 3rd Asia Pacific Conference on Plant Physiology (p. 53), November 3-7, Shanghai, China.

Widholm, J.M. (1995). Leguminous plants. In *Transformation of Plants and Soil Microorganisms* (pp. 101-124), Eds. Wang, K., A. Herrera-Estrella, and M. Van Montagu. Cambridge, UK: Cambridge University Press.

Widholm, J.M. (1999). Status of soybean transformation methods. In *World Soybean Research Conference VI, Proceedings* (pp. 62-67), Ed. Kauffman, H.E. Champaign, IL: Superior Printing.

Xing, A., Z. Zhang, S. Sato, P. Staswick, and T. Clemente (2000). The use of the two T-DNA binary system to derive marker-free transgenic soybeans. *In Vitro Cellular Development Biology—Plant* 36:456-463.

Yan, B., M.S.S. Reddy, and G.B. Collins (2000). *Agrobacterium tumefaciens*-mediated transformation of soybean [*Glycine max* (L.) Merrill] using immature zygotic cotyledon explants. *Plant Cell Reports* 19:1090-1097.

Zhang, X.-H., A.R. Portis, and J.M. Widholm (2001). Plastid transformation of soybean suspension cultures. *Journal of Plant Biotechnology* 3:39-44.

Zhang, X.-H., J.M. Widholm, and A.R. Portis (2001). Photosynthetic properties of two different soybean suspension cultures. *Journal of Plant Physiology* 158:357-365.

Zhang, Z., A. Xing, P. Staswick, and T.E. Clemente (1999). The use of glufosinate as a selective agent in *Agrobacterium*-mediated transformation of soybean. *Plant Cell, Tissue and Organ Culture* 56:37-46.

Chapter 12

Progress in Vegetable Crop Transformation and Future Prospects and Challenges

Zamir K. Punja
Mistianne Feeney

INTRODUCTION

Vegetable crop species are grown worldwide and provide an important source of fiber, nutrients, and vitamins in the human diet. They are consumed fresh or may be eaten after cooking, processing, or pickling and constitute an important part of the meals of billions of people. The crops may be grown under field conditions or under controlled environment conditions, e.g., in greenhouses. A large number of vegetable crop species have been genetically transformed, from at least nine different taxonomic families. Most crops are annual or, infrequently, biennial plants (such as carrot and chicory); a few species are perennial (such as asparagus and watercress). The edible portions of these plants represent the complete spectrum of botanical features, including root (beet, carrot), stem (asparagus), leaf (cabbage, chicory, lettuce, spinach, watercress), flower (broccoli, cauliflower), and fruit (cucumber, eggplant, pepper, tomato).

Significant progress has been made using breeding strategies to produce horticulturally improved, high-yielding, and nutritionally enhanced cultivars of virtually all of the vegetable crops presently grown under cultivation. In addition, resistance to insect pests and diseases and tolerance to environmental stresses have been incorporated using conventional breeding methods. This has resulted in vegetable crop species being grown in a wide range of environments and niches throughout the world.

With the advent of recent techniques in genetic engineering that permit the introduction and transformation of plants with foreign genes (Hansen and Wright, 1999; Newell, 2000), various methods have been utilized to introduce additional genes to potentially enhance the horticultural quality of vegetable crops. In this chapter, the general approaches used to transform vegetable crop species and examples of crops with the specific traits introduced are described. A total of 23 crop species are discussed and the recent literature (from 1995 onward) is included. Issues that need to be addressed and future prospects of this technology are discussed with regard to vegetable crops.

TRANSFORMATION TECHNOLOGIES USED FOR VEGETABLE CROPS

Most of the vegetable crop plants considered in this chapter are dicotyledenous species (with the exception of asparagus and onion), and the majority have been transformed using *Agrobacterium tumefaciens* (Gelvin, 2000) and to a lesser extent with *A. rhizogenes*. However, there are reports of foreign gene introduction using techniques such as microprojectile bombardment (biolistic) (Southgate et al., 1995; Christou, 1996), as well as electroporation (Barcelo and Lazzeri, 1998). A summary of the different approaches used for each crop species is presented in Table 12.1. For many crops, the introduction and expression of genes used as selectable markers [neomycin phosphotransferase *(nptII),* hygromycin phosphotransferase *(hpt),* and phosphinothricin acetyltransferase *(pat, bar)*] or reporter genes [β-glucuronidase *(uidA, gus)*] was achieved as a part of research leading to the development of efficient transformation protocols. For other crops, the introduction of potentially useful genes for resistance to insect pests, virus diseases, fungal pathogens, herbicides, as well as other traits such as stress tolerance and improved horticultural characteristics (enhanced quality, shelf life, increased nutrients) has been achieved (Table 12.2). These latter reports are described in more detail in the following section.

TABLE 12.1. Summary of vegetable crop species that have been genetically transformed (1995-2001)

Family	Crop	Transformation method	Novel protein introduced	Reference
Asteraceae	Chicory (*Cichorium intybus* L. var. *foliosum*)	*A. tumefaciens*	neomycin phosphotransferase, β-glucuronidase	Abid et al. (1995); Frulleux et al. (1997)
		A. tumefaciens	6G-fructosyltransferase, neomycin phosphotransferase	Vijn et al. (1997)
	Lettuce (*Lactuca sativa* L.)	*A. tumefaciens*	neomycin phosphotransferase, auxin synthesis	Curtis et al. (1996)
		A. tumefaciens	virus coat protein, neomycin phosphotransferase	Pang et al. (1996); Dinant et al. (1997)
		A. tumefaciens	nitrate reductase, neomycin phosphotransferase	Curtis et al. (1999)
		A. tumefaciens	phosphinothricin, acetyltransferase, neomycin phosphotransferase	McCabe et al. (1999); Mohapatra et al. (1999)
		A. tumefaciens	isopentenyl phosphotransferase, neomycin phosphotransferase, β-glucuronidase	McCabe et al. (2001)
		A. tumefaciens	gibberellin 20-oxidase, neomycin phosphotransferase	Niki et al. (2001)
Chenopodiaceae	Red beet (*Beta vulgaris* L.)	*A. tumefaciens*	rolB protein	Xing et al. (1996)
	Spinach (*Spinacia oleracea* L.)	*A. tumefaciens*	virus coat protein, neomycin phosphotransferase	Yang et al. (1997)
Convolvulaceae	Sweet potato [*Ipomoea batatas* (L.) Lam.]	*A. tumefaciens*	trypsin inhibitor, snowdrop lectin, neomycin phosphotransferase, β-glucuronidase	Newell et al. (1995)

TABLE 12.1 *(continued)*

Family	Crop	Transformation method	Novel protein introduced	Reference
		A. tumefaciens	neomycin phosphotransferase, β-glucuronidase	Gama et al. (1996)
		Electroporation	virus coat protein, hygromycin phosphotransferase	Nishiguchi et al. (1998)
		A. tumefaciens	hygromicin phosphotransferase, β-glucuronidase	Otani et al. (1998)
		A. tumefaciens	fatty acid desaturase, hygromycin phosphotransferase	Wakita et al. (2001)
Cruciferaceae	Bok choy (*Brassica rapa* L. ssp. *chinensis*)	*A. tumefaciens*	phosphinothricin acetyltransferase	Quing et al. (2000)
	Broccoli (*Brassica oleracea* L. var. *italica*)	*A. tumefaciens*	*B. thuringiensis* CryIA, neomycin phosphotransferase	Metz, Dixit, and Earle (1995)
		A. rhizogenes	*B. thuringiensis* CryIA, neomycin phosphotransferase, β-glucuronidase	Christey et al. (1997)
		A. tumefaciens	*B. thuringiensis* CryIC, hygromycin phosphotransferase	Cao et al. (1999)
		A. rhizogenes	antisense ACC oxidase, neomycin phosphotransferase	Henzi et al. (1999, 2000b)
		A. rhizogenes	neomycin phosphotransferase, β-glucuronidase	Henzi et al. (2000a)
		A. tumefaciens	phosphinothricin acetyltransferase, neomycin phosphotransferase	Waterer et al. (2000)

Cabbage (*Brassica oleraceae* L. var. *capitata*)	*A. tumefaciens*	*B. thuringiensis* CryIA, neomycin phosphotransferase	Metz, Dixit, and Earle (1995)
	A. rhizogenes	*B. thuringiensis* CryIA, neomycin phosphotransferase, β-glucuronidase	Christey et al. (1997)
	A. tumefaciens	*B. thuringiensis* Cry1Ab3, neomycin phosphotransferase	Jin et al. (2000)
	A. tumefaciens	neomycin phosphotransferase, β-glucuronidase	Pius and Achar (2000)
Cauliflower (*Brassica oleracea* var. *botrytis*)	*A. tumefaciens*	virus coat protein, hygromycin phosphotransferase, neomycin phosphotransferase, β-glucuronidase	Passelègue and Kerlan (1996)
	A. rhizogenes	*B. thuringiensis* Cry IA, neomycin phosphotransferase, β-glucuronidase	Christey et al. (1997)
	A. tumefaciens	trypsin inhibitor, neomycin phosphotransferase	Ding et al. (1998)
	A. tumefaciens	antibacterial peptides, neomycin phosphotransferase	Braun et al. (2000)
	A. tumefaciens	*B. thuringiensis* Cry9Aa, hygromycin phosphotransferase	Kuvshinov et al. (2001)
	A. rhizogenes	β-glucuronidase	Puddephat et al. (2001)
Chinese cabbage (*Brassica campestris* L.)	*A. tumefaciens*	virus coat protein, neomycin phosphotransferase	Jun et al. (1995)
	A. tumefaciens	*B. thuringiensis* CryIAb, CryIAc, neomycin phosphotransferase	Xiang et al. (2002)

TABLE 12.1 *(continued)*

Family	Crop	Transformation method	Novel protein introduced	Reference
		A. tumefaciens	*B. thuringiensis* CryIC, hygromycin phosphotransferase	Cho et al. (2001)
	Rutabaga (*Brassica napobrassica* L.)	*A. tumefaciens*	*B. thuringiensis* Cry1A, neomycin phosphotransferase	Li et al. (1995)
	Watercress (*Rorippa nasturtium-aquaticum* L.)	*A. tumefaciens*	*B. thuringiensis* CryIIa3, neomycin phosphotransferase	Jin et al. (1999)
Cucurbitaceae	Cucumber (*Cucumis sativus* L.)	Biolistic	neomycin phosphotransferase, β-glucuronidase	Schulze et al. (1995)
		A. tumefaciens	hygromycin phosphotransferase, β-glucuronidase	Nishibayashi et al. (1996)
		A. tumefaciens	chitinase, neomycin phosphotransferase	Raharjo et al. (1996)
		A. tumefaciens	chitinase, neomycin phosphotransferase	Tabei et al. (1998)
	Squash (*Cucurbita pepo* L.)	*A. tumefaciens*	virus coat protein, neomycin phosphotransferase	Clough and Hamm (1995)
Leguminosae	Bean (*Phaseolus vulgaris* L.)	Electroporation	β-glucuronidase	Dillen et al. (1995)
		Biolistic	antisense viral RNA, neomycin phosphotransferase, β-glucuronidase	Aragão et al. (1996, 1998)
		Biolistic, *A. tumefaciens*	β-glucuronidase	Brasileiro et al. (1996)
		Biolistic	β-glucuronidase	Kim and Minamikawa (1996, 1997)
		Biolistic	β-glucuronidase	Aragão and Rech (1997)

		A. tumefaciens	phosphinothricin acetyltransferase, neomycin phosphotransferase, β-glucuronidase	Nagl et al. (1997)
	Pea (*Pisum sativum* L.)	Electroporation	β-glucuronidase	Chowrira et al. (1995)
		A. tumefaciens	phosphinothricin acetyltransferase, neomycin phosphotransferase	Grant et al. (1995)
		A. tumefaciens	α-amylase, phosphinothricin acetyltransferase	Schroeder et al. (1995)
		A. tumefaciens	phosphinothricin acetyltransferase	Bean et al. (1997)
		A. tumefaciens	neomycin phosphotransferase	Grant et al. (1998)
		A. tumefaciens	β-glucuronidase	Lurquin et al. (1998)
Liliaceae	Asparagus (*Asparagus officinalis* L.)	Biolistic	hygromycin phosphotransferase, phosphinothricin acetyltransferase, β-glucuronidase	Cabrera-Ponce et al. (1997)
		Biolistic	neomycin phosphotransferase, β-glucuronidase	Li and Wolyn (1997)
		A. tumefaciens	neomycin phosphotransferase, β-glucuronidase	Limanton-Grevet and Julien (2001)
	Onion (*Allium cepa* L.)	Biolistic	β-glucuronidase	Barandiaran et al. (1998)
		A. tumefaciens	neomycin phosphotransferase, green fluorescent protein	Eady et al. (2000)
		A. tumefaciens	hygromycin phosphotransferase, β-glucuronidase	Zheng et al. (2001)

TABLE 12.1 *(continued)*

Family	Crop	Transformation method	Novel protein introduced	Reference
Solanaceae	Eggplant (*Solanum melongena* L.)	*A. tumefaciens*	*B. thuringiensis* CryIIIB, β-glucuronidase	Chen et al. (1995)
		A. tumefaciens	neomycin phosphotransferase, β-glucuronidase	Fari et al. (1995)
		A. tumefaciens	*B. thuringiensis* CryIIIB, neomycin phosphotransferase	Iannacone et al. (1995); Arapaia et al. (1997)
		A. tumefaciens	*B. thuringiensis* CryIIIB, neomycin phosphotransferase, β-glucuronidase	Billings et al. (1997)
		A. tumefaciens	tryptophan monoxigenase, neomycin phosphotransferase	Rotino et al. (1997); Donzella et al. (2000)
		A. tumefaciens	*B. thuringiensis* CryIIIA, neomycin phosphotransferase, β-glucuronidase	Jelenkovic et al. (1998)
		A. tumefaciens	*B. thuringiensis* CryIAb, neomycin phosphotransferase	Kumar et al. (1998)
	Pepper (sweet) (*Capsicum annum* L.)	*A. tumefaciens*	phosphinothricin acetyltransferase, neomycin phosphotransferase	Tsaftaris (1996)
		A. tumefaciens	virus coat protein, neomycin phosphotransferase	Zhu et al. (1996)
	Pepper (hot) (*C. annuum* L.)	*A. tumefaciens*	viral satellite RNA, neomycin phosphotransferase	Kim et al. (1997)
		A. tumefaciens	neomycin phosphotransferase, β-glucuronidase	Manoharen et al. (1998); Ochoa-Alejo and Ramirez-Malagon (2001)
	Tomato (*Lycopersicon esculentum* Mill.)	*A. tumefaciens*	phytoene synthase, neomycin phosphotransferase	Fray et al. (1995)

	A. tumefaciens	chitinase, glucanase, neomycin phosphotransferase	Jongedijk et al. (1995)
	Biolistic	neomycin phosphotransferase, β-glucuronidase	Van Eck et al. (1995)
	A. tumefaciens	neomycin phosphotransferase, β-glucuronidase	Frary and Earle (1996)
	A. tumefaciens	viral coat protein, neomycin phosphotransferase	Gielen et al. (1996)
	A. tumefaciens	antisense β-fructosidase, neomycin phosphotransferase	Klann et al. (1996)
	A. tumefaciens	yeast desaturase, neomycin phosphotransferase	Wang et al. (1996)
	Electroporation	β-glucuronidase	Lin et al. (1997)
	A. tumefaciens	stilbene synthase, neomycin phosphotransferase	Thomzik et al. (1997)
	A. tumefaciens	viral replicase, neomycin phosphotransferase	Gal-On et al. (1998)
	A. tumefaciens	alcohol dehydrogenase, neomycin phosphotransferase	Speirs et al. (1998)
	A. tumefaciens	expansin, chlorsulfuron acetyltransferase	Brummell et al. (1999)
	A. tumefaciens	tryptophan monoxigenase, neomycin phosphotransferase	Ficcadenti et al. (1999)
	A. tumefaciens	viral coat protein, neomycin phosphotransferase	Kaniewski et al. (1999)

TABLE 12.1 *(continued)*

Family	Crop	Transformation method	Novel protein introduced	Reference
		A. tumefaciens	chitinase, neomycin phosphotransferase	Tabaeizadeh et al. (1999)
		A. tumefaciens	*B. thuringiensis* CryIAc, neomycin phosphotransferase	Mandaokar et al. (2000)
		A. tumefaciens	defensin, neomycin phosphotransferase	Parashina et al. (2000)
		A. tumefaciens	polygalacturonase-inhibiting protein, neomycin phosphotransferase	Powell et al. (2000)
		A. tumefaciens	phytoene desaturase, phytoene synthase, neomycin phosphotransferase	Römer et al. (2000); Fraser et al. (2001)
		A. tumefaciens	β-galactosidase, neomycin phosphotransferase	Carey et al. (2001)
		Biolistic	spectinomycin resistance	Ruf et al. (2001)
		A. rhizogenes	ACC deaminase	Grichko and Glick (2001)
		A. tumefaciens	chitinase, neomycin phosphotransferase	Gongora et al. (2001)
		A. tumefaciens	chalcone isomerase, neomycin phosphotransferase	Muir et al. (2001)
		A. tumefaciens	vacuolar Na^+/H^+ antiport, neomycin phosphotransferase	Zhang and Blumwald (2001)
Umbelliferae	Carrot (*Daucus carota* L.)	Polythylene-glycol	neomycin phosphotransferase	Dirks et al. (1996)
		A. tumefaciens	chitinase, neomycin phosphotransferase	Gilbert et al. (1996)

		A. tumefaciens	neomycin phosphotransferase, β-glucuronidase	Hardeggar and Sturm (1998)
		A. tumefaciens	lysozyme, neomycin phosphotransferase	Takaichi and Oeda (2000)
		A. tumefaciens	thaumatin-like protein, hygromycin phosphotransferase, phosphinothricin acetyltransferase	Chen and Punja (2002)

STRESS-RESISTANT VEGETABLE CROPS

Insect-Resistant Vegetable Crops

The most common approach that has been used to achieve enhanced resistance to insect pests in vegetable crops has been through the introduction and expression of a wide range of *Bacillus thuringiensis* crystal delta-endotoxins specific against different insect pests (Table 12.1). Presently at least 140 *B. thuringiensis* crystal protein genes are described based on differences at the nucleotide level (Jouanin et al., 1998; Hilder and Boulter, 1999). The major groups of proteins effective against insects are encoded by specific genes represented by *cryI, cryII, cryIII,* and *cryIV* families. Expression of a number of different synthetic *B. thuringiensis cry* genes has been reported in several vegetable crops, including broccoli, cabbage, cauliflower, rutabaga, eggplant, and tomato (Table 12.1), with enhanced resistance to insect pests observed in many cases (Table 12.2). Expression of trypsin-inhibitor proteins has been reported for cauliflower and sweet potato (Table 12.1). Other strategies being explored for non-vegetable crops could have applications for insect control on vegetables in the future (Gatehouse and Gatehouse, 1998; Hilder and Boulter, 1999).

Virus-Resistant Vegetable Crops

The successful development of virus-resistant transgenic vegetable crops has primarily utilized the coat protein-mediated strategy to

TABLE 12.2. Examples of agronomically useful traits that have been introduced into vegetable crop species through genetic engineering

Crop species	Trait	Reference
Asparagus	Resistance to the herbicide phosphinothricin	Cabrera-Ponce et al. (1997)
Bean	Delayed development and reduced symptoms due to virus infection	Aragão et al. (1998)
Bok choy	Resistance to the herbicide phosphinothricin	Qing et al. (2000)
Broccoli	Enhanced resistance to insect pests	Metz, Roush, et al. (1995); Cao et al. (1999)
	Resistance to the herbicide glufosinate ammonium	Waterer et al. (2000)
	Reduced senescence, enhanced flower head color	Henzi et al. (1999, 2000b)
Cabbage	Enhanced resistance to insect pests	Jin et al. (2000)
Carrot	Enhanced resistance to fungal pathogens	Punja and Raharjo (1996); Melchers and Stuiver (2000); Takaichi and Oeda (2000); Chen and Punja (2002)
Cauliflower	Enhanced resistance to several insect pests	Ding et al. (1998); Kuvshinov et al. (2001)
Chinese cabbage	Enhanced protection against insect feeding damage	Xiang et al. (2000); Cho et al. (2001)
Cucumber	Enhanced resistance to fungal pathogens	Tabei et al. (1998)
Eggplant	Enhanced resistance to insect attack	Arpaia et al. (1997); Jelenkovic et al. (1998); Kumar et al. (1998)
	Enhanced production of IAA, resulting in parthenocarpic fruit and enhanced fruit set	Rotino et al. (1997); Donzella et al. (2000)
Lettuce	Resistance to the herbicide bialaphos	McCabe et al. (1999); Mohapatra et al. (1999)
	Enhanced resistance to virus infection	Pang et al. (1996); Dinant et al. (1997)
	Reduced nitrate accumulation in leaves	Curtis et al. (1999)

	Reduced leaf senescence	McCabe et al. (2001)
Pea	Enhanced resistance to insect attack	Schroeder et al. (1995)
Pepper (sweet)	Resistance to the herbicide phosphinothricin	Tsaftaris (1996)
Pepper (hot)	Reduced symptom development due to virus infection	Kim et al. (1997)
Rutabaga	Enhanced tolerance to insect damage	Li et al. (1995)
Squash	Reduced virus development, enhanced fruit yield	Clough and Hamm (1995); Fuchs et al. (1999)
Sweet potato	Enhanced resistance to virus infection	Nishiguchi et al. (1998)
	Enhanced linolenic acid content	Wakita et al. (2001)
Tomato	Enhanced protection against insect feeding damage	Mandaokar et al. (2000); Gongora et al. (2001)
	Enhanced resistance to virus infection	Fuchs et al. (1996); Gielen et al. (1996); Gal-On et al. (1998); Kaniewski et al. (1999)
	Enhanced resistance to fungal pathogens	Jongedijk et al. (1995); Thomzik et al. (1997); Tabaeizadeh et al. (1999); Parashina et al. (2000); Powell et al. (2000)
	Enhanced levels of flavonols in fruit and in tomato paste	Muir et al. (2001)
	Enhanced fruit flavor	Wang et al. (1996); Speirs et al. (1998)
	Increased sucrose in fruit	Klann et al. (1996)
	Delayed fruit ripening	Picton et al. (1995)
	Parthenocarpic fruit	Ficcadenti et al. (1999)
	Enhanced fruit firmness, thicker tomato paste	Brummell and Lebavitch (1997); Brummell et al. (1999)

TABLE 12. 2 *(continued)*

Crop species	Trait	Reference
	Enhanced tolerance to salt stress	Zhang and Blumwald (2001)
	Increased tolerance to flooding stress	Grichko and Glick (2001)
	Enhanced levels of β-carotene in fruit	Römer et al. (2000); Fraser et al. (2001)
	Increased quality of fruit juice	Thakur et al. (1996)

engineer pathogen resistance (Grumet, 1995; Kavanagah and Spillane, 1995). In some cases, satellite RNA, replicase, and antisense RNA have been expressed in transgenic vegetables to provide protection against virus infection. The crops displaying enhanced virus resistance include bean, lettuce, hot pepper, sweet potato, squash, and tomato (Table 12.2).

Fungal-Resistant Vegetable Crops

Enhanced resistance to fungal pathogen infection in transgenic vegetable crop species has been demonstrated through the expression of antifungal pathogenesis-related proteins, particularly chitinases (Table 12.2). The crops which have been engineered include carrot, cucumber, and tomato. In one report, the expression of a lysozyme enhanced resistance of carrot to fungal infection (Takaichi and Oeda, 2000).

Herbicide-Resistant Vegetable Crops

Herbicide resistance is a frequently used selectable marker in transformation studies, and the genes *pat* or *bar* that provide resistance to the related herbicides phosphinothricin, bialaphos, and glufosinate ammonium have been expressed in asparagus, bean, bok choy, broccoli, lettuce, pea, and sweet pepper (Table 12.1).

IMPROVEMENT OF HORTICULTURAL TRAITS THROUGH GENETIC ENGINEERING

Among all of the vegetable crop species that have been genetically engineered, tomato has been the target of much of the work to enhance horticulturally desirable traits, such as improved shelf life, nutritional value, and processing characteristics (Picton et al., 1995; Thakur et al., 1996). The expression of antisense RNA to fruit-ripening-related genes, such as those encoding for the production of polygalacturonase, pectin esterase, and ethylene, and for carotenoid biosynthesis (phytoene synthase), has shown that the rate of ripening-related processes can be substantially reduced and fruit quality enhanced or altered (Picton et al., 1995; Thakur et al., 1996). A reduction in ethylene production in broccoli by expression of antisense RNA to 1-aminocyclopropane-1-carboxylic acid oxidase also reduced the rate of senescence and enhanced flower head color (Henzi et al., 1999, 2000b).

Expression of a chalcone isomerase gene from petunia in transgenic tomato resulted in significantly higher accumulation of flavonols in the fruit peel as well as in processed paste (Muir et al., 2001). Flavonols are potent antioxidants which can have health-promoting properties. The level of ß-carotene, another compound with demonstrated health benefits, was enhanced in tomato fruit by expression of a bacterial phytoene saturase gene (Römer et al., 2000). These studies demonstrate the potential for engineering enhanced vitamin and nutrient levels, as well as quality, in vegetable crops such as tomato and broccoli.

Transgenic lettuce that was engineered to express a nitrate reductase gene from tobacco resulted in lower nitrate accumulation in the leaves (Curtis et al., 1999). Because lettuce can accumulate high levels of nitrate compared to other vegetables, nitrate concentration of leaves is an important quality determinant, with lower levels being desirable.

Transgenic eggplant expressing a bacterial tryptophan monoxigenase gene resulted in enhanced auxin (indole-3-acetic acid) levels in fruit, which in turn resulted in parthenocarpic fruit development (Rotino et al., 1997; Donzella et al., 2000), a desirable characteristic. Fruits developed on transgenic plants under conditions that were not conducive for development of nontransgenic fruit, suggesting that overall yield of transgenic plants could be higher (Donzella et al., 2000).

Transgenic tomato expressing a bacterial 1-aminocyclopropane-1-carboxylic acid deaminase gene withstood flooding stress (low oxygen) better than untransformed plants and were less subject to the deleterious effects of root hypoxia on plant growth (Grichko and Glick, 2001). The transgenic plants had greater shoot fresh and dry weight, produced lower amounts of ethylene, and had higher amounts of leaf chlorophyll content.

Transgenic sweet potato expressing a tobacco microsomal ω-3 fatty acid desaturase gene had enhanced linolenic acid content and lower linoleic acid content (Wakita et al., 2001), suggesting that changes to the fatty acid composition of the tubers could improve their nutritional value.

ISSUES TO BE ADDRESSED

The commercial deployment of the transgenic vegetable crop species described in this chapter has not yet been undertaken to any great extent. Some of the issues that likely will need to be addressed prior to widespread consumer and producer acceptance of these genetically engineered vegetables are described here. These issues are not unique to vegetable crops and have been identified for other transgenic crop species (Daniell, 1999; Halford and Shewry, 2000). They include the avoidance of antibiotic resistance markers, determining the potential impact of spread of the transgenes through pollen movement, assessing the potential for resistance buildup on insect and pathogen populations, and ensuring minimal allergenicity/toxigenicity due to proteins in the engineered vegetable foods, especially those that are consumed fresh.

In almost all of the reports summarized in Table 12.1, either kanamycin or hygromycin was used as the antibiotic selection agent for transformation. Alternative selectable markers that do not rely on antibiotic selection have been developed, such as mannose selection (Zhang, Potrykus, and Puonti-Kaerlas, 2000; Joersbo, 2001) or cytokinin expression (Kunkel et al., 1999). These should be viewed more favorably by consumers of vegetable foods, especially in light of the recent public concerns over genetically modified foods (Ferber, 1999; Halford and Shewry, 2000). In addition, antibiotic selection marker genes can be eliminated from engineered crops using several approaches (Zubko et al., 2000; Hohn et al., 2001). Most of these alternative approaches have not yet been explored for vegetable crop species, with the exception of lettuce (Kunkel et al., 1999).

The potential for spread of transgenes from cultivated vegetable crop species to weedy relatives through pollen is likely to occur in areas where the cultivated and wild species are growing in close proximity to one another, e.g., lettuce and wild lettuce. An assessment of the impact of transgene introgression into weedy species or other organisms should be conducted to determine if any potential ecological or environmental consequence exists (Malik and Saroha, 1999). The engineering of transgene expression into chloroplasts is one approach that can minimize gene flow (Daniell, 1999; Khan and Maliga, 1999; Iamthan and Day, 2000), although this has been achieved for only one of the vegetable crop species described here (tomato) (Ruf et al., 2001).

The greatest concern regarding the utilization of *B. thuringiensis cry* gene expression in transgenic vegetable crops is the potential for development of resistance in the insect species to the specific toxins as a result of increased selection pressure. Several strategies have been described to minimize the rate of development of resistance (Jouanin et al., 1998; Daniell, 1999; Hilder and Boulter, 1999; Hoy, 1999). These as well as other approaches should similarly be implemented for transgenic vegetable crop species when they begin to be cultivated extensively.

The testing for allergenicity/toxigenicity due to proteins in engineered vegetable crop species (Franck-Oberaspach and Keller, 1997) would require that the guidelines set forth by the appropriate regulatory agencies be followed before the vegetable can be approved for use as food (Kaeppler, 2000).

FUTURE PROSPECTS

Although genetically engineered tomato with enhanced fruit quality characteristics was the first example of a transgenic product to be marketed in the United States and the United Kingdom in the 1990s, transgenic vegetable crop species have not reached the market as rapidly as, for example, transgenic canola, corn, cotton, potato, and soybean. Transgenic squash with virus resistance is currently one of the few examples of a transgenic vegetable on the U.S. market. This chapter has summarized reports on the potential benefits of engineering specific traits in certain vegetable crops. Some of these would result in direct benefits to the grower, e.g., insect or herbicide resistance, while others would have benefits for the consumer, e.g., enhanced vitamins, flavor and shelf life. The fact that 23 vegetable crop species have been genetically engineered (excluding potato and sweet corn) attests to the rapidly growing interest in the applications of this technology for improvement of vegetable crops.

Future applications of transgenic technology to vegetable crops should be able to capitalize on the existing knowledge available for other crops. For example, engineering resistance to bacterial diseases (During, 1996; Mourgues et al., 1998), increasing tolerance to a range of abiotic stresses (Grover et al., 1998; Holmberg and Bülow, 1998; Bajaj et al., 1999; Rathinasabapathi, 2000; Zhang, Klueva, et al., 2000), and targeting enhancement of micronutrient levels for human health (DellaPenna, 1999) should be achievable within the next ten years. Additional applications of transgenic crops are summarized in Dunwell (1999), some of which could also be applicable to vegetable crops in the future.

REFERENCES

Abid, M., B. Palms, R. Derycke, J.P. Tissier, and S. Rambour (1995). Transformation of chicory and expression of the bacterial *uidA* and *nptII* genes in the transgenic regenerants. *Journal of Experimental Botany* 46:337-346.

Aragão, F.J.L., L.M.G. Barros, A.C.M. Brasileiro, S.G. Ribeiro, F.D. Smith, J.C. Sanford, J.C. Faria, and E.L. Rech (1996). Inheritance of foreign genes in transgenic bean (*Phaseolus vulgaris* L.) co-transformed via particle bombardment. *Theoretical and Applied Genetics* 93:142-150.

Aragão, F.J.L. and E.L. Rech (1997). Morphological factors influencing recovery of transgenic bean plants (*Phaseolus vulgaris* L.) of a carioca cultivar. *International Journal of Plant Science* 158:157-163.

Aragão, F.J.L., S.G. Ribeiro, L.M.G. Barros, A.C.M. Brasileiro, D.P. Maxwell, E.L. Rech, and J.C. Faria (1998). Transgenic beans (*Phaseolus vulgaris* L.) engineered to express viral antisense RNAs show delayed and attenuated symptoms to bean golden mosaic geminivirus. *Molecular Breeding* 4:491-499.

Arpaia, S., G. Mennella, V. Onofaro, E. Perri, F. Sunseri, and G.L. Rotino (1997). Production of transgenic eggplant (*Solanum melanogena* L.) resistant to Colorado potato beetle (*Leptinotarsa decemlineata* Say). *Theoretical and Applied Genetics* 95:329-334.

Bajaj, S., J. Targolli, L.-F. Liu, T.-H.D. Ho, and R. Wu (1999). Transgenic approaches to increase dehydration-stress tolerance in plants. *Molecular Breeding* 5:493-503.

Barandiaran, X., A. DiPietro, and J. Martin (1998). Biolistic transfer and expression of a *uidA* reporter gene in different tissues of *Allium sativum* L. *Plant Cell Reports* 17:737-741.

Barcelo, P. and P.A. Lazzeri (1998). Direct gene transfer: Chemical, electrical and physical methods. In *Transgenic Plant Research* (pp. 35-55), Ed. Lindsey, K. Reading, UK: Harwood Academic Publishers.

Bean, S.J., P.S. Gooding, P.M. Mullineaux, and D.R. Davies (1997). A simple system for pea transformation. *Plant Cell Reports* 16:513-519.

Billings, S., G. Jelenkovic, C.-K. Chin, and J. Eberhardt (1997). The effect of growth regulators and antibiotics on eggplant transformation. *Journal of the American Society for Horticultural Science* 122:158-162.

Brasileiro, A.C.M., F.J.L. Aragão, S. Rossi, D.M.A. Dusi, L.M.G. Barros, and E.L. Rech (1996). Susceptibility of common and tepary beans to *Agrobacterium* spp. strains and improvement of *Agrobacterium*-mediated transformation using microprojectile bombardment. *Journal of the American Society for Horticultural Science* 121:810-815.

Braun, R.H., J.K. Reader, and M.C. Christey (2000). Evaluation of cauliflower transgenic for resistance to *Xanthomonas campestris* pv. *campestris*. *Acta Horticulturae* 539:137-143.

Brummell, D.A., M.H. Harpster, P.M. Civello, J.M. Palys, A.B. Bennett, and P. Dunsmuir (1999). Modification of expansion protein abundance in tomato fruit alters softening and cell wall polymer metabolism during ripening. *The Plant Cell* 11:2203-2216.

Brummell, D.A. and J.M. Labavitch (1997). Effect of antisense suppression of endopolygalacturonase activity on polyuronide molecular weight in ripening tomato fruit and in fruit homogenates. *Plant Physiology* 115:717-725.

Cabrera-Ponce, J.L., L. López, N. Assad-Garcia, C. Medina-Arevalo, A.M. Bailey, and L. Herrera-Estrella (1997). An efficient particle bombardment system for the genetic transformation of asparagus (*Asparagus officinalis* L.). *Plant Cell Reports* 16:255-260.

Cao, J., J.D. Tang, N. Strizhov, A.M. Shelton, and E.D. Earle (1999). Transgenic broccoli with high levels of *Bacillus thuringiensis* Cry1C protein control diamondback moth larvae resistant to Cry1A or Cry1C. *Molecular Breeding* 5:131-141.

Carey, A.T., D.L. Smith, E. Harrison, C.R. Bird, K.G. Gross, G.B. Seymour, and G.A. Tucker (2001). Down-regulation of a ripening-related 13-galactosidase gene (TBG 1) in transgenic tomato fruits. *Journal of Experimental Botany* 52:663-668.

Chen, Q., G. Jelenkovic, C.-K. Chin, S. Billings, J. Eberhardt, J.C. Goffreda, and P. Day (1995). Transfer and transcriptional expression of coleopteran *cryIIIB* endotoxin gene of *Bacillus thuringiensis* in eggplant. *Journal of the American Society for Horticultural Science* 120:921-927.

Chen, W.P. and Z.K. Punja (2002). Transgenic herbicide- and disease-tolerant carrot (*Daucus carota* L.) plants obtained through *Agrobacterium*-mediated transformation. *Plant Cell Reports* 20:929-935.

Cho, H.S., J. Cao, J.P. Ren, and E.D. Earle (2001). Control of lepidopteran insect pests in transgenic Chinese cabbage (*Brassica rapa* ssp. *pekinensis*) transformed with a synthetic *Bacillus thuringiensis cry1C* gene. *Plant Cell Reports* 20:1-7.

Chowrira, G.M., V. Akella, and P.F. Lurquin (1995). Electroporation-mediated gene transfer into intact nodal meristems *in planta:* Generating transgenic plants without in vitro tissue culture. *Molecular Biotechnology* 3:17-23.

Christey, M.C., B.K. Sinclair, R.H. Braun, and L. Wyke (1997). Regeneration of transgenic vegetable brassicas (*Brassica oleracea* and *B. campestris*) via Ri-mediated transformation. *Plant Cell Reports* 16:587-593.

Christou, P. (1996) *Particle Bombardment for Genetic Engineering of Plants.* Austin, TX: R.G. Landes Company.

Clough, G.H. and P.B. Hamm (1995). Coat protein transgenic resistance to watermelon mosaic and zucchini yellows mosaic virus in squash and cantaloupe. *Plant Disease* 79:1107-1109.

Curtis, I.S., C. He, J.B. Power, D. Mariotti, A. de Laat, and M.R. Davey (1996). The effects of *Agrobacterium rhizogenes rolAB* genes in lettuce. *Plant Science* 115:123-135.

Curtis, I.S., J.B. Power, A.M.M. de Laat, M. Caboche, and M.R. Davey (1999). Expression of a chimeric nitrate reductase gene in transgenic lettuce reduces nitrate in leaves. *Plant Cell Reports* 18:889-896.

Daniell, H. (1999). Environmentally friendly approaches to genetic engineering. *In Vitro Cellular and Developmental Biology—Plant* 35:361-368.

DellaPenna, D. (1999). Nutritional genomics: Manipulating plant micronutrients to improve human health. *Science* 285:375-379.

Dillen, W., G. Engler, M. Van Montagu, and G. Angenon (1995). Electroporation-mediated DNA delivery to seedling tissues of *Phaseolus vulgaris* L. (common bean). *Plant Cell Reports* 15:119-124.

Dinant, S., B. Maisonneuve, J. Albouy, Y. Chupeau, M.-C. Chupeau, Y. Bellec, F. Gaudefroy, C. Kusiak, S. Souche, C. Robaglia, and H. Lot (1997). Coat protein gene-mediated protection in *Lactuca sativa* against lettuce mosaic potyvirus strains. *Molecular Breeding* 3:75-86.

Ding, L.-C., C.-Y. Hu, K.-W. Yeh, and P.-J. Wang (1998). Development of insect-resistant transgenic cauliflower plants expressing the trypsin inhibitor gene isolated from local sweet potato. *Plant Cell Reports* 17:854-860.

Dirks, R., V. Sidorov, and C. Tulmans (1996). A new protoplast culture system in *Daucus carota* L. and its applications for mutant selection and transformation. *Theoretical and Applied Genetics* 93:809-815.

Donzella, G., A. Spena, and G.L. Rotino (2000). Transgenic parthenocarpic eggplants: Superior germplasm for increased winter production. *Molecular Breeding* 6:79-86.

Dunwell, J.M. (1999). Transgenic crops: The next generation, or an example of 2020 vision. *Annals of Botany* 84:269-277.

During, K. (1996). Genetic engineering for resistance to bacteria in transgenic plants by introduction of foreign genes. *Molecular Breeding* 2:297-305.

Eady, C.C., R.J. Weld, and C.E. Lister (2000). *Agrobacterium tumefaciens*-mediated transformation and transgenic-plant regeneration of onion (*Allium cepa* L.). *Plant Cell Reports* 19:376-381.

Fari, M., I. Nagy, M. Csanyl, J. Mityko, and A. Andrasfalvy (1995). *Agrobacterium* mediated genetic transformation and plant regeneration via organogenesis and somatic embryogenesis from cotyledon leaves in eggplant (*Solanum melongena* L. cv. 'Kecskenmefi lila'). *Plant Cell Reports* 15:82-86.

Ferber, D. (1999). GM crops in the cross hairs. *Science* 286:1662-1666.

Ficcadenti, N., S. Sestili, T. Pandolfini, C. Cirillo, G.L. Rotino, and A. Spena (1999). Genetic engineering of parthenocarpic fruit development in tomato. *Molecular Breeding* 5:463-470.

Franck-Oberaspach, S.L. and B. Keller (1997). Consequences of classical and biotechnological resistance breeding for food toxicology and allergenicity. *Plant Breeding* 116:1-17.

Frary, A. and E.D. Earle (1996). An examination of factors affecting the efficiency of *Agrobacterium*-mediated transformation of tomato. *Plant Cell Reports* 16:235-240.

Fraser, P.D., S. Romer, J.W. Kiano, C.A. Shipton, P.B. Mills, R. Drake, W. Schuch, and P.M. Bramley (2001). Elevation of carotenoids in tomato by genetic manipulation. *Journal of the Science of Food and Agriculture* 81:822-827.

Fray, R.G., A. Wallace, P.D. Fraser, D. Valero, P. Hedden, P.M. Bramley, and D. Grierson (1995). Constitutive expression of a fruit phytoene synthase gene in transgenic tomatoes causes dwarfism by redirecting metabolites from the gibberellin pathway. *The Plant Journal* 8:693-701.

Frulleux, F., G. Weyens, and M. Jacobs (1997). *Agrobacterium tumefaciens*-mediated transformation of shoot-buds of chicory. *Plant Cell, Tissue and Organ Culture* 50:107-112.

Fuchs, M., A. Gal-On, B. Raccah, and D. Gonzalves (1999). Epidemiology of an aphid nontransmissible potyvirus in fields of nontransgenic and coat protein transgenic squash. *Transgenic Research* 8:429-439.

Fuchs, M., R. Provvidenti, J.L. Slightom, and D. Gonsalves (1996). Evaluation of transgenic tomato plants expressing the coat protein gene of cucumber mosaic virus strain WL under field conditions. *Plant Disease* 80:270-275.

Gal-On, A., D. Wolf, Y. Wang, J.-E. Faure, M. Pilowski, and A. Zelcer (1998). Transgenic resistance to cucumber mosaic virus in tomato: Blocking of long-distance movement of the virus in lines harboring a defective viral replicase gene. *Phytopathology* 88:1101-1107.

Gama, M.I.C.S., R.P. Leite Jr., A.R. Cordeiro, and D.J. Cantliffe (1996). Transgenic sweet potato plants obtained by *Agrobacterium tumefaciens*-mediated transformation. *Plant Cell, Tissue and Organ Culture* 46:237-244.

Gatehouse, A.M.R. and J.A. Gatehouse (1998). Identifying proteins with insecticidal activity: Use of encoding genes to produce insect-resistant transgenic crops. *Pesticide Science* 52:165-175.

Gelvin, S.B. (2000). *Agrobacterium* and plant genes involved in T-DNA transfer and integration. *Annual Review of Plant Physiology and Plant Molecular Biology* 51:223-256.

Gielen, J., T. Ultzen, S. Bontems, W. Loots, A. van Schepen, A. Westerbroek, P. de Haan, and M. van Grinsven (1996). Coat protein-mediated protection to cucumber mosaic virus infections in cultivated tomato. *Euphytica* 88:139-149.

Gilbert, M.O., Y.Y. Zhang, and Z.K. Punja (1996). Introduction and expression of chitinase encoding genes in carrot following *Agrobacterium*-mediated transformation. *In Vitro Cellular and Developmental Biology—Plant* 32:171-178.

Gongora, C.E., S. Wang, R.V. Barbehenn, and R.M. Broadway (2001). Chitinolytic enzymes from *Streptomyces albidoflavus* expressed in tomato plants: Effects on *Trichoplusia ni. Entomologia Experimentalis et Applicata* 99:193-204.

Grant, J.E., P.A. Cooper, B.J. Gilpin, S.J. Hoglund, J.K. Reader, M.D. Pither-Joyce, and G.M. Timmerman-Vaughan (1998). Kanamycin is effective for selecting transformed peas. *Plant Science* 139:159-164.

Grant, J.E., P.A. Cooper, A.E. McAra, and T.J. Frew (1995). Transformation of peas (*Pisum sativum* L.) using immature cotyledons. *Plant Cell Reports* 15:254-258.

Grichko, V.P. and B.R. Glick (2001). Flooding tolerance of transgenic tomato plants expressing the bacterial enzyme ACC deaminase controlled by the 35S, *rolD* or PRB-1b promoter. *Plant Physiology and Biochemistry* 39:19-25.

Grover, A., A. Pareek, S.L. Singla, D. Minhas, S. Katiyar, S. Ghawana, H. Dubey, M. Agarwal, G.U. Rao, J. Rathee, and A. Grover (1998). Engineering crops for tolerance against abiotic stresses through gene manipulation. *Current Science* 75:689-696.

Grumet, R. (1995). Genetic engineering for crop virus resistance. *HortScience* 30:449-456.

Halford, N.G. and P.R. Shewry (2000). Genetically modified crops: Methodology, benefits, regulation and public concerns. *British Medical Bulletin* 56:62-73.

Hansen, G. and M.S. Wright (1999). Recent advances in the transformation of plants. *Trends in Plant Science* 4:226-231.

Hardegger, M. and A. Sturm (1998). Transformation and regeneration of carrot (*Daucus carota* L.). *Molecular Breeding* 4:119-127.

Henzi, M.X., M.C. Christey, and D.L. McNeil (2000a). Factors that influence *Agrobacterium rhizogenes*-mediated transformation of broccoli (*Brassica oleracea* L. var. *italica*). *Plant Cell Reports* 19:994-999.

Henzi, M.X., M.C. Christey, and D.L. McNeil (2000b). Morphological characterisation and agronomic evaluation of transgenic broccoli (*Brassica oleraceae* L. var. *italica*) containing an antisense ACC oxidase gene. *Euphytica* 113:9-18.

Henzi, M.X., M.C. Christey, D.L. McNeil, and K.M. Davies (1999). *Agrobacterium rhizogenes*-mediated transformation of broccoli (*Brassica oleracea* L. var. *italica*) with an antisense 1-aminocyclopropane-1-carboxylic acid oxidase gene. *Plant Science* 143:55-62.

Hilder, V.A. and D. Boulter (1999). Genetic engineering of crop plants for insect resistance: A critical review. *Crop Protection* 18:177-191.

Hohn, B., A.A. Levy, and H. Puchta (2001). Elimination of selection markers from transgenic plants. *Current Opinion in Biotechnology* 12:139-143.

Holmberg, N. and L. Bülow (1998). Improving stress tolerance in plants by gene transfer. *Trends in Plant Science* 3:61-66.

Hoy, C.W. (1999). Colorado potato beetle resistance management strategies for transgenic potatoes. *American Journal of Potato Research* 76:215-219.

Iamtham, S. and A. Day (2000). Removal of antibiotic resistance genes from transgenic tobacco plastids. *Nature Biotechnology* 18:1172-1176.

Iannacone, R., M.C. Fiore, A. Macchi, P.D. Grieco, S. Arpaia, D. Perrone, G. Mennella, F. Sunseri, F. Cellini, and G.L. Rotino (1995). Genetic engineering of eggplant (*Solanum melanogena* L.). *Acta Horticulturae* 392:227-233.

Jelenkovic, G., S. Billings, Q. Chen, J. Lashomb, G. Hamilton, and G. Ghidiu (1998). Transformation of eggplant with synthetic *cryIIIA* gene produces a high level of resistance to the Colorado potato beetle. *Journal of the American Society for Horticultural Science* 123:19-25.

Jin, R.-G., Y.-B. Liu, B.E. Tabashnik, and D. Borthakur (1999). Tissue culture and *Agrobacterium*-mediated transformation of watercress. *Plant Cell, Tissue and Organ Culture* 58:171-176.

Jin, R.-G., Y.-B. Liu, B.E. Tabashnik, and D. Borthakur (2000). Development of transgenic cabbage (*Brassica oleracea* var. *capitata*) for insect resistance by *Agrobacterium tumefaciens*-mediated transformation. *In Vitro Cellular and Developmental Biology—Plant* 36:231-237.

Joersbo, M. (2001). Advances in the selection of transgenic plants using non-antibiotic marker genes. *Physiologia Plantarum* 111:269-272.

Jongedijk, E., H. Tigelaar, J.S.C. van Roekel, S.A. Bres-Vloemans, I. Dekker, P.J.M. van den Elzen, B.J.C. Cornelissen, and L.S. Melchers (1995). Synergistic activity of chitinases and ß-1,3-glucanases enhances fungal resistance in transgenic tomato plants. *Euphytica* 85:173-180.

Jouanin, L., M. Bonadé-Bottino, C. Girard, G. Morrot, and M. Giband (1998). Transgenic plants for insect resistance. *Plant Science* 131:1-11.

Jun, S.I., S.Y. Kwon, K.Y. Paek, and K.-H. Paek (1995). *Agrobacterium*-mediated transformation and regeneration of fertile transgenic plants of Chinese cabbage (*Brassica campestris* ssp. *pekinensis* cv. 'spring flavor'). *Plant Cell Reports* 14:620-625.

Kaeppler, H.F. (2000). Food safety assessment of genetically modified crops. *Agronomy Journal* 92:793-797.

Kaniewski, W., V. Ilardi, L. Tomassoli, T. Mitsky, J. Layton, and M. Barba (1999). Extreme resistance to cucumber mosaic virus (CMV) in transgenic tomato expressing one or two viral coat proteins. *Molecular Breeding* 5:111-119.

Kavanagh, T.A. and C. Spillane (1995). Strategies for engineering virus resistance in transgenic plants. *Euphytica* 85:149-158.

Khan, M.S. and P. Maliga (1999). Fluorescent antibiotic resistance marker for tracking plastid transformation in higher plants. *Nature Biotechnology* 17:910-915.

Kim, J.W. and T. Minamikawa (1996). Transformation and regeneration of French bean plants by the particle bombardment process. *Plant Science* 117:131-138.

Kim, J.W. and T. Minamikawa (1997). Stable delivery of a canavalin promoter-β-glucuronidase gene fusion into French bean by particle bombardment. *Plant Cell Physiology* 38:70-75.

Kim, S.J., S.J. Lee, B.-D. Kim, and K.-H. Paek (1997). Satellite-RNA-mediated resistance to cucumber mosaic virus in transgenic plants of hot pepper (*Capsicum annuum* cv. Golden Tower). *Plant Cell Reports* 16:825-830.

Klann, E.M., B. Hall, and A.B. Bennett (1996). Antisense acid invertase *(TIV 1)* gene alters soluble sugar composition and size in transgenic tomato fruit. *Plant Physiology* 112:1321-1330.

Kumar, P.A., A. Mandaokar, K. Sreenivasu, S.K. Chakrabarti, S. Bisaria, S.R. Sharma, S. Kaur, and R.P. Sharma (1998). Insect-resistant transgenic brinjal plants. *Molecular Breeding* 4:33-37.

Kunkel, T., Q.-W. Niu, Y.-S. Chan, and N.-H. Chua (1999). Inducible isopentenyl transferase as a high-efficiency marker for plant transformation. *Nature Biotechnology* 17:916-919.

Kuvshinov, V., K. Koivu, A. Kanerva, and E. Pehu (2001). Transgenic crop plants expressing synthetic *cry9Aa* gene are protected against insect damage. *Plant Science* 160:341-353.

Li, B. and D.J. Wolyn (1997). Recovery of transgenic asparagus plants by particle gun bombardment of somatic cells. *Plant Science* 126:59-68.

Li, X.-B., H.-Z. Mao, and Y.-Y. Bai (1995). Transgenic plants of rutabaga *(Brassica napobrassica)* tolerant to pest insects. *Plant Cell Reports* 15:97-101.

Limanton-Grevet, A. and M. Jullien (2001). *Agrobacterium*-mediated transformation of *Asparagus officinalis* L.: Molecular and genetic analysis of transgenic plants. *Molecular Breeding* 7:141-150.

Lin, C.H., L. Xiao, B.H. Hou, S.-B. Ha, and J.A. Saunders (1997). Optimization of electroporation conditions for expression of GUS activity in electroporated protoplasts and intact plant cells. *Plant Physiology and Biochemistry* 35:959-968.

Lurquin, P.F., Z. Cai, C.M. Stiff, and E.P. Fuerst (1998). Half-embryo cocultivation technique for estimating the susceptibility of pea (*Pisum sativum* L.) and lentil (*Lens culinaris* Medik.) cultivars to *Agrobacterium tumefaciens*. *Molecular Biotechnology* 9:175-179.

Malik, V.S. and M.K. Saroha (1999). Marker gene controversy in transgenic plants. *Journal of Plant Biochemistry and Biotechnology* 8:1-13.

Mandaokar, A.D., R.K. Goyal, A. Shukla, S. Bisaria, R. Bhalla, V.S. Reddy, A. Chaurasia, R.P. Sharma, I. Altosaar, and P.A. Kumar (2000). Transgenic tomato plants resistant to fruit borer (*Helicoverpa armigera* Hubner). *Crop Protection* 19:307-312.

Manoharan, M., C.S. Sree Vidya, and G.L. Sita (1998). *Agrobacterium*-mediated genetic transformation in hot chili (*Capsicum annuum* L. var. Pusa jwala). *Plant Science* 131:77-83.

McCabe, M.S., L.C. Garratt, F. Schepers, W.J.R.M. Jordi, G.M. Stoopen, E. Davelaar, J.H.A. van Rhijn, J.B. Power, and M.R. Davey (2001). Effects of *PSAG12-IPT* gene expression on development and senescence in transgenic lettuce. *Plant Physiology* 127:505-516.

McCabe, M.S., F. Schepers, A. van der Arend, U. Mohapatra, A.M.M. de Laat, J.B. Power, and M.R. Davies (1999). Increased stable inheritance of herbicide resistance in transgenic lettuce carrying a *petE* promoter-*bar* gene compared with CaMV 35S-*bar* gene. *Theoretical and Applied Genetics* 99:587-592.

Melchers, L.S. and M.H. Stuiver (2000). Novel genes for disease-resistance breeding. *Current Opinion in Plant Biology* 3:147-152.

Metz, T.D., R. Dixit, and E.D. Earle (1995). *Agrobacterium tumefaciens*-mediated transformation of broccoli (*Brassica oleracea* var. *italica*) and cabbage (*B. oleracea* var. *capitata*). *Plant Cell Reports* 15:287-292.

Metz, T.D., R.T. Roush, J.D. Tang, A.M. Shelton, and E.D. Earle (1995). Transgenic broccoli expressing a *Bacillus thuringiensis* insecticidal crystal protein: Implications for pest resistance management strategies. *Molecular Breeding* 1:309-317.

Mohapatra, U., M.S. McCabe, J.B. Power, F. Schepers, A. Van Der Arend, and M.R. Davey (1999). Expression of the *bar* gene confers herbicide resistance in transgenic lettuce. *Transgenic Research* 8:33-44.

Mourgues, F., M.-N. Brisset, and E. Chevreau (1998). Strategies to improve plant resistance to bacterial diseases through genetic engineering. *Trends in Biotechnology* 16:203-210.

Muir, S.R., G.J. Collins, S. Robinson, S. Hughes, A. Bovy, C.H. Ric De Vos, A.J. van Tunen, and M.E. Verhoeyen (2001). Overexpression of petunia chalcone isomerase in tomato results in fruit containing increased levels of flavonols. *Nature Biotechnology* 19:470-474.

Nagl, W., S. Ignacimuthu, and J. Becker (1997). Genetic engineering and regeneration of *Phaseolus* and *Vigna:* State of the art and new attempts. *Journal of Plant Physiology* 150:625-644.

Newell, C.A. (2000). Plant transformation technology: Developments and applications. *Molecular Biotechnology* 16:53-65.

Newell, C.A., J.M. Lowe, A. Merryweather, L.M. Rooke, and W.D.O. Hamilton (1995). Transformation of sweet potato [*Ipomoea batatas* (L.) Lam.] with *Agrobacterium tumefaciens* and regeneration of plants expressing cowpea trypsin inhibitor and snowdrop lectin. *Plant Science* 107:215-227.

Niki, T., T. Nishijima, M. Nakayama, T. Hisamatsu, N. Oyama-Okubo, H. Yamazaki, P. Hedden, T. Lange, L.N. Mander, and M. Koshioka (2001). Production of dwarf lettuce by overexpressing a pumpkin gibberellin 20-oxidase gene. *Plant Physiology* 126:965-972.

Nishibayashi, S., H. Kaneko, and T. Hayakawa (1996). Transformation of cucumber (*Cucumis sativus* L.) plants using *Agrobacterium tumefaciens* and regeneration from hypocotyl explants. *Plant Cell Reports* 15:809-814.

Nishiguchi, M., M. Mori, Y. Okada, T. Murata, T. Kimura, J.-I. Sakai, K. Hanada, C. Miyazaki, and A. Saito (1998). Virus resistant transgenic sweet potato with the CP gene: Current challenge and perspective of its use. *Phytoprotection* 79 (Suppl.):112-116.

Ochao-Alejo, N. and R. Ramirez-Malagon (2001). In vitro chili pepper biotechnology. *In Vitro Cellular and Developmental Biology—Plant* 37:701-729.

Otani, M., T. Shimada, T. Kimura, and A. Saito (1998). Transgenic plant production from embryogenic callus of sweet potato [*Ipomoea batatas* (L.) Lam.] using *Agrobacterium tumefaciens. Plant Biotechnology* 15:11-16.

Pang, S.-Z., F.-J. Jan, K. Carney, J. Stout, D.M. Tricoli, H.D. Quemada, and D. Gonsalves (1996). Post-transcriptional transgene silencing and consequent tospovirus resistance in transgenic lettuce are affected by transgene dosage and plant development. *The Plant Journal* 9:899-909.

Parashina, E.V., L.A. Serdobinskii, E.G., Kalle, N.V. Lavrova, V.A. Avetisov, V.G. Lunin, and B.S. Naroditskii (2000). Genetic engineering of oilseed rape and tomato plants expressing a radish defensin gene. *Russian Journal of Plant Physiology* 47:417-423.

Passelègue, E. and C. Kerlan (1996). Transformation of cauliflower (*Brassica oleracea* var. *botrytis*) by transfer of cauliflower mosaic virus genes through combined co-cultivation with virulent and avirulent strains of *Agrobacterium. Plant Science* 113:79-89.

Picton, S., J.E. Gray, and D. Grierson (1995). The manipulation and modification of tomato fruit ripening by expression of antisense RNA in transgenic plants. *Euphytica* 85:193-202.

Pius, P.K. and P.N. Achar (2000). *Agrobacterium*-mediated transformation and plant regeneration of *Brassica oleracea* var. *capitata. Plant Cell Reports* 19:888-892.

Powell, A.L.T., J. van Kan, A. ten Have, J. Visser, L.C. Greve, A.B. Bennett, and J.M. Labavitch (2000). Transgenic expression of pear PGIP in tomato limits fungal colonization. *Molecular Plant-Microbe Interactions* 13:942-950.

Puddephat, I.J., H.T. Robinson, T.M. Fenning, D.J. Barbara, A. Morton, and D.A.C. Pink (2001). Recovery of phenotypically normal transgenic plants of *Brassica oleracea* upon *Agrobacterium* rhizogenes-mediated co-transformation and selection of transformed hairy roots by GUS assay. *Molecular Breeding* 7:229-242.

Punja, Z.K. and S.H.T. Raharjo (1996). Response of transgenic cucumber and carrot plants expressing different chitinase enzymes to inoculation with fungal pathogens. *Plant Disease* 80:999-1005.

Qing, C.M., L. Fan, Y. Lei, D. Bouchez, C. Tourneur, L. Yan, and C. Robaglia (2000). Transformation of pakchoi (*Brassica rapa* L. *chinensis*) by *Agrobacterium* infiltration. *Molecular Breeding* 6:67-72.

Raharjo, S.H.T., M. Hernandez, Y.Y. Zhang, and Z.K. Punja (1996). Transformation of pickling cucumber with chitinase encoding genes using *Agrobacterium tumefaciens. Plant Cell Reports* 15:591-596.

Rathinasabapathi, B. (2000). Metabolic engineering for stress tolerance: Installing osmoprotectant synthesis pathways. *Annals of Botany* 86:709-716.

Römer, S., P.D. Fraser, J.W. Kiano, C.A. Shipton, N. Misawa, W. Schuch, and P.M. Bramley (2000). Elevation of the provitamin A content of transgenic tomato plants. *Nature Biotechnology* 18:666-669.

Rotino, G.L., E. Perri, M. Zottini, H. Sommer, and A. Spena (1997). Genetic engineering of parthenocarpic plants. *Nature Biotechnology* 15:1398-1401.

Ruf, S., M. Hermann, I.J. Berger, H. Carrer, and R. Bock (2001). Stable genetic transformation of tomato plastids and expression of a foreign protein in fruit. *Nature Biotechnology* 19:870-875.

Schroeder, H.E., S. Gollasch, A. Moore, L.M. Tabe, S. Craig, D.C. Hardie, M.J. Chrispeels, D. Spencer, and T.J.V. Higgins (1995). Bean α-amylase inhibitor confers resistance to the pea weevil *(Bruchus pisorum)* in transgenic peas (*Pisum sativum* L.). *Plant Physiology* 107:1233-1239.

Schulze, J., C. Balko, B. Zellner, T. Koprek, R. Hänsch, A. Nerlich, and R.R. Mendel (1995). Biolistic transformation of cucumber using embryogenic suspension cultures: Long-term expression of reporter genes. *Plant Science* 112:197-206.

Southgate, E.M., M.R. Davey, J.B. Power, and R. Marchant (1995). Factors affecting the genetic engineering of plants by microprojectile bombardment. *Biotechnology Advances* 13:631-651.

Speirs, J., E. Lee, K. Holt, K. Yong-Duk, N. Steele Scott, B. Loveys, and W. Schuch (1998). Genetic manipulation of alcohol dehydrogenase levels in ripening tomato fruit affects the balance of some flavor aldehydes and alcohols. *Plant Physiology* 117:1047-1058.

Tabaeizadeh, Z., Z. Agharbaoui, H. Harrak, and V. Poysa (1999). Transgenic tomato plants expressing a *Lycopersicon chilense* chitinase gene demonstrate improved resistance to *Verticillium dahliae* race 2. *Plant Cell Reports* 19:197-202.

Tabei, Y., S. Kitade, Y. Nishizawa, N. Kikuchi, T. Kayano, T. Hibi, and K. Akutsu (1998). Transgenic cucumber plants harboring a rice chitinase gene exhibit enhanced resistance to gray mold *(Botrytis cinerea). Plant Cell Reports* 17:159-164.

Takaichi, M. and K. Oeda (2000). Transgenic carrots with enhanced resistance against two major pathogens, *Erysiphe heraclei* and *Alternaria dauci. Plant Science* 153:135-144.

Thakur, B.R., R.K. Singh, and A.K. Handa (1996). Effect of an antisense pectin methylesterase gene on the chemistry of pectin in tomato *(Lycopersicon esculentum)* juice. *Journal of Agricultural and Food Chemistry* 44:628-630.

Thomzik, J.E., K. Stenzel, R. Stöcker, P.H. Schreier, R. Hain, and D.J. Stahl (1997). Synthesis of a grapevine phytoalexin in transgenic tomatoes (*Lycopersicon esculentum* Mill.) conditions resistance against *Phytophthora infestans. Physiological and Molecular Plant Pathology* 51:265-278.

Tsaftaris, A. (1996). The development of herbicide-tolerant transgenic crops. *Field Crop Research* 45:115-123.

Van Eck, J.M., A.D. Blowers, and E.D. Earle (1995). Stable transformation of tomato cell cultures after bombardment with plasmid and YAC DNA. *Plant Cell Reports* 14:299-304.

Vijn, I., A. van Dijken, N. Sprenger, K. van Dun, P. Weisbeek, A. Wiemken, and S. Smeekens (1997). Fructan of the inulin neoseries is synthesized in transgenic chicory plants (*Chicorium intybus* L.) harbouring onion (*Allium cepa* L.) fructan:fructan 6G-fructosyltransferase. *The Plant Journal* 11:387-398.

Wakita, Y., M. Otani, T. Hamada, M. Mori, K. Iba, and T. Shimada (2001). A tobacco microsomal ω-3 fatty acid desaturase gene increases the linolenic acid content in transgenic sweet potato *(Ipomoea batatas). Plant Cell Reports* 20:244-249.

Wang, C., C.-K. Chin, C.-T. Ho, C.-F. Hwang, J.J. Polashock, and C.E. Martin (1996). Changes in fatty acids and fatty acid-derived flavor compounds by expressing the yeast Δ–9 desaturase gene in tomato. *Journal of Agricultural Food Chemistry* 44:3399-3402.

Waterer, D., S. Lee, G. Scoles, and W. Keller (2000). Field evaluation of herbicide-resistant transgenic broccoli. *HortScience* 35:930-932.

Xiang, Y., W.-K.R. Wong, M.C. Ma, and R.S.C. Wong (2000). *Agrobacterium*-mediated transformation of *Brassica campestris* ssp. *parachinensis* with synthetic *Bacillus thuringiensis cry1Ab* and *cry1Ac* genes. *Plant Cell Reports* 19:251-256.

Xing, T., D.-Y. Zhang, J.F. Hall, and E. Blumwald (1996). Auxin levels and auxin binding protein availability in *rolB* transformed *Beta vulgaris* cells. *Biologia Plantarum* 38:351-362.

Yang, Y., J.M. Al-Khayri, and E.J. Anderson (1997). Transgenic spinach plants expressing the coat protein of cucumber mosaic virus. *In Vitro Cellular and Developmental Biology—Plant* 33:200-204.

Zhang, H.X. and E. Blumwald (2001). Transgenic salt-tolerant tomato plants accumulate salt in foliage but not in fruit. *Nature Biotechnology* 19:765-768.

Zhang, J., N.Y. Klueva, Z. Wang, R. Wu, T.-H.D. Ho, and H.T. Nguyen (2000). Genetic engineering for abiotic stress resistance in crop plants. *In Vitro Cellular and Developmental Biology—Plant* 36:108-114.

Zhang, P., I. Potrykus, and J. Puonti-Kaerlas (2000). Efficient production of transgenic cassava using negative and positive selection. *Transgenic Research* 9:405-415.

Zheng, S.-J., L. Khrustaleva, B. Henken, E. Sofiari, E. Jacobsen, C. Kik, and F.A. Krens (2001). *Agrobacterium tumefaciens*-mediated transformation of *Allium cepa* L.: The production of transgenic onions and shallots. *Molecular Breeding* 7:101-115.

Zhu, Y.-X., W.-J. Ou-Yang, Y.-F. Zhang, and Z.-L. Chen (1996). Transgenic sweet pepper plants from *Agrobacterium*-mediated transformation. *Plant Cell Reports* 16:71-75.

Zubko, E., C. Scutt, and P. Meyer (2000). Intrachromosomal recombination between attP regions as a tool to remove selectable marker genes from tobacco transgenes. *Nature Biotechnology* 18:442-445.

Chapter 13

Genetic Transformation of Turfgrass

Barbara A. Zilinskas
Xiaoling Wang

INTRODUCTION

Turfgrasses commonly fall into two major groups: cool-season turfgrass and warm-season turfgrass. The cool-season turfgrasses are well adapted to a temperature of 60 to 75°F and are widely distributed throughout the cool humid and cool subhumid regions. Most of them originated from the fringe forest species of Eurasia. The cultivated cool-season turfgrasses include bluegrass (*Poa* spp.), fescue (*Festuca* spp.), ryegrass (*Lolium* spp.), and bentgrass (*Agrostis* spp.). The warm-season turfgrasses refer to turfgrasses having a temperature optimum of 80 to 95°F, which are widely distributed throughout the warm humid and warm subhumid climates. As compared with the common European origin of cool-season turfgrass, the warm-season turfgrasses have varied origins such as Africa, South America, and Asia. The more important and widely adopted warm-season turfgrasses are bermudagrass (*Cynodon dactylon* Rich.), zoysiagrass [*Zoysia japonica* (Willd.) Steud.], bahiagrass (*Paspalum notatum* Flugge.), St. Augustinegrass [*Stenotaphrum secundatum* (Walt.) Kuntze], and centipedegrass [*Eremochloa ophiuroides* (Munro.) Hack.]. The low-growing warm-season turfgrasses with their deeper roots are more tolerant to heat, drought, and close mowing, but they are more likely to discolor at low temperatures (Beard, 1973).

Two groups of cultivars are identified according to their scope of genetic variation. Those that are genetically homogeneous are developed by vegetative propagation, self-pollination, or an apomictic breeding system. Genetic variation within this group is limited to differences between cultivars, while those that are developed by outcrossing are genetically heterogeneous, and their genetic variation resides both between and within cultivars. This characteristic not only complicates the identification, analysis, and selection of germplasm for breeding purposes, but it also imposes the prerequisite to use cultures derived from single genotypes for genetic engineering.

Traditionally, turfgrass improvement has relied on conventional breeding, by which more than 245 warm-season and cool-season cultivars have been produced since 1946 in the United States (Lee et al., 1996). However, inaccessibility of genetic material due to barriers to sexual reproduction, as well as the relatively long time needed in breeding programs, has restricted improvement by conventional breeding. Plant biotechnology has permitted more efficient improvement of turfgrass by introducing useful traits from a broader range of sources within an economically viable time frame. Establishment of regenerable cultures from explants that are meristemic and undifferentiated has provided reliable targets for genetic manipulation. Development of various transformation technologies will render more efficient the introduction of useful genes into a larger number of plant species. Advances in tissue culture and transformation technology have resulted in an increased number of successful transformations of turfgrasses to date, and soon it will be possible to introduce beneficial genes directly into elite cultivars.

The turfgrass industry has become a big business in terms of dollars, labor, and land acreage. In 1965, $4 billion were spent on maintaining lawns, golf courses, parks, roadsides, and other turfgrass areas (Aldous, 1999); however, three decades later the expenditure has reached $45 billion. It has been estimated that the industry growth rate will be very rapid, and turfgrass maintenance expenditures will soon approach $90 billion (Duble, 1996). The annual retail sales of turfgrass seed are second only to hybrid corn (Kidd, 1993; Lee et al., 1996). At the same time, the acreage of land covered by turfgrass was 35 million acres in the United States in 1995, and it continues to increase. In 1993, more than 1 million persons worked in the turfgrass industry in the United States (Duble, 1996).

With the rapid growth of the turfgrass industry, especially the expansion in the number of golf courses in the United States, more and more environmental issues are put into the calendar of turfgrass industry development. Besides the positive impact of turfgrass on environmental quality, great concerns are raised mainly with regard to water shortage and application of pesticides and fertilizer (Duble, 1996; Snow and Kenna, 1999). For example, the severe drought during the 1970s and early 1980s resulted in the extreme restriction on use of water for turfgrass (Duble, 1996). Similarly, the application of fertilizer and pesticides in golf course maintenance is under attack because people are growing concerned over potential risks of their use (Duble, 1996; Snow and Kenna, 1999). In 1993, 73 million pounds of pesticide and 500 million pounds of fertilizer were used for maintaining turfgrass in lawns and gardens, which contributes to the pollution of water sources and poses a potential risk to both human health and the environment (Duble, 1996).

The turfgrass industry needs to address the problem not only for the benefit of human beings but also for its own safety. Until recently, the turfgrass industry has employed traditional breeding programs to create new varieties with improved characteristics. Plant biotechnology offers the opportunity to overcome limitations of sexual reproduction and to modify plants with useful traits within an economically viable time frame (Hansen and Wright, 1999). Introduction of genetic traits from unrelated plants and even other kingdoms becomes possible. Based on the use of cell culture and molecular biology techniques, plant biotechnology provides the potential to produce plants with enhanced resistance to biotic and abiotic stresses and with numerous value-added qualities.

The focus of this chapter will be on genetic transformation of turfgrass, although it must be mentioned that other biotechnological approaches to turfgrass improvement are available. These include the use of molecular markers in assisting conventional breeding programs and in protecting breeders' rights; somatic hybridization; and selection of somaclonal variants, produced in tissue culture, which may have improved qualities. Readers are referred to three comprehensive reviews on these subjects (Lee et al., 1996; Chai and Sticklen, 1998; Spangenberg et al., 1998).

REGENERABLE TISSUE CULTURE PREREQUISITE

General Considerations

Current transformation systems in use for turfgrasses always involve tissue culture, even though it may not be necessary. As a result, establishment of a system for highly efficient regeneration becomes one of the most important factors that determines the success and efficiency of transformation.

As in other graminaceous monocots, grasses are among the most recalcitrant species to be genetically manipulated in vitro (Vasil, 1988). In early work, mature and differentiated tissues were used as explant materials in turfgrass in vitro culture, but this was not a wise choice as these grass tissues lack the ability for secondary growth by cambium.

The research summarized by Vasil (1995) on establishing regenerable cell suspension cell cultures for grass species formed the basis for the successful establishment of a regeneration system in all major turfgrasses (Chai and Sticklen, 1998). The major considerations include (1) choice of meristem and undifferentiated explants such as immature embryos or seeds; (2) supplementation of the culture medium with high concentrations of potent auxins, such as 2,4-dichlorophenoxyacetic acid (2,4-D) and 3,6-dichloro-2-methoxy-benzoic acid (dicamba) to induce embryogenic calli; (3) selection of tissues in culture that regenerate predominantly by formation of somatic embryos; and (4) isolation of protoplasts from cell suspensions derived from embryogenic callus.

Explants

The genotype and explant type are considered the most critical factors in determining the in vitro response of the tissue. Although genetic differences in regeneration potential have been reported in many species, success in plant regeneration from a wide range of genotypes and species suggests that genetic differences can be circumvented. Commonly employed techniques to improve regenerative ability are to use suitable explants at specific physiological and developmental stages or to modify the culture medium in which the explants are grown (Vasil, 1988).

The most suitable explants for tissue culture of grasses are those in which a large proportion of the tissues or cells are still meristematic and undifferentiated, such as immature embryos or seeds, young inflorescences, and young leaves (Vasil, 1987). The endogenous pool of plant growth regulators and nutrients is related to the embryogenic competency in these explants of grass species; the level of endogenous growth regulators in these tissues is higher than that in mature tissue (Rajasekaran et al., 1987; Vasil, 1987). Various explants have been used to establish regenerable cultures of turfgrasses, and these explants have been used to produce embryogenic callus, suspensions derived from embryogenic callus, and protoplasts isolated from embryogenic cell suspensions that have been used in transformation (Table 13.1).

Selection and Maintenance of Embryogenic Callus and Cell Suspensions

Two kinds of calli are formed in monocot tissue culture: embryogenic and nonembryogenic. Embryogenic calli (type I) are compact, organized, and yellow/white in color. They grow slowly but maintain their regenerability in culture for a long time if one takes care to select and subculture only embryogenic calli. Nonembryogenic calli (type II) are soft and friable and have low regenerability. Most cultures are a mixture of the two kinds of calli in which the embryogenic callus consists of a small portion. Initially the surrounding nonembryogenic calli may act as nurse tissue (Taylor and Vasil, 1985), but gradually they overgrow the embryogenic callus. Thus for maintenance of embryogenic potential and regeneration of callus, it is essential to select embryogenic callus for subculture (Vasil, 1987).

It has been demonstrated that embryogenic cell suspensions may provide an efficient way to retain the embryogenic potential of tissues in several major turfgrasses. However, in maintenance of embryogenic cell suspensions, several problems are encountered. Maintenance of embryogenic cell suspension by weekly subculture is time-consuming and expensive, and it also may cause the loss of valuable material by contamination and genetic changes. Embryogenic cell suspensions also tend to lose their regenerability with prolonged culture, and new embryogenic callus must be initiated frequently. Continuous initiation of new embryogenic cell suspensions is difficult in practice, especially with genetically heterogeneous species, such as *Festuca* and *Lolium,* because different seeds of these species may have different genotypes both within and between cultivars, and the establishment of these cultures is genotype dependent (Wang, Nagel, et al., 1993; Wang, Valles, et al., 1993). Efficient maintenance of

TABLE 13.1. Diverse turfgrass explants that have been used to establish regenerable cultures

Species	Explant	References
Festuca arundinacea (tall fescue)	Mature seed	Lowe and Conger, 1979; Dalton, 1988a,b; Rajoelina et al., 1990; Ha et al., 1992; Wang et al., 1995; Chai and Niu, 1997; Bai and Qu, 2000a,b
	Immature inflorescence	Rajoelina et al., 1990; Bai and Qu, 2000b
	Leaf base	Takamizo et al., 1990
	Shoot tip	Dalton et al., 1998
Festuca rubra L. (red fescue)	Mature seed	Torello et al., 1984; Zaghmout and Torello, 1989; Spangenberg et al., 1994; Wang et al., 1995
Lolium perenne L. (perennial ryegrass)	Mature seed	Creemers-Molenaar et al., 1988
	Mature embryo	Dalton, 1988a,b; Wang, Valles, 1993; Wang, Nagel, et al., 1993
	Shoot tip	Dalton et al., 1998, 1999
Lolium multiflorum Lam. (Italian ryegrass)	Mature seed	Creemers-Molenaar et al., 1988; Rajoelina et al., 1990; Wang, Nagel, 1993; Ye et al., 1997; Dalton et al., 1999
	Shoot tip	Dalton et al., 1998, 1999
Agrostis palustris Huds. (creeping bentgrass)	Mature seed	Krans et al., 1982; Tarakawa et al., 1992; Zhong et al., 1991; Hartman et al.,1994; Asano and Ugaki, 1994; Xiao and Ha, 1997
Agrostis alba L. (redtop)	Mature seed	Shetty and Yoshito, 1991
Poa pratensis L. (Kentucky bluegrass)	Mature seed	McDonnell and Conger, 1984; van der Valk et al., 1989

	Mature embryo	Nielsen and Knudsen, 1993; Nielsen et al., 1993
	Immature inflorescence	van der Valk et al., 1989
Dactylis glomerata (orchardgrass)	Young leaf tissue	Denchev et al., 1997
	Mature seed	Cho et al., 2001
Cynodon dactylon L. (bermudagrass)	Immature inflorescence	Ahn et al., 1985; Artunduaga et al., 1988; Chaudhury and Qu, 2000
Buchloe dactyloides (Nutt.) Engelm (buffalograss)	Immature inflorescence	Fei, 2000

regenerability of in vitro cultures has challenged and interested many turfgrass researchers.

Cryopreservation of embryogenic cell suspensions is considered to be a good alternative for retaining high-level green plant regeneration frequency without changing genetic stability. For different species and cultivars of *Festuca* and *Lolium,* a reproducible method for cryopreservation of embryogenic cell suspension cultures derived from single genotypes has been established (Wang et al., 1994). This procedure ensured the availability of regenerable embryogenic suspension cultures for a reasonably long time period. The in vitro performance of reestablished embryogenic cells from long-term cryopreserved cultures retained high regenerability, while unfrozen cultures had low regenerability.

Another in vitro system was recently described that eliminated the cell suspension step by proliferating highly regenerable green tissues from embryogenic callus derived from mature seed of tall fescue and red fescue (Cho et al., 2000), orchardgrass (Cho et al., 2001), and Kentucky bluegrass (Ha et al., 2001). Callus was initiated on 2,4-D and cupric sulfate with or without benzylaminopurine (BAP), and it was then maintained on 2,4-D and cupric sulfate with a low concentration of BAP. Through this intermediate step, a light-green, shoot-like structure was developed on the same medium under dim light, which could be maintained for more than one year with minimal reduction of regenerability. This is the first report wherein protoplasts or embryogenic cell suspensions were not used as transformation targets for *Festuca* and *Poa*. Similar systems had previously been established in barley (Cho et al., 1998) and oat (Cho et al., 1999).

Culture Medium and Regeneration

Plants are regenerated from tissue culture via two ways, organogenesis (or shoot morphogenesis) and somatic embryogenesis. In most cases, organogenesis does not occur de novo but rather by derepression of preexisting shoot meristems whose multicellular origin may give rise to chimeras unsuitable as breeding material. In contrast, plant regeneration via somatic embryogenesis is nonchimeric in most cases because of the assumed single-cell origin of somatic embryos (Vasil, 1987). This makes embryogenic cultures more suitable for genetic manipulation.

Much effort has been focused on varying the components of the culture medium to obtain an optimum medium for regeneration. The culture medium used in turfgrasses is very similar to that for cereal crops. Murashige and Skoog (MS) medium has been used as the basic medium. High concentrations of 2,4-D or dicamba are critical for the initiation and production of regenerable callus. However, the addition of low concentrations of cytokinins, such as 6-benzylaminopurine, can be significant in promoting callus initiation of turfgrasses (Chai and Sticklen, 1998). Other essential elements such as carbohydrate source, amino acids, and vitamins also play a role in callus initiation and regeneration, which vary according to different explants, genotype, and species used. The fact is, however, that many efficient regeneration systems have been established through the use of explants from immature organs or inflorescences rather than through the optimization of the components of culture medium. However, modification of the medium composition may extend the range of genotypes possible for establishment of regenerable tissue culture, or it may improve the in vitro response of explants.

In several major turfgrass species, the establishment of embryogenic cell suspensions or protoplasts derived from them may yield better regeneration potential. In general, embryogenic cell suspensions are established from embryogenic calli. Embryogenic calli are selectively transferred to liquid medium, such as AA (Muller and

Grafe, 1978) or MS (Murashige and Skoog, 1962) medium, to establish the finely dispersed and fast-growing cell suspensions. The cell suspension is maintained in liquid medium with high levels of auxins, such as 2 to 3 mg·L^{-1} 2,4-D, and it is subcultured every three to seven days. Prior to transfer to regeneration medium, the cell suspension must be plated on semisolid medium for embryo development. For many turfgrasses species, embryogenic cell suspensions have been established and subsequent plant regeneration has been achieved. Embryogenic cell suspension cultures provide the sole source of totipotent cells from which protoplasts are isolated (Chai and Sticklen, 1998).

Ever since protoplasts have been isolated from cell suspensions derived from embryogenic callus, plating efficiency and formation of protoplast-derived colonies have been greatly improved. Most of the early advances with in vitro manipulation of turfgrass were made using protoplasts as explants. However, protoplasts are still considered the most difficult explants in terms of plant regeneration. Protoplast culture is also thought to be difficult in practice because of the laborious worked involved, poor longevity of donor-suspension cells, low regeneration rate from protoplasts, and genotype dependency in culture establishment.

The main factors involved in establishment of protoplast culture in turfgrasses are donor material and culture conditions. Finely dispersed suspension cells derived from embryogenic callus produce protoplasts more capable of regeneration. The age of the embryogenic suspension cells is critical to the regeneration potential of the protoplasts (Takamizo et al., 1990; Nielsen et al., 1993). In addition to the basic requirements for protoplast culture, including the basic medium, growth regulators, and osmoticum, several culture conditions are important, such as the use of the bead-type culture (Takamizo et al., 1990; Wang et al., 1992, 1995) and the use of "nurse" or "feeder" cells (Takamizo et al., 1990; Ha et al., 1992; Wang et al., 1992, 1995). Embedding the protoplasts in agarose improves their regeneration. The feeder-layer system increases plating efficiency and is effective in the formation of protoplast-derived callus and subsequent regeneration. Manipulation of culture conditions makes the establishment of protoplast culture of recalcitrant species possible.

TRANSFORMATION SYSTEMS

General Considerations

Plant transformation technology is now a very useful tool for crop improvement and also for studying gene function in plants. Many diverse plant species have been transformed successfully, including some plant species that were previously considered recalcitrant to transformation (Birch, 1997). Three major issues must be addressed to produce stable transgenic plants. These include (1) a DNA delivery method; (2) integration of DNA into germline cells such that the transgene will be inherited by the progeny; and (3) a means to regenerate fertile plants from explants, callus, or cells into which the foreign DNA was introduced.

The commonly used targets for heritable transformation of turfgrass are protoplasts (cells stripped of their cell walls), embryogenic callus (dedifferentiated cells with the potential to produce somatic embryos), and suspension cell cultures (which are usually derived from embryogenic callus). Conditions to establish appropriate materials which serve as targets are described, as are considerations in achieving regeneration of transformed cells. Even though tissue culture is not a prerequisite for plant transformation, it is used in almost all current transformation systems. It is also the reason why transformation remains an art, because of the unique culture conditions for different species. Many factors should be considered such as genotype, explant, and culture medium, especially the combination of different hormones contained in the culture medium. Whatever is considered good target tissue, it should have the capability of proliferation and should retain the capability for regeneration after the duration of necessary proliferation and selection treatments (Birch, 1997). Microscope-assisted selection of embryogenic callus improves the success rate in transformation and subsequent regeneration of healthy, fertile transgenic plants. It is also important to minimize the time that plant material is maintained in tissue culture in order to decrease the chances of introduction of somaclonal mutations.

The two major ways that DNA can be successfully delivered into turfgrasses include direct DNA transfer and transformation mediated by *Agrobacterium tumefaciens.* Transformation of turf species was first achieved by direct DNA uptake by protoplasts of orchardgrass (*Dactylis glomerata* L.) (Horn et al., 1988). Cell walls are removed with cell wall-degrading enzymes to produce protoplasts which are subsequently made competent to take up DNA by osmotic treatment or by subjecting the protoplasts to electric shock (electroporation). Several turf species have been transformed by direct DNA delivery into protoplasts as described in detail in the next section; however, as it is difficult to regenerate fertile plants from protoplasts, this method is not commonly used today.

Biolistic delivery of DNA, modified after the invention described by Sanford and colleagues (1987), is now the preferred method of DNA transfer into turfgrass and other plant species. Also known as particle bombardment, plant tissue or callus is bombarded with DNA-coated gold particles that are accelerated with considerable force by use of a "gene gun" or, more commonly today, by a stream of compressed He gas or by an acceleration system based on an airgun. This method has been used successfully for several turfgrass species to produce fertile transgenic plants.

More recently, *Agrobacterium tumefaciens* has been used to mediate DNA transfer into turfgrass (Lakkaraju et al., 1999; Yu, Skinner, et al., 2000). *Agrobacterium*-mediated transformation offers several advantages over particle gun bombardment or other means of direct gene transfer. These include stable transgene integration without rearrangement of either host or transgene DNA; preferential integration of the transgene into transcriptionally active regions of the chromosome; ability to transfer large segments of DNA; and integration of low numbers of gene copies into plant nuclear DNA, which is particularly important to minimize possible cosuppression of the transgene in later generations.

Until recently, *Agrobacterium*-mediated transformation was thought to be limited to dicotyledonous plants. However, Hiei and colleagues (1994) described efficient transformation of rice by *Agrobacterium* and subsequently there have been convincing reports for other agronomically important cereal crops. Numerous factors are of critical importance in *Agrobacterium*-mediated transformation of monocots, including the type of tissue that is infected, the vector and bacterial strains used, plant genotype, tissue culture conditions, and the actual infection process.

DNA Uptake by Protoplasts

Since the establishment of the protoplast-to-plant regeneration system, direct delivery of DNA into protoplasts has been used rather widely. In this method, protoplasts could be pretreated to take up plasmid DNA from the surrounding culture medium either by chemical treatment with membrane-altering compounds such as PEG (polyethylene glycol) or by the application of electric pulses (electroporation). The use of protoplasts as recipients of foreign DNA has several advantages, the most important of which is that the removal of the cell wall eliminates the barrier for transformation (Songstad et al., 1995). Protoplast culture could produce a large number of identical cells in an optimum medium, thus providing sufficient material for transformation. Protoplast-derived tissue is usually colonial in origin due to the nature of protoplasts as individual cells; therefore, selection of transformants is more efficient and formation of chimeras is minimized (Songstad et al., 1995).

The commonly used procedure for direct DNA delivery into plant protoplasts is PEG-mediated DNA uptake by protoplasts. In this process, DNA is incubated with protoplasts in the presence of PEG together with divalent cations such as Ca^{2+} or Mg^{2+}. They are then diluted with a buffer which also contains the divalent cation. PEG causes reversible permeability of the plasma membrane and enables exogenous DNA to enter the cytoplasm. PEG also affects the conformation of DNA, thus allowing DNA association with the plasma membrane. In general, 8 to 20 percent PEG with a molecular weight of 4000 to 8000 is used; typically, Ca^{2+} promotes better transient expression, and Mg^{2+} increases transformation frequency. Electroporation induces the uptake of DNA by protoplasts by promoting and increasing the stability of micropores in the cell membrane. The opening of the pore is related to the voltage and duration of the electrical pulse (Shillito, 1999).

Festuca

Transgenic plants were recovered from protoplasts obtained from different forage-type cultivars of tall fescue by using PEG-mediated transformation (Wang et al., 1992). Chimeric *hph* and *bar* genes, providing resistance to hygromycin and phosphinothricin (PPT) or bialaphos, respectively, were used as selectable markers; both markers were under the control of the CaMV 35S promoter. A dose response experiment was performed in order to determine the optimum selection concentration for transformation. Hygromycin at 200 $mg{\cdot}L^{-1}$ provided efficient selection on bead-type culture followed by plate selection, and 50 $mg{\cdot}L^{-1}$ PPT was lethal to nontransformed callus. Southern blot analysis showed the integration of multiple full-length transgenes and the occurrence of transgene rearrangement. Expression of the transgenes in mature plants was determined by enzyme assay or by herbicide spray. The regeneration of plants transformed with *hph* and *gus* genes was also reported after electroporation of protoplasts of *Festuca arundinacea* (Ha et al., 1992).

The selection scheme is generally determined first for the sake of efficient and stable transformation. The selection scheme has impacted on the nature of the transgenic plants and transformation frequency (Dalton et al., 1995). PEG-treated protoplasts of tall fescue were transformed with a chimeric CaMV 35S-*hph* gene. Discontinuous and continuous selection schemes with 50 to 200 $mg{\cdot}L^{-1}$ hygromycin were tested. Plant regeneration was high with discontinuous selection. In certain selection stages, no selection agent was used, but a significant number of escapes was obtained. Continuous selection at 50 $mg{\cdot}L^{-1}$ hygromycin produced a high number of transgenic plants without escapes. These two selection schemes produced plants with two or fewer copies of the transgene. High concentrations of hygromycin favored recovery of plants with multiple-copy transgenes.

The physiological state (both the stage of the growth cycle and the age of the cell culture) of suspension-cultured cells used to isolate protoplasts affected both transient and stable expression of the transgene in transgenic tall fescue (Kuai and Morris, 1995). A rice *act1* promoter-*gus* chimeric gene was used. Transient expression, but not stable expression, of GUS was dependent on the cell growth cycle, and prolonged subculturing of cells resulted in a gradual decline of both transient and stable expression. A high level of transient GUS expression and a 5.7 percent frequency of stable GUS expression were obtained with protoplasts isolated from suspension cells 10 to 20 weeks after initiation and three to four days after subculturing. The osmolarity of the protoplast culture medium was critical for the stable expression of GUS. The highest rate of transformation was obtained when a relatively high osmolarity (768 mOsm) was used in the culture medium after transformation.

Subsequently, Kuai and Morris (1996) obtained GUS-positive callus by directly screening for gene expression cell colonies derived from protoplasts of tall fescue *(Festuca arundinacea).* Of 320 protoplast-derived cell colonies screened, only seven were found to be GUS positive. Two types of callus clones with different levels of GUS expression were derived from these. GUS expression was stable over repeated culture in one type of callus which was initially light blue after staining; however, GUS expression was unstable with the other culture which initially stained intensely blue. The integration of the *gus* gene in the genome was demonstrated by Southern blot analysis. Following the screening procedure, only albino plants were regenerated.

Using PEG-mediated DNA uptake by protoplasts, transgenic plants of tall fescue were produced by introducing the *bar* gene as selectable marker on one plasmid and the *gus* gene as reporter on another plasmid, both genes are driven by the rice *act1* promoter (Kuai et al., 1999). Phosphinothricin at 1.5 to 3 $mg{\cdot}L^{-1}$ was used for selection. For all 12 plants randomly selected from 117 putative transgenic plants, none of them had GUS activity using the GUS histochemical assay. The *gus* gene was not detected in transgenic plants by polymerase chain reaction (PCR), which indicated that cotransformation was not successful in this experiment, whereas the presence of the *bar* gene was confirmed by Southern blot hybridization.

The stability of expression of a nonselected gene during vegetative propagation of protoplast-derived tall fescue (*Festuca arundinacea* Schreb.) was first studied by Bettany and colleagues (1998). The *gus* gene, under the control of the rice *act1* promoter, was introduced with the CaMV 35S-*hph* chimeric gene via PEG-mediated cotransformation of protoplasts. Selection was with 50 $mg{\cdot}L^{-1}$ hygromycin. Expression of GUS was unstable during the early generations of tillering but was more stable in the fourth or fifth tiller generation.

Lolium

Although a successful system for protoplast culture and plant regeneration was reported earlier, the first transgenic perennial ryegrass and Italian ryegrass plants were obtained by direct gene transfer into protoplasts only recently (Wang et al., 1997). The chimeric CaMV 35S-neomycin phosphotransferase II *(nptII)* and *nos-gus* genes were introduced into *Lolium* protoplasts via PEG-mediated transformation. Mild selection was used with 25 mg·L^{-1} kanamycin at an early stage of protoplast culture, and a stepwise selection from 25 to 40 mg·L^{-1} with G418 was applied later. The *gus* gene was confirmed to be integrated into the plant genome and transmitted to progeny of all of the transgenic plants tested.

Agrostis

Asano and Ugaki (1994) reported the first successful genetic transformation of *Agrostis*. Protoplasts of redtop (*Agrostis alba* L.) were isolated from an embryogenic suspension culture derived from callus which was initiated from seed. A CaMV 35S-*nptII* chimeric gene was introduced into protoplasts of redtop plants by electroporation. Among the antibiotics tested, G418 was chosen as selective agent due to its total inhibition of the growth of cell clones at a low concentration of 20 mg·L^{-1} G418 at 20 mg·L^{-1} was continuously included in the medium to select resistant calli and resistant plants from *A. alba* transformed cell clones. PCR analysis and Southern hybridization showed that all randomly chosen G418-resistant plants contained the *nptII* gene. Enzymatic analysis showed that the transgene was expressed in the transgenic plants.

An efficient plant regeneration system was established for protoplasts isolated from embryogenic suspension cultures of creeping bentgrass (*Agrostis palustris* Huds.) using a feeder cell method (Lee et al., 1996). All seven cultivars tested were regenerated using a simple system. The *bar* gene under the control of the CaMV 35S promoter was transformed into protoplasts of these creeping bentgrass cultivars by means of either electroporation or PEG treatment. Colonies resistant to 4 mg·L^{-1} bialaphos were obtained from five of the seven creeping bentgrass cultivars. Resistant calli of the cultivar Cobra regenerated into plants. All 153 'Cobra' transgenic plants analyzed survived 2 mg·L^{-1} of the bialaphos-containing herbicide, Herbiace. The stable integration of the *bar* gene into the genome and the presence of the *bar* transcript were demonstrated by Southern and Northern blot hybridization, respectively.

The first published field evaluation of herbicide resistance in creeping bentgrass and inheritance in the progeny was conducted in the United States (Lee et al., 1997). The herbicide-resistant transgenic creeping bentgrass lines were obtained using biolistics and protoplast transformation to introduce a chimeric *bar* gene (Hartman et al., 1994; Lee et al., 1996), and the transgenic plants were used in 1994 and 1995 field trials. In both trials, all plants that were completely resistant to herbicide treatment in the greenhouse survived 1× and 3× field rate herbicide applications (0.8 $kg \cdot ha^{-1}$ and 2.52 $kg \cdot ha^{-1}$). Seeds were produced from transgenic plants pollinated with seed-grown control plants. In most transgenic lines, the ratio of herbicide-resistant seedlings to herbicide-sensitive seedlings was about one to one, indicating Mendelian inheritance of herbicide resistance as a dominant locus.

Asano and colleagues (1997) reported the development of herbicide-resistant transgenic creeping bentgrass plants (*Agrostis stolonifera* 'Penncross') using a transformation method similar to that developed previously for protoplasts of redtop (Asano and Ugaki, 1994). Protoplasts were isolated from embryogenic suspension cells initiated from seed-derived callus of creeping bentgrass. Plasmids containing the *bar* gene under the control of either the CaMV 35S or enhanced CaMV 35S promoter were introduced into protoplasts. More than 500 resistant calli were selected using bialaphos at 2 to 5 $mg \cdot L^{-1}$, a high percentage of which were shown by PCR-Southern hybridization to contain the *bar* gene under the control of the enhanced CaMV 35S promoter. PCR and PCR-Southern hybridization demonstrated the presence of the *bar* gene in all test plants. Application of 1 percent Herbiace herbicide showed that all of the transgenic plants tested were herbicide resistant. The level of resistance varied between individuals. More of the plants transformed with the *bar* gene downstream from the enhanced CaMV 35S promoter showed stronger resistance.

Herbicide-resistant transgenic creeping bentgrass was obtained by electroporation using an alternative buffer (Asano et al., 1998). Modification of an electroporation buffer using $Ca(NO_3)_2$, instead of $CaCl_2$, and elevation of the pH (9-10), increased the transformation frequency about twofold. Bialaphos at a concentration of 2 mg·L^{-1} was used initially in selection, and the concentration was subsequently increased to 5 mg·L^{-1}. A total of 278 bialaphos-resistant cell colonies were obtained, and 50 percent of them regenerated. All of the randomly chosen herbicide-resistant transgenic plants showed the presence of the *bar* gene by molecular analysis, and herbicide resistance was demonstrated by spraying herbicide at a field rate of 0.5 to 1 percent.

Zoysia

The successful transformation of Japanese lawngrass was achieved by means of PEG-mediated direct gene transfer into protoplasts (Inokuma et al., 1997, 1998). The chimeric CaMV 35S-*pat* gene (1997) or maize alcohol dehydrogenase-1 *(Adh1)-gus* gene (1998) was cotransformed into protoplasts of Japanese lawngrass (*Zoysia japonica* Steud.) along with the CaMV 35S-*hph* gene. In both reports, a high concentration of hygromycin at 400 mg·L^{-1} was used in selection of resistant calli. Hundreds of transgenic plants were regenerated from resistant calli on regeneration medium without hygromycin. Southern blot and PCR analysis revealed that all plants tested contained one or more genes conferring hygromycin resistance, in addition to the *gus* gene (Inokuma et al., 1998). However, only half of them contained the *pat* gene (Inokuma et al., 1997). The *gus* gene regulated by the *adh1* promoter was expressed in both leaves and roots of transgenic Japanese lawngrass plants (Inokuma et al., 1998).

Dactylis

The very first transgenic turfgrass plants were obtained from protoplasts of orchardgrass (*Dactylis glomerata* L.) (Horn et al., 1988). Using heat shock followed by electroporation or PEG treatment, a chimeric CaMV 35S-*hph* gene was transferred into protoplasts isolated from embryogenic suspension cultures. Resistant cell colonies were selected in liquid medium containing 20 mg·L^{-1} hygromycin. Callus lines resistant to hygromycin were analyzed by Southern blot hybridization; of 15 callus lines tested, five of them showed the presence of the *hph* gene. Plants were subsequently regenerated from transformed calli. Some of them did not contain the *hph* gene. The authors suggested that the hygromycin-resistant calli may have been chimeric.

Biolistic Transformation

Biolistics is a process in which DNA carried on metal particles is delivered into targets by gunpowder, compressed gasses, or other means to accelerate particles at high velocity. It was invented by Sanford and colleagues at Cornell University (Sanford et al., 1987). Utilizing a macroprojectile and a stopping plate, tungsten coated with plasmid DNA was carried into targets by a mechanical impulse. Subsequently, the original gunpowder version of the particle gun went through several modifications. The gunpowder was replaced first by compressed helium gas and then by an acceleration system based on an airgun. Finally, it was developed further into the present gene delivery device, such as the Bio-Rad biolistic PDS-1000. Though this instrument can regulate the gas pressure, the depth of delivery of DNA-coated particles into the explants is still random (Christou, 1996).

Klein and colleagues (1987) first reported biolistic transformation of plant tissue where DNA was delivered into epidermal cells of onion. Following that first report, the method has evolved into a reliable process, and success has been achieved in numerous monocot species including corn, rice, barley, and wheat (Christou, 1992). Turfgrasses have also been transformed using biolistics. It has been used to transform embryogenic calli of creeping bentgrass (Zhong et al., 1993; Warkentin et al., 1998; Xiao and Ha, 1997), cell suspensions of creeping bentgrass (Hartman et al., 1994), embryogenic suspension cells (Spangenberg et al., 1995b) or highly regenerative tissue from embryogenic callus of tall fescue and red fescue (Cho et al., 2000), and embryogenic suspension cells of perennial ryegrass (Spangenberg et al., 1995a; Dalton et al., 1999) and Italian ryegrass (Ye et al., 1997; Dalton et al., 1999).

Festuca

Production of transgenic *Festuca* plants was first reported by Spangenberg and colleagues (1995b). Seed-derived embryogenic cell suspensions from *Festuca arundinacea* and *F. rubra* were used as targets. Different treatments of target cells and parameters for DNA particle delivery were evaluated by using transient expression assays of a chimeric *gusA* gene whose expression was driven by the CaMV 35S promoter. Various biolistic conditions were optimized including bombardment pressure and distance, baffle mesh size and baffle distance, the number of bombardments, and the concentration of plasmid DNA and microprojectiles. Osmotic treatment of the suspension cells prior to bombardment had a significant effect on increasing the number of transient GUS expression loci. Stable transformation of these two species was performed with a plasmid vector containing the *hph* gene driven by the CaMV 35S or rice *act1* 5' regulatory sequences. After bombardment, the transformed suspension cells were selected with hygromycin in two selection schemes using either 250 $mg \cdot L^{-1}$ hygromycin for tall fescue and 150 $mg \cdot L^{-1}$ hygromycin for red fescue in solid media or liquid selection medium containing 50 $mg \cdot L^{-1}$ hygromycin. The regeneration frequencies from antibiotic-resistant calli were 35 percent and 85 percent for tall fescue and red fescue, respectively. The integration and expression of the two genes in the transformed plants were demonstrated by Southern and Northern hybridization, respectively.

In another study (Cho et al., 2000), suspension cells were replaced with highly regenerative tissues of tall fescue and red fescue as targets. These highly regenerative tissues contained multiple, light green, shoot meristem-like structures, which developed from mature seed-derived callus under dim light. Stable transformation of tall fescue was obtained by bombardment with three plasmids containing the *hph, bar,* and *gus* genes, and red fescue with three plasmids containing *hph, gus,* and a synthetic gene encoding green florescent protein *(sgfp).* Prior to bombardment, tissues were treated with osmoticum (0.2 M mannitol and 0.2 M sorbitol). Gold particles, coated with a mixture of the three plasmids at a molar ratio of 1:1:1 or a similar ratio for red fescue, were used for bombardment with a Bio-Rad PDS-1000 He biolistic device. After bombardment, tissues were selected on hygromycin in a stepwise way (30 $mg \cdot L^{-1}$ to 50 $mg \cdot L^{-1}$ to 100 $mg \cdot L^{-1}$). The regeneration frequency from hygromycin-resistant calli was 6.8 percent for tall fescue and 4.5 percent for red fescue. The transgenic nature of the turfgrass was demonstrated by PCR and

Southern blot hybridization. The coexpression frequency of all three transgenes in transgenic tall fescue *(hph/bar/gus)* and red fescue *(hph/gus/sgfp)* was 25 to 27 percent.

Altpeter and Xu (2000) first reported transformation of turf-type *Festuca rubra* L. with the *nptII* gene, where expression was controlled by the maize ubiquitin *(ubi1)* promoter and first intron. They showed that 100 mg·L^{-1} paramomycin gave tight selection, and no escapes were found. Of the bombarded calli 4 to 8 percent were resistant to paramomycin. Eight independent transgenic plants were regenerated from 100 bombarded calli in two independent experiments. Transgene integration was variable from line to line, as confirmed by Southern blot analysis.

Lolium

Biolistic transformation of *Lolium* was first reported in 1993 (Hensgens et al., 1993; Perez-Vicente et al., 1993); other papers soon followed (van der Maas et al., 1994; Spangenberg et al., 1995a; Ye et al., 1997; Dalton et al., 1999; Ye et al., 2001).

In early works, no transgenic *Lolium* plants were obtained. Hensgens and colleagues (1993) studied transient expression of GUS driven by the 5' regulatory sequence of the constitutively expressed rice *gos2* gene or *gos5* gene, CaMV 35S promoter, and the mannopine synthase *(mas)* promoter. These plasmids were delivered into seedlings and nonmorphogenic suspension cells of *Lolium perenne* by biolistics. The chimeric *gos2-gus* and CaMV 35S-*gus* were more effective in expressing GUS. The stable transformation of the *gos2-gus* gene was obtained in a callus line of *L. perenne* in which a CaMV 35S-driven *hph* gene was used as the selectable marker (van der Maas et al., 1994). The stable integration of the *gusA* and *hph* genes in the genome of this callus line was demonstrated by Southern hybridization, but no transgenic plants were regenerated. The direct delivery of DNA particles carrying a chimeric *act1-gus* gene into floral and vegetative apices of *Lolium multiflorum* and *L. perenne* was reported; transient expression of GUS was detected in the first (L1) and second (L2) cell layers (Perez-Vicente et al., 1993).

Transgenic *Lolium* plants were obtained for the first time by biolistic transformation from single genotype-derived embryogenic suspension cells of *L. perenne* (Spangenberg et al., 1995a) and later of *L. multiflorum* (Ye et al., 1997). In these studies, parameters for the bombardment of embryogenic suspension cells were evaluated by transient expression assays of a chimeric *gusA* gene driven by either the CaMV 35S promoter (Spangenberg et al., 1995a) or the maize *ubi1* promoter (Ye et al., 1997). These parameters included different pressures and distances for bombardment, baffle mesh sizes, particles per bombardment, and DNA concentration per bombardment. The optimized conditions used for the stable transformation of an *act1-hph* gene were similar to those described for biolistic transformation of *Festuca* (Spangenberg et al., 1995b). Stably transformed clones were recovered with a selection scheme involving a stepwise increase in hygromycin concentration in liquid medium (from 25 to 50 $mg \cdot L^{-1}$ for *L. multiflorum* and from 50 to 200 $mg \cdot L^{-1}$ for *L. perenne*) followed by plate selection (200 $mg \cdot L^{-1}$ hygromycin for *L. perenne* and 100 $mg \cdot L^{-1}$ for *L. multiflorum*). Plants were regenerated from 23 percent of the hygromycin-resistant calli of *L. perenne* (Spangenberg et al., 1995a) and 33 percent of the hygromycin-resistant calli of *L. multiflorum* (Ye et al., 1997). The transgenic nature of regenerated plants was demonstrated in both *Lolium* species by Southern hybridization; both single and multiple copies of the *hph* gene in the plant genome were observed. Expression of the transgene in both species of transgenic plants was confirmed by northern hybridization and a hygromycin phosphotransferase enzyme assay.

Dalton and colleagues (1999) later described the cotransformation of *Lolium perenne* and *Lolium multiflorum* using two separate plasmids: one containing the *hph* gene and the other containing the *gus* gene. In their studies, transformation was performed according to the method of Spangenberg and colleagues (1995a). Cell suspensions from shoot tip-derived embryogenic callus were cobombarded with one plasmid containing the *hph* gene under the control of the CaMV 35S promoter and another plasmid containing the *gus* gene downstream from either the truncated rice *act1* promoter and its first intron or the maize *ubi1* promoter and its first intron. Resistant calli were selected using a liquid medium containing hygromycin, and the concentration of hygromycin was increased from 50 to 75 $mg \cdot L^{-1}$ with each round of selection. Southern hybridization showed that the nonselected *(gus)* gene was present at a higher copy number than the *hph* gene. Histochemical staining of leaf tissues showed that the co-expression frequency varied from 37 to 50 percent.

A chimeric bacterial levansucrase *sacB* gene was introduced into *Lolium multiflorum* via biolistics in order to study the regulation and role of fructan metabolism, and the physiological consequence of its manipulation in native fructan-accumulating plants (Ye et al., 2001). A *Bacillus subtilis* levansucrase *sacB* gene driven by maize *ubi1* promoter was transferred into embryogenic suspension cells of Italian ryegrass by biolistics. A low level of steady-state *sacB* transcripts in 16 out of 42 of the transgenic Italian ryegrass plants was demonstrated by RT-PCR. Three of these 16 *sacB*-positive transgenic plants were found to accumulate a low level of bacterial levan-type fructan. Growth of *sacB*-transgenic ryegrass slowed down with the onset of the reproductive state, and flowering plants had an abnormal phenotype.

Agrostis

Transgenic *Agrostis palustris* plants were first obtained from embryogenic callus bombarded by tungsten particles coated with a chimeric *act1-gus* gene using the PDS 1000/He biolistic device (Zhong et al., 1993). To evaluate the transient expression of the *gus* gene, one-half of the bombarded embryogenic calli from each sample was assayed for GUS activity. Data were collected from 10 out of the 19 bombarded callus lines. Among the calli tested, 15 percent of them showed GUS activity. Of 15 putative transgenic plants chosen randomly from 500 regenerated plants, integration of the *gus* gene was demonstrated in four plants by Southern hybridization. However, only one of the four expressed a high level of GUS. GUS staining was visible in all tested plant organs but with different intensities. Though the rapid photochemical GUS assay could be used for isolating transformants in the absence of genes encoding antibiotic or herbicide resistance as a selectable marker, the frequency of transformation was significantly lower (Zhong et al., 1993).

Hartman and colleagues (1994) reported the reproducible generation of transgenic creeping bentgrass using the PDS 1000/He biolistic device. A chimeric *bar* gene under the control of a CaMV 35S promoter was delivered on gold particles instead of tungsten to reduce toxicity. Transgenic plants were recovered from embryogenic suspension cultures, which were subjected to selection on either plate or liquid medium containing 2 to 4 mg·L^{-1} bialaphos. Of 900 plantlets recovered from the transformation, 55 survived the higher spray rate (2 mg·mL^{-1}) of the bialaphos-containing herbicide, Herbiace. The transformation frequency in different experiments ranged from 0 to 13.7 percent. The presence of the *bar* gene in bialaphos-resistant plants was confirmed by PCR and Southern blots, and expression of the transgene was further demonstrated by Northern blot hybridization.

Following the procedure described by Zhong and colleagues (1993), Warkentin and colleagues (1997) tried to produce transgenic creeping bentgrass with enhanced resistance to fungal disease using particle bombardment. Two plasmids carrying either a chimeric CaMV 35S-*bar* gene or chimeric CaMV 35S-chitinase gene were cotransformed into embryogenic calli of creeping bentgrass. After bombardment, calli were selected on a medium containing increasing concentrations of bialaphos (from 3 to 10 mg·L^{-1}). Many putative transformants were recovered which survived two spray applications of 1 percent Ignite containing 180 mg·L^{-1} phosphinothricin. However, no molecular data were available to confirm gene integration into the plant genome; nor was any bioassay reported for fungal disease resistance. Xiao and Ha (1997) established an efficient transformation system for creeping bentgrass by particle bombardment using gold particles coated with plasmid DNA containing a reporter *gus* gene and a selectable marker *hph* gene. Embryogenic calli or embryogenic suspension cells of creeping bentgrass were bombarded using the PDS 1000/He biolistic device. Efficient selection of transformants using 200 mg·L^{-1} hygromycin resulted in an average of 4.6 resistant colonies per bombardment. Thirteen putative transgenic plants were regenerated from a total of 124 resistant colonies; of these, 11 were tested and showed the presence of the *gus* and *hph* genes by PCR and Southern hybridization analysis, indicating that selection by hygromycin was highly effective.

Poa

The first reported success in producing transgenic plants of Kentucky bluegrass was achieved using highly regenerative tissues derived from mature seed-derived callus via biolistics (Ha et al., 2001).

Similar systems were established in tall fescue and red fescue (Cho et al., 2000) and orchardgrass (Cho et al., 2001). These tissues contained multiple, light green, shoot meristem-like structures, and their regenerability lasted for 26 months without occurrence of albino plants. These tissues were established under dim light, using a medium supplemented with 2,4-D (4.5 to 9.0 μM), 6BAP (0.44 or 2.2 μM), and cupric sulfate (5.0 μM). They were cotransformed with three plasmids containing the *hph* gene under the control of rice *act1* promoter, or *gus* and *gfp* genes under the control of the maize *ubi1* promoter. Of 463 pieces of green tissue bombarded, ten independent transgenic lines (2.2 percent) were obtained, of which 70 percent were regenerable. During a three- to four-month selection period, a high concentration of hygromycin (100 $mg \cdot L^{-1}$) was used in order to eliminate the early growth of nontransgenic cells; 30 $mg \cdot L^{-1}$ was used during the later stages of selection. Stable integration of the transgenes was confirmed in transgenic plants by PCR and Southern hybridization. Transformation frequency for all three genes *(hph/gus/gfp)* was 20 percent, and for two transgenes, either *hph/gus* or *hph/gfp,* was 30 to 40 percent.

The gene encoding betaine aldehyde dehydrogenase (BADH) was cotransformed via biolistics with *hph* into seed-derived callus of Kentucky bluegrass for improving salt and drought tolerance. Of 81 hygromycin-resistant callus lines, 65 transgenic plants were regenerated. A PCR assay showed that 45 of them contained the gene encoding BADH. Three of the transgenic lines were reported to be more tolerant to saline conditions and drought in greenhouse studies (Meyer et al., 2000).

Dactylis

Since direct embryogenesis from mesophyll cells in cultured leaf segments of orchardgrass was reported (Conger et al., 1983), the leaf culture system in this species became more attractive for genetic transformation. Young leaf tissue of orchardgrass has been used in biolistic experiments to obtain transgenic plants transformed with both the *bar* and *gusA* genes (Denchev et al., 1997). Bombardment parameters were optimized using transient expression of GUS. It showed that 48 to 96 hour preculture time and 4 hour osmotic treatment before bombardment increased transformation rate greatly. Bombarded leaf segments were cultured on SH medium for three weeks, and somatic embryos were developed from them. These embryos were subjected to 3 $mg{\cdot}L^{-1}$ bialaphos. Of 67 plants tested for Basta tolerence, ten showed complete tolerance to Basta, and six showed a localized response. GUS histochemical staining was observed in embryogenic tissue cultures cultured from leaf segments of T_0 transgenic plants, but not in the T_0 leaf tissue itself. The presence of the transgenes in T_0 was confirmed by PCR and Southern hybridization. Three of the plants that were resistant to Basta were tested by Southern hybridization, and all contained the *bar* sequence. Of six plants analyzed by PCR for the presence of the *bar* and *gus* genes, three of them showed the presence of both genes, two of them only showed the presence of *gus* gene, and one plant showed neither.

An efficient and reproducible transformation system for orchardgrass was established using the same kind of tissue as described earlier for *Festuca* and Kentucky bluegrass (Cho et al., 2001). These tissues were developed from embryogenic callus under dim light which could be maintained in culture for two years with minimal loss in regenerability. These tissues were bombarded with a mixture of three plasmids containing the *act1-hph, ubi1-bar,* and *ubi1-gus* chimeric genes. A stepwise concentration of hygromycin was used for selection (30 to 50 $mg{\cdot}L^{-1}$). Eleven independent hygromycin-resistant lines were obtained from 147 individual explants bombarded, giving a 7.5 percent transformation frequency. Of 11 independent lines, ten of them were regenerable. The transgenic nature of regenerated plants was determined by PCR and Southern hybridization. Coexpression frequency of all three genes *(hph/bar/gus)* was 20 percent, and two of the three transgenes (either *hph/bar* or *hph/gusA*) were 45 percent and 60 percent, respectively. A high frequency of change in ploidy level (70 percent) was observed in transformed T_0 orchardgrasses, while none of the nontransformed plants obtained from tissue culture or seed had abnormal ploidy. Furthermore, phenotypic variation was observed in transgenic plants, which correlated with changes in ploidy level. The authors speculated that the increased stress of the transformation process, in addition to the in vitro culture step, contributed to chromosomal abnormalities.

Switchgrass

The first report of genetic transformation in switchgrass came from the work by Richard and colleagues (2001). A dual marker plasmid containing the *gfp* gene under the control of the rice *act1* promoter and the *bar* gene under the control of the maize *ubi1* promoter was introduced via biolistics into embryogenic callus derived from immature inflorescence of switchgrass. A two-step selection procedure of initial culture on 5 mg·L^{-1} bialaphos, followed by culture on 10 mg·L^{-1} bialaphos, was used for selection of transformants. A total of 97 Basta-tolerant plants were obtained. GFP expression was observed in leaf tissue and pollen of transgenic plants. Southern blot hybridization confirmed the presence of both genes. Some of the T_1 progeny resulting from a cross between transgenic and control plants were tolerant to Basta, indicating inheritance of the *bar* gene. However, a high frequency of nontolerant to tolerant T_1 plants was observed, which the author suggested resulted either from gene silencing or less than 50 percent of the pollen participating in pollination possessed the *bar* gene.

Agrobacterium-*Mediated Transformation*

Agrobacterium tumefaciens, a soil bacterium, is a plant pathogen of susceptible plants, causing a disease called crown gall. Since its role in tumor development of plants was revealed, the detailed process and mechanism of bacterial infection and transfer of bacterial T-DNA into the plant have been studied extensively, which led to its utilization in plant transformation since the 1980s. The successful transformation of many dicots and several important crop species demonstrates its potential as a convenient, efficient method for foreign gene transfer.

Agrobacterium-mediated transformation has been well established in dicotyledonous plants (Smith and Hood, 1995). The advantages of the method over other transformation methods include high efficiency of transformation, one or few copies of transferred DNA with defined ends, the transfer of large segments of DNA, and the absence of a protoplast-culture step (Hiei et al., 1994). However, transformation of monocots via *Agrobacterium* is relatively difficult because monocot species are not natural hosts for *Agrobacterium tumefaciens*. As a consequence, alternative techniques were developed for monocots, including protoplast-mediated transformation and biolistics. In the meantime, many investigations continued on the subject, attempting to manipulate *Agrobacteria* to genetically transform monocots.

Early studies of *Agrobacterium*-mediated transformation in monocots were controversial (Raineri et al., 1990; Mooney et al., 1991). Raineri and colleagues (1990) obtained transformed japonica rice cells by cocultivation of mature embryos with *Agrobacteria*. Cells transformed with the *nptII* gene and the *gus* gene were selected on a medium containing kanamycin (200 $mg{\cdot}L^{-1}$). Southern hybridization showed the integration of T-DNA into the plant genome, but no transgenic plants were regenerated. *Agrobacterium*-mediated transformation of wheat was demonstrated at very low transformation frequency (Mooney et al., 1991). Pretreatment of embryos with partial enzyme digestion was essential for successful transfer. They detected the presence of T-DNA carrying the *nptII* gene in callus tissue by Southern hybridization, but they also could not regenerate plants from transformed callus. All these studies lack unequivocal evidence for stable transformation of monocots with *Agrobacterium* (Potrykus, 1990).

The controversy seemed to be resolved as a result of the studies by Chan and colleagues (1993) and Hiei and colleagues (1994). In 1993, Chan and colleagues obtained a few transgenic japonica rice plants by inoculation of immature embryos with *Agrobacterium tumefaciens*. They clearly proved the integration of foreign genes *nptII* and *gus* into the rice genome and inheritance of the transferred DNA to progeny by Southern hybridization, although the progeny of only a single transformed plant was analyzed. For the first time in monocots, it was clearly demonstrated that the DNA was integrated and stably inherited in the transgenic plant.

Soon thereafter, the most important and convincing demonstration of *Agrobacterium*-mediated transformation of monocots was made by Hiei and colleagues (1994). A large number of morphologically normal, fertile transgenic japonica rice plants were obtained after cocultivation of rice tissue with a strain of *A. tumefaciens* which carried a supervirulent vector having *virB, virC,* and *virG* genes from plasmid pTiBo542. Among various tissues, calli induced from scutella were the best starter material. Transformation frequency of rice was as high as that of transformation of dicots with this *Agrobacterium* strain. Stable integration of T-DNA into the rice genome was confirmed by Southern hybridization and sequencing of the T-DNA-genome junction. Clear Mendelian segregation of hygromycin resistance and GUS expression was also observed in the T_1 and T_2 progeny.

Since then, successful transformation has been achieved in japonica (Aldemita and Hodges, 1996) and javanica (Dong et al., 1996) rice cultivars, as well as the relatively recalcitrant indica cultivar (Aldemita and Hodges, 1996; Khanna and Raina, 1999). Aldemita and Hodges (1996) showed that immature embryos were also good starter materials for *Agrobacterium*-mediated transformation. Immature embryos were inoculated with an *Agrobacterium* strain containing either an ordinary vector or supervirulent vector based on the same basic protocol by Hiei and colleagues (1994); inoculation of immature embryos with only the supervirulent strain resulted in efficient transformation. Studies by both Hiei and colleagues (1994) and Aldemita and Hodges (1996) showed that at least four factors contributed to the success in the high-efficiency induction of transformants either from scutellar calli of mature embryos or from immature embryos: the choice of explants; the use of supervirulent binary vectors; the maintenance of acetosyringone during cocultivation; and the use of a selectable agent which did not impair the regenerability of callus.

Since these initial studies with rice, the efficient *Agrobacterium*-initiated transformation of several other agronomically important monocots has been demonstrated. These include maize (Ishida et al., 1996), wheat (Cheng et al., 1997), barley (Tingay et al., 1997), sugarcane (Enriquez-Obregon, 1998), and sorghum (Zhao et al., 2000). These successes indicate that the monocot nature of plants should not prevent the application of *Agrobacterium*-mediated transformation to other monocot species. By optimizing methodological parameters, gene transfer mediated by *Agrobacterium* should be possible for other recalcitrant monocots (Hiei et al., 1997).

Few reports are available on *Agrobacterium*-mediated transformation of turfgrass. In our laboratory at Rutgers University we have developed an efficient *Agrobacterium*-mediated transformation protocol for several turfgrass species including creeping bentgrass (Lakkaraju et al., 1999), tall fescue, and velvet bentgrass (Pitcher et al., 2000; Lakkaraju et al., 2001). Transformation with the *hph* gene and the *gus* gene has been achieved. Calli initiated from mature seeds of varied turfgrass species were cocultivated with a supervirulent strain of *A. tumefaciens,* which contains the *hph* gene under the control of rice *act1* promoter, and the *gus* gene under the control of the maize *ubi1* promoter within the T-DNA borders of its intermediate vector. Transformed calli were selected on a medium containing 200 $mg \cdot L^{-1}$ hygromycin. Transgenic creeping bentgrass plants were regenerated, whose transgenic nature was demonstrated by Southern hybridization (Lakkaraju et al., 1999). The presence of the *hph* and *gus* genes in transformed calli of tall fescue and velvet bentgrass was demonstrated by Southern hybridization, but initially only albino regenerants were recovered. Of 17 transformed calli tested, half of them showed one or two integration sites, and the remainder had several (Pitcher et al., 2000). After modification of culture conditions in subsequent experiments, green transgenic tall fescue and velvet bentgrass plants were regenerated.

Yu, Skinner, and colleagues (2000) used *Agrobacterium* to transform calli obtained from mature seeds of creeping bentgrass with the *sgfp* gene encoding a synthetic green fluorescent protein. They did not use a selectable marker but relied on the green florescence produced by putative transformants to identify both the transformed calli and the plants that subsequently regenerated from these calli. They demonstrated integration of the transgene by Southern hybridization and the relative efficiency of the transformation procedure.

DEVELOPMENT OF VALUE-ADDED TRANSGENIC TURFGRASS

The methodology for genetic manipulation of the key turfgrass species has been established, as described previously. Reproducible and efficient regeneration systems are available for most widely used grass species, which allows for recovery of fertile and genetically stable plants. Efficient transformation systems have been developed to transfer foreign genes into various turfgrass species. They include direct gene transfer to protoplasts, as well as *Agrobacterium*-mediated and biolistic-mediated transformation. These techniques are available to create new turfgrass cultivars with value-added traits (Spangenberg et al., 1998; Chai and Sticklen, 1998). Some of the desirable value-added traits for turfgrass are as follows.

1. *Enhanced resistance to pathogens and pests through regulated expression of antifungal and pesticide proteins.* A number of genes have been identified and strategies developed for engineering disease and pest resistance in plants. These include chitinase, glucanase, enzymes involved in phytoalexin synthesis, ribosome-inactivating proteins, and Bt toxin. Transgenic turf fescue expressing a chimeric endochitinase gene has been developed by biolistic transformation, and an increased level of chitinase activity was observed which may confer resistance to fungal disease (Spangenberg et al., 1998). The chitinase gene has also been transferred into creeping bentgrass (Belanger et al., 1996; Chai et al., 2002; Yu, Huang, et al., 2000). Brown patch resistance in transgenic creeping bentgrass engineered with an elm chitinase was demonstrated in two out of five independent transgenic lines in laboratory and greenhouse inoculation studies (Chai et al., 2002). Qu and colleagues (2001) reported that tall fescue engineered to express both glucanase and chitinase were resistant to brown patch. A number of other genes with disease resistance potential have been introduced into creeping bentgrass via biolistics (Belanger et al., 1996). These include genes encoding bacterio-opsin (Mittler et al., 1995), pokeweed antiviral protein (Zoubenko et al., 1997; Dai et al., 2003), and glucose oxidase (Wu et al., 1995). Belanger and colleagues (personal communication) have introduced the *Arabidopsis thaliana* gene encoding the PR5K receptor protein kinase (Wang et al., 1996) into creeping bentgrass by particle bombardment. In field tests, three of the transgenic lines showed delays in dollar spot disease expression of 36 to 45 days, relative to the control plants. Recently, Fu and colleagues (2001) used *Agrobacterium* to mediate transformation of creeping bentgrass with the rice *tlp* gene encoding the thaumatin-like protein (PR-5) and the *bar* gene. Calli were selected on bialaphos, and regenerated seedlings were shown to be tolerant to painting with glufosinate-ammonium. Resistance to fungal pathogens is yet to be evaluated.

2. *Improved tolerance to abiotic stress such as salt, drought, heat, and shade.* Salt and drought tolerance in Kentucky bluegrass and

bermudagrass have been enhanced by introducing the gene encoding betaine aldehyde dehydrogenase (BADH) via biolistic cotransformation with the *hph* gene encoding hygromycin resistance. After selection of transgenic plants with hygromycin, several plants expressing BADH were shown to have significantly higher tolerance to drought and salt stress than control plants under greenhouse conditions (Meyer et al., 2000; Zhang et al., 2001).

3. *Manipulation of protein content.* Pollen allergens from grasses contribute 25 percent of the allergies experienced by patients in cool, temperate climates (Tamborini et al., 1995). Transgenic *L. perenne* has been engineered to express an antisense pollen allergen cDNA under the control of a pollen-specific promoter for the purpose of reducing the allergen level (Spangenberg et al., 1998). In an attempt to improve the protein quality of forage grass for ruminant nutrition, Wang and colleagues (2001) produced transgenic tall fescue plants that expressed sunflower seed albumin 8 (SFA8) protein which is rich in sulfur-containing amino acids. The SFA8 protein accumulated up to 0.2 percent of the total soluble protein when expressed under the control of the CaMV 35S promoter. Use of monocot promoters that are stronger will be necessary to achieve higher expression levels of the *sfa8* gene that are nutritionally useful. Nonetheless, this work is the first example of production of transgenic forage grasses which express genes of agronomic interest.

4. *Manipulation of growth and development.* There is interest in producing turfgrasses with low maintenance requirements, for example a dwarf phenotype requiring less frequent mowing. Genes that might be manipulated include those that affect the synthesis or steady-state levels of certain plant hormones, such as gibberellins or brassinosteroids. Candidate genes include the *GA5* gene which encodes gibberellin 20-oxidase (Xu et al., 1995) and *BAS1* which encodes a cytochrome P450 (CYP7281) that inactivates the steroid hormone, brassinolide, by hydroxylation (Neff et al., 1999). Also receiving attention is the production of male sterile plants because male-sterile transgenic turfgrass may minimize or prevent pollen dispersal from transgenic turfgrass plants, thus reducing the potential risk to the environment (Wipff and Fricker, 2000).

5. *Production of herbicide-tolerant turfgrasses.* Several turfgrass species with resistance to phosphinothricin herbicides have been ob-

tained by introducing the *bar* gene into turfgrasses. These include creeping bentgrass (Hartman et al., 1994; Lee et al., 1996; Asano et al., 1997), tall fescue (Wang et al., 1992; Kuai et al., 1999), red fescue (Cho et al., 2000), and orchardgrass (Denchev et al., 1997; Cho et al., 2001). More recently, glyphosate-resistant creeping bentgrass has been produced by introduction of the *cp4 epsps* gene which encodes an altered 5-enolpyruvylshikimate-3-phosphate synthase. This gene has been used previously in producing several Roundup Ready agronomic crops (Zhou et al., 1995). At glyphosate application rates sufficient to provide complete control of susceptible bentgrasses, no injury or retardation of growth was observed in glyphosate-resistant lines (Hart et al., 2002).

FUTURE PROSPECTS

Prior to commercial release of transgenic turfgrass, certain questions must be addressed, the most important of which is the likelihood of transgene spread to related species. Few studies have been conducted in turfgrasses, and those have concerned *Agrostis* spp. Wipff and Fricker (2000) reported that herbicide-resistant interspecific hybrids were obtained between transgenic glufosinate-resistant creeping bentgrass *(Agrostis palustris)* and *A. canina, A. capillaries, A. castellana, A. gigantea,* and *A. pallens* when accessions of these species were placed in the transgenic nursery prior to flowering. No information, however, was reported on the frequency of interspecific hybridization.

Belanger and colleagues (2003) have also addressed the issue of transgene spread to related species by undertaking a field study to evaluate the frequency of interspecific hybridization between creeping bentgrass and four related species. The species used were *A. canina* L. ssp. *canina* (velvet bentgrass), *A. tenuis* (colonial bentgrass), *A. gigantea* (redtop), and *A. castellana* (dryland bentgrass). In this study transgenic creeping bentgrass expressing the *bar* gene which confers glufosinate resistance were used as the pollen source. Two fields were established with the transgenic plants in the center and individuals of each related species placed around the center in all directions at 3 m increments. There were a total of 90 sample points in each field, and the maximum distance from the center was 15 m. At each sample point was one plant of each of the four related species. The plants were vernalized and allowed to flower. Seed was harvested from each of the plants and the progeny screened for glufosinate resistance to identify any hybrids.

Interspecific hybrids were recovered with *A. tenuis* and *A. castellana* at overall frequencies of 0.04 percent and 0.001 percent, respectively. In both cases the percentage of hybrids recovered was greatest at the 3 m distance from the center and dropped dramatically at greater distances. No hybrids were recovered with *A. gigantea* or *A. canina*.

The conditions of this field study were similar to what would be encountered in nature. The plants were flowering at their normal times and competing pollen was available from other plants of the same species. Under these conditions the frequency of interspecific hybridization was found to be low.

Additional work must be done to resolve the question of transgene escape. However, as with most new technologies, potential risks must be weighed relative to expected benefits. These include lower costs and improved efficiency in operations and the environmental benefit forthcoming from use of smaller amounts of chemicals as pesticides, herbicides, and possibly also fertilizers.

REFERENCES

Ahn, B.J., F.H. Huang, and J.W. King (1985). Plant regeneration through somatic embryogenesis in common bluegrass tissue culture. *Crop Science* 25:1107-1109.

Aldemita, R.R. and T.K. Hodges (1996). *Agrobacterium tumefaciens*-mediated transformation of *japonica* and *indica* rice varieties. *Planta* 199:612-617.

Aldous, D.E. (1999). Introduction in turfgrass science and management. *International Turf Management* (pp.1-18), Ed. Aldous, D.E. Boca Raton, FL: CRC Press.

Altpeter, F. and J. Xu (2000). Rapid production of transgenic turfgrass (*Festuca rubra* L.) plants. *Journal of Plant Physiology* 157:441-448.

Artunduaga, I.R., C.M. Taliaferro, and B.B. Johnson (1988). Effects of auxin concentrations on embryogenic callus induction of cultured young inflorescence of old world blue stems (*Bothriochloa* spp.) and bermudagrass (*Cynodon* spp.). *Plant Cell, Tissue, and Organ Culture* 12:13-19.

Asano, Y., Y. Ito, M. Fukami, M. Morifuji, and A. Fujie (1997). Production of herbicide resistant, transgenic creeping bent plants. *International Turfgrass Society Research Journal* 8:261-266.

Asano, Y., Y. Ito, M. Fukami, K. Sugiura, and A. Fujie (1998). Herbicide-resistant transgenic creeping bentgrass plants obtained by electroporation using an altered buffer. *Plant Cell Reports* 17:963-967.

Asano, Y. and M. Ugaki (1994). Transgenic plants of *Agrostis alba* obtained by electroporation-mediated direct gene transfer into protoplasts. *Plant Cell Reports* 13:243-246.

Bai, Y. and R. Qu (2000a). An evaluation of callus induction and plant regeneration in twenty-five turf-type tall fescue (*Festuca arundinacea* Scherb.) cultivars. *Grass Forage Science* 55:326-330.

Bai, Y. and R. Qu (2000b). Factors influencing tissue culture responses of mature seeds and immature embryos in turf-type tall fescue (*Festuca arundinacea* Scherb.). *Plant Breeding* 120:239-242.

Beard, J.B. (1973). *Turfgrass: Science and Culture*. Englewood Cliffs, NJ: Prentice-Hall.

Belanger, F., C. Laramore, and P. Day (1996). Turfgrass biotechnology. *Rutgers Turfgrass Proceedings* 28:1-3.

Belanger, F.C., T.R. Meagher, P.R. Day, K. Plumley, and W.A. Meyer (2003). Interspecific hybridization between *Agrostis stolonifera* and related *Agrostis* species under field conditions. *Crop Science* 43:240-246.

Bettany, A.J.E., S.J. Dalton, E. Timms, and P. Morris (1998). Stability of transgene expression during vegetative propagation of protoplast-derived tall fescue (*Festuca arundinacea* Scherb.) plants. *Journal of Experimental Botany* 49:1797-1804.

Birch, R.G. (1997). Plant transformation: Problems and strategies for practical application. *Annual Review of Plant Physiology and Plant Molecular Biology* 48:297-326.

Chai, B., S.B. Maqbool, R.K. Hajela, D. Green, J.M. Vargas, D. Warkentin, R. Sabzihar, and M.B. Sticklen (2002). Cloning of chitinase-like cDNA (hs2), its transfer to creeping bentgrass (*Agrostis palustris* Huds.) and development of brown patch *(Rhizoctonia solani)* disease resistant transgenic lines. *Plant Science* 163:183-193.

Chai, B. and M.B. Sticklen (1998). Application of biotechnology in turfgrass genetic improvement. *Crop Science* 38:1320-1338.

Chai, M.L. and Y.M. Niu (1997). Variant plantlets regenerated from callus differentiation of *Festuca arundinacea* cv. Tempo. *International Turfgrass Society Research Journal* 8:279-281.

Chan, M.-T., H.-H. Chang, S.-L. Ho, W.-F. Tong, and S.M. Yu (1993). *Agrobacterium*-mediated production of transgenic rice plants expressing a chimeric α-amylase promoter/β-glucuronidase gene. *Plant Molecular Biology* 22:491-506.

Chaudhury, A. and R. Qu (2000). Somatic embryogenesis and plant regeneration of turf-type bermudagrass: Effect of 6-benzyladenine in callus induction medium. *Plant Cell Tissue and Organ Culture* 60:113-120.

Cheng, M., J.E. Fry, S. Pang, H. Zhou, C.M. Hironaka, D.R. Duncan, T.W. Conner, and Y. Wan (1997). Genetic transformation of wheat mediated by *Agrobacterium tumefaciens*. *Plant Physiology* 115:971-980.

Cho, M.-J., H.W. Choi, and P.G. Lemaux (2001). Transformed T_0 orchardgrass (*Dactylis glomerata* L.) plants produced from highly regenerative tissues derived from mature seeds. *Plant Cell Reports* 20:28-33.

Cho, M.-J., C.D. Ha, and P.G. Lemaux (2000). Production of transgenic tall fescue and red fescue plants by particle bombardment of mature seed-derived highly regenerative tissues. *Plant Cell Reports* 19:1084-1089.

Cho, M.-J., W. Jiang, and P.G. Lemaux (1998). Transformation of recalcitrant barley cultivars through improvement of regenerability and decreased albinism. *Plant Science* 138:229-244.

Cho, M.-J., W. Jiang, and P.G. Lemaux (1999). High-frequency transformation of oat via microinjectile bombardment of seed-derived highly regenerative cultures. *Plant Science* 148:9-17.

Christou, P. (1992). Genetic transformation of crop plants using microprojectile bombardment. *Plant Journal* 2:275-281.

Christou, P. (1996). Electric discharge particle acceleration technology for the creation of transgenic plants with altered characteristics. *Field Crops Research* 45:143-151.

Conger, B.V., G.E. Hanning, D.J. Gray, and R.E. McDaniel (1983). Direct embryogenesis from mesophyll cells of orchardgrass. *Science* 221:850-851.

Creemers-Molenaar, T., J.P.M. Loeffe, and M.A.C.M. Zaal (1988). Isolation, culture and regeneration of *Lolium perenne* and *Lolium multiflorum* protoplasts. *Current Plant Science and Biotechnology in Agriculture* 7:53-54.

Dai, W.D., S. Bonos, Z. Guo, W.A. Meyer, P.R. Day, and F.C. Belanger (2003). Expression of pokeweed antiviral proteins in creeping bentgrass. *Plant Cell Reports* 21:497-502.

Dalton, S.J. (1988a). Plant regeneration from cell suspension protoplasts of *Festuca arundinacea* Schreb., *Lolium perenne* L., and *L. multiflorum* Lam. *Plant Cell Tissue and Organ Culture* 12:137-140.

Dalton, S.J. (1988b). Plant regeneration from cell suspension protoplasts of *Festuca arundinacea* Schreb. (tall fescue) and *Lolium perenne* L. (perennial ryegrass). *Journal of Plant Physiology* 132:170-175.

Dalton, S.J., A.J.E. Bettany, E. Timms, and P. Morris (1995). The effects of selection pressure on transformation frequency and copy number in transgenic plants of tall fescue (*Festuca arundinecea* Schreb.) *Plant Science* 108:63-70.

Dalton, S.J., A.J.E. Bettany, E. Timms, and P. Morris (1998). Transgenic plants of *Lolium perenne, Lolium multiflorium, Festuca arundinacea,* and *Agrostis stolonifera* by silicon carbide fiber-mediated transformation of cell suspension cultures. *Plant Science* 132:31-42.

Dalton, S.J., A.J.E. Bettany, E. Timms, and P. Morris (1999). Co-transformed, diploid *Lolium perenne* (perennial ryegrass), *Lolium multiflorium* (Italian ryegrass) and *Lolium temulentum* (darnel) plants produced by microprojectile bombardment. *Plant Cell Reports* 18:721-726.

Denchev, P.D., D.D. Songstad, J.K. McDaniel, and B.V. Conger (1997). Transgenic orchardgrass *(Dactylis glomerata)* plants by direct embryogenesis from microprojectile bombarded leaf cells. *Plant Cell Reports* 16:813-819.

Dong, J., W.G. Teng, and T.C. Hall (1996). *Agrobacterium*-mediated transformation of *javanica* rice. *Molecular Breeding* 2:267-276.

Duble, R.L. (1996). *Turfgrasses: Their Management and Use in the Southern Zone*. College Station, TX: Texas A&M University Press.

Enriquez-Obregon, G.A., R.I. Vazquez-Padron, D.L. Prieto-Sanmsonov, G.A. de la Riva, and G. Selman-Housein (1998). Herbicide-resistant sugarcane (*Saccharum officinarum* L.) plants by *Agrobacterium*-mediated transformation. *Planta* 206:20-27.

Fei, S. (2000). Improvement of embryogenic callus induction and shoot regeneration of buffalograss by silver nitrate. *Plant Cell Tissue and Organ Culture* 60:196-203.

Fu, D., S. Muthukrishnan, and G.H. Liang (2001). *Agrobacterium*-mediated transformation of creeping bentgrass with TLP gene. Annual Meeting Abstracts:328. Crop Science Society of America, Madison, WI.

Ha, C.D., P.G. Lemaux, and M. Cho (2001). Stable transformation of recalcitrant Kentucky bluegrass (*Poa pratensis* L.) cultivars using mature seed-derived highly regenerative tissues. *In Vitro Cellular and Developmental Biology—Plant: Journal of the Tissue Culture Association* 37:6-11.

Ha, S.B., F.S. Wu, and T.K. Throne (1992). Transgenic turf-type tall fescue (*Festuca arundinacea* Schreb.) plants regenerated from protoplasts. *Plant Cell Reports* 11:601-604.

Hansen, G. and M.S. Wright (1999). Recent advances in the transformation of plants. *Trends in Plant Science* 4:226-231.

Hart, S.E., D.W. Lycan, M. Faletti, E.K. Nelson, and G. Marquez (2002). Response of glyphosate resistant and susceptible bentgrass (*Agrostis* spp.) to postemergence herbicides. *Weed Science Society of America (WSSA) Abstracts* 42:79-80.

Hartman, C.L., L. Lee, P.R. Day, and N.E. Tumer (1994). Herbicide resistant turfgrass (*Agrostis palustris* Huds.) by biolistic transformation. *Bio/Technology* 12:919-923.

Hensgens, L.A.M., E.P.H.M. de Bakker, E.P. van Os-Ruygrok, S. Rueb, F. van de Mark, H. van der Mass, S. van der Veen, M. Kooman-Gersmann, L. Hart, and R.A. Schilperoot (1993). Transient and stable expression of gus fusion with rice genes in rice, barley and perennial ryegrass. *Plant Molecular Biology* 22:1101-1127.

Hiei, Y., T. Komari, and T. Kubo (1997). Transformation of rice mediated by *Agrobacterium tumefaciens*. *Plant Molecular Biology* 35:205-218.

Hiei, Y., S. Ohta, T. Komari, and T. Kumashiro (1994). Efficient transformation of rice (*Oryza sativa* L.) mediated by *Agrobacterium* and sequence analysis of the boundaries of the T-DNA. *Plant Journal* 6:271-282.

Horn, M.E., R.D. Shillito, B.V. Conger, and C.T. Harms (1988). Transgenic plants of orchardgrass (*Dactylis glomerata* L.) from protoplasts. *Plant Cell Reports* 7: 469-472.

Inokuma, C., K. Sugiura, N. Imaizumi, and C. Cho (1998). Transgenic Japanese lawngrass (*Zoysia japonica* Steud.) plants regenerated from protoplasts. *Plant Cell Reports* 17:334-338.

Inokuma, C., K. Sugiura, N. Imaizumi, C. Cho, and S. Kaneko (1997). Transgenic *Zoysia* (*Zoysia japonica* Steud.) plants regenerated from protoplasts. *International Turfgrass Society Research Journal* 8:297-303.

Ishida, Y., H. Saito, S. Ohta, Y. Hiei, T. Komari, and T. Kumashiro (1996). High efficiency transformation of maize (*Zea mays* L.) mediated by *Agrobacterium tumefaciens*. *Nature Biotechnology* 14:745-750.

Khanna, H.K. and S.K. Raina (1999). *Agrobacterium*-mediated transformation of indica rice cultivars using binary and superbinary vectors. *Australian Journal of Plant Physiology* 26:311-324.

Kidd, G. (1993). Why do agbiotech firms neglect turfgrasses? *Bio/Technology* 11:268.

Klein, T.M., E.D. Wolf, R. Wu, and J.C. Sanford (1987). High velocity microprojectiles for delivering nucleic acids into living cells. *Nature* 327:70-73.

Krans, J.V., V.T. Henning, and K.C. Torres (1982). Callus induction, maintenance and plantlet regeneration in creeping bentgrass. *Crop Science* 22:1193-1197.

Kuai, B., S.J. Dalton, A.J.E Bettany, and P. Morris (1999). Regeneration of fertile transgenic tall fescue plants with a stable highly expressed foreign gene. *Plant Cell Tissue and Organ Culture* 58:149-154.

Kuai, B. and P. Morris (1995). The physiological state of suspension cultured cells affects the expression of the β-glucuronidase gene following transformation of tall fescue *(Festuca arundinacea)* protoplasts. *Plant Science* 110:235-247.

Kuai, B. and P. Morris (1996). Screening for stable transformation and stability of β-glucuronidase gene expression in suspension cultured cells of tall fescue *(Festuca arundinacea)*. *Plant Cell Reports* 15:804-808.

Lakkaraju, S., L.H. Pitcher, X.L. Wang, and B.A. Zilinskas (2001). *Agrobacterium*-mediated transformation of turfgrass. In *Proceedings of the Tenth Annual Rutgers Turfgrass Symposium* (p. 20). New Brunswick, NJ: Rutgers University.

Lakkaraju, S., L.H. Pitcher, and B.A. Zilinskas (1999). Turfgrass transformation mediated by *Agrobacterium tumefaciens*. In *Proceedings of the Eighth Annual Rutgers Turfgrass Symposium* (p. 23). New Brunswick, NJ: Rutgers University.

Lee, L., C. Laramore, P.R. Day, and N.E. Tumer (1996). Transformation and regeneration of creeping bentgrass (*Agrostis palustris* Huds.) protoplasts. *Crop Science* 36:401-406.

Lee, L., C. Laramore, C.L. Hartman, L. Yang, C.R. Funk, J. Grande, J.A. Murphy, S.A. Johnston, B.A. Majek, N.E. Tumer, and P.R. Day (1997). Field evaluation of herbicide resistance in transgenic *Agrostis stolonifera* L. and inheritance in the progeny. *International Turfgrass Society Research Journal* 8:337-344.

Lowe, K.W. and B.V. Conger (1979). Root and shoot formation from callus cultures of tall fescue. *Crop Science* 19:397-400.

McDonnell, R.E. and B.V. Conger (1984). Callus induction and plantlet formation from mature embryo explants of Kentucky bluegrass. *Crop Science* 24:573-578.

Meyer, W., G. Zhang, S. Lu, S. Chen, T.A. Chen, and R. Funk (2000). Transformation of Kentucky bluegrass (*Poa pratensis* L) with betaine aldehyde dehydrogenase gene for salt and drought tolerance. Annual Meeting Abstracts:167. Crop Science Society of America, Madison, WI.

Mittler, R., V. Shulaev, and E. Lam (1995). Coordinated activation of programmed cell death and defense mechanisms in transgenic tobacco plants expressing a bacterial proton pump. *Plant Cell* 7:29-42.

Mooney, P.A., P.B. Goodwin, E.S. Dennis, and D.J. Llewellyn (1991). *Agrobacterium tumefaciens* gene transfer into wheat tissues. *Plant Cell Tissue and Organ Culture* 25:209-218.

Muller, A.J. and R. Grafe (1978). Isolation and characterization of cell lines of *Nicotiana tabacum* lacking nitrate reductase. *Molecular and General Genetics* 161:67-76.

Murashige, T. and F. Skoog (1962). A revised medium for rapid growth and bioassays with tobacco tissue cultures. *Physiologia Plantarum* 15:473-493.

Neff, M.M., S.M. Nguyen, E.J. Malancharuvil, S. Fujioka, T. Noguchi, H. Seto, M. Tsubuki, T. Honda, S. Takatsuto, S. Yoshida, and J. Chory (1999). *BAS1:* A gene regulating brassinosteriod levels and light responsiveness in *Arabidopsis. Proceedings of the National Academy of Sciences of the United States of America* 96:15316-15323.

Nielsen, K.A. and E. Knudsen (1993). Regeneration of green plants from embryogenic suspension cultures of Kentucky bluegrass (*Poa pratensis* L.). *Journal of Plant Physiology* 14:589-595.

Nielsen, K.A., E. Larsen, and E. Knudsen (1993). Regeneration of protoplast-derived green plants of Kentucky bluegrass (*Poa pratensis* L). *Plant Cell Reports* 12:537-540.

Perez-Vicente, R., X.D. Wen, Z.Y. Wang, N. Leduc, C. Sautter, E. Wehrli, I. Portrykus, and G. Spangenberg (1993). Culture of vegetative and floral meristems in ryegrass: Potential targets for microballistic transformation. *Journal of Plant Physiology* 142:610-617.

Pitcher, L.H., S. Lakkaraju, and B.A. Zilinskas (2000). Transformation of tall fescue and velvet bentgrass using *Agrobacterium tumefaciens.* In *Proceedings of the Ninth Annual Rutgers Turfgrass Symposium* (p. 39). New Brunswick, NJ: Rutgers University.

Potrykus, I. (1990). Gene transfer to cereals: An assessment. *Bio/Technology* 8:535-542.

Qu, R., Y. Bai, L. Li, D. Bradley, X. Chen, H.D. Shew, and A.N. Bruneau (2001). Turfgrass transformation. Annual Meeting Abstracts. Crop Science Society of America, Madison, WI.

Raineri, D.M., P. Bottino, M.P. Gordon, and E.W. Nester (1990). *Agrobacterium*-mediated transformation of rice (*Oryza sativa* L.). *Bio/Technology* 8:33-38.

Rajasekaran, K., M.B. Hein, G.C. Davis, and I.K. Vasil (1987). Endogenous plant growth regulators in leaves and tissue cultures of napier grass (*Pennisetum purpureum* Schum). *Journal of Plant Physiology* 130:13-25.

Rajeolina, S.R., G. Alibert, and C. Planchon (1990). Continuous plant regeneration from established embryogenic cell suspension culture of Italian ryegrass and tall fescue. *Journal of Plant Breeding* 104:265-271.

Richard, H.A., V.A. Rudas, H. Sun, J.K. McDaniel, Z. Tomaszewski, and B.V. Conger (2001). Construction of a GFP-BAR plasmid and its use for switchgrass transformation. *Plant Cell Reports* 20:48-54.

Sanford, J.C., T.M. Klein, E.D. Wolf, and N. Allen (1987). Delivery of substances into cells and tissue using a particle bombardment process. *Journal of Particle Science and Technology* 5:27-37.

Shetty, K. and A. Yoshito (1991). The influence of organic nitrogen source on the induction of embryogenic callus in *Agrostis alba* L. *Journal of Plant Physiology* 149:82-85.

Shillito, R. (1999). Methods of genetic transformation: Electroporation and polyethylene glycol treatment. In *Molecular Improvement of Cereal Crops* (pp. 9-20), Ed. Vasil, I.K.). Dordrecht, the Netherlands: Kluwer Academic Publishers.

Smith, R.H. and E.E. Hood (1995). *Agrobacterium tumefaciens* transformation of moncots. *Crop Science* 35:301-309.

Snow, J.T. and M.P. Kenna (1999). Environmental issues in turf management. In *International Turf Management* (pp. 327-343), Ed. Aldous, D.E. Boca Raton, FL: CRC Press.

Songstad, D.D., D.A. Somers, and R.J. Griesbach (1995). Advances in alternative DNA delivery techniques. *Plant Cell Tissue and Organ Culture* 40:1-15.

Spangenberg, G., Z.Y. Wang, J. Nagel, and I. Potrykus (1994). Protoplast culture and generation of transgenic plants in red fescue (*Festuca rubra* L). *Plant Science* 97:83-94.

Spangenberg, G., Z.-Y. Wang, and I. Potrykus (1998). *Biotechnology in Forage and Turfgrass Improvement*. Berlin: Springer-Verlag.

Spangenberg, G., Z.-Y. Wang, X.L. Wu, J. Nagel, and I. Potrykus (1995a). Transgenic perennial ryegrass *(Lolium perenne)* plants from microprojectile bombardment of embryogenic suspension cells. *Plant Science* 108:209-217.

Spangenberg, G., Z.-Y. Wang, X.L. Wu, J. Nagel, and I. Potrykus (1995b). Transgenic tall fescue *(Festuca arundinacea)* and red fescue *(Festuca rubra L.)* plants from microprojectile bombardment of embryogenic suspension cells. *Journal of Plant Physiology* 145:693-701.

Sticklen, M.B. (2001). Genetic engineering: An ultimate solution to the turfgrass problems. *Michigan State University Turfgrass Team Reports,* pp. 40-43.

Takamizo, T., K. Suginobu, and R. Ohsugi (1990). Plant regeneration from suspension culture derived protoplasts of tall fescue of a single genotype. *Plant Science* 72:125-131.

Tamborini, E., A. Brandazza, C. de Lalla, G. Musco, A.G. Siccardi, P. Arosio, and A. Sidoli (1995). Recombinant allergen *Lol p II:* Expression, purification and characterization. *Molecular Immunology* 32:505-513.

Tarakawa, T., T. Sato, and M. Koike (1992). Plant regeneration from protoplasts isolated from embryogenic suspension cultures of creeping bentgrass (*Agrostis palustris* Huds). *Plant Cell Reports* 11:457-461.

Taylor, M.G. and I.K. Vasil (1985). Ultrastructural characterization of embryogenic callus formation in cultured embryos of *Pennisetum americanum. American Journal of Botany* 72:833-834.

Tingay, S., D. McElroy, R. Kalla, S. Fieg, M. Wang, S. Thorton, and R. Brettell (1997). *Agrobacterium*-mediated barley transformation. *Plant Journal* 11:1369-1376.

Torello, W.A., A.J. Symington, and R. Rufner (1984). Callus induction, initiation, plant regeneration and evidence of somatic embryogenesis in red fescue. *Crop Science* 24:1037-1040.

van der Mass, H.M., E.R. de Jong, S. Rueb, L.A.M. Hengnes, and F.A. Krens (1994). Stable transformation and long-term expression of the gusA reporter gene in callus lines of perennial ryegrass (*Lolium perenne* L.). *Plant Molecular Biology* 24:401-405.

van der Valk, P., M.A.C.M. Zaal, and J. Creemers-Molenaar (1989). Somatic embryogenesis and plant regeneration in inflorescence and seed derived callus of *Poa pratensis* L (Kentucky bluegrass). *Plant Cell Reports* 7:644-647.

Vasil, I.K. (1987). Developing cell and tissue culture systems for the improvement of cereal and grass crops. *Journal of Plant Physiology* 128:193-218.

Vasil, I.K. (1988). Progress in the regeneration and genetic manipulation of cereal crops. *Bio/Technology* 6:397-402.

Vasil, I.K. (1995). Cellular and molecular genetic improvement of cereals. In *Current Issues in Plant Molecular and Cellular Biology* (pp. 5-18), Eds. Terzi, M., A. Cella, and A. Falaugina). Dordrecht, the Netherlands: Kluwer Academic Publishers.

Wang, G.R., H. Binding, and U.K. Posselt (1997). Fertile transgenic plants from direct gene transfer to protoplasts of *Lolium perenne* L. and *Lolium multiflorum* Lam. *Journal of Plant Physiology* 151:83-90.

Wang, X., P. Zafian, M. Choudhary, and M. Lawton (1996). The PR5K receptor protein kinase from *Aradidopsis thaliana* is structurally related to a family of plant defense proteins. *Proceedings of the National Academy of Sciences of the United States of America* 93:2598-2602.

Wang, Z.Y., G. Legris, J. Nagel, I. Potrykus, and G. Spangenberg (1994). Cryopreservation of embryogenic cell suspensions in *Festuca* and *Lolium* species. *Plant Science* 103:93-106.

Wang, Z.Y., G. Legris, M.P. Valles, I. Potrykus, and G. Spangenberg (1995). Plant regeneration from suspension and protoplast cultures in the temperate grasses *Festuca* and *Lolium*. In *Current Issues in Plant Molecular and Cellular Biology* (pp. 81-86), Eds. Terzi, M., A. Cella, and A. Falaugina. Dordrecht, the Netherlands: Kluwer Academic Publishers.

Wang, Z.Y., J. Nagel, I. Potrykus, and G. Spangenberg (1993). Plants from cell suspension-derived protoplasts in *Lolium* species. *Plant Science* 94:179-193.

Wang, Z.Y., T. Takamizo, V. Iglesias, M. Osusky, J. Nagel, I. Potrykus, and G. Spangenberg (1992). Transgenic plants of tall fescue (*Festuca arunidinacea* Schreb) obtained by direct gene transfer to protoplasts. *Bio/Technology* 10:691-696.

Wang, Z.Y., M.P. Valles, P. Montavon, I. Potrykus, and G. Spangenberg (1993). Fertile plant regeneration from protoplasts of meadow fescue (*Festuca pratensis* Huds). *Plant Cell Reports* 12:95-100.

Wang, Z.Y., X.D. Ye, J. Nagel, I. Potrykus, and G. Spangenberg (2001). Expression of a sulfur-rich sunflower albumin gene in transgenic tall fescue (*Festuca arundinacea* Schreb.) plants. *Plant Cell Reports* 20:213-219.

Warkentin, D., B. Chai, R.K. Hajela, H. Zhong, and M.B. Sticklen (1997). Development of transgenic creeping bentgrass (*Agrostis palustris* Huds.) for fungal disease resistance. In *Turfgrass Biotechnology: Cell and Molecular Genetic Approaches to Turfgrass Improvement* (pp. 153-161), Eds. Sticklen, M.B. and M.P. Kenna. Chelsea, MI: Ann Arbor Press.

Wipff, J. and C.R. Fricker (2000). Determining gene flow of transgenic creeping bentgrass and gene transfer to other bentgrass species. *Diversity* 16:36-39.

Wu, G., B.J. Shortt, E.B. Lawrence, E.B. Levine, K.C. Fitzsimmons, and D.M. Shah (1995). Disease resistance conferred by expression of a gene encoding H_2O_2-generating glucose oxidase in transgenic potato plants. *Plant Cell* 7:1357-1368.

Xiao, H. and S.B. Ha (1997). Efficient selection and regeneration of creeping bentgrass transformation following particle bombardment. *Plant Cell Reports* 16:874-878.

Xu, Y., L. Li, K. Wu, A.J.M. Peeters, D.A. Gage, and A.D. Zeevaart (1995). The GA5 locus of *Arabidopsis thaliana* encodes a multifunctional gibberellin 20-oxidase: Molecular cloning and functional expression. *Proceedings of the National Academy of Sciences of the United States of America* 92:6640-6644.

Ye, X., Z.Y. Wang, X. Wu, and I. Potrykus (1997). Transgenic Italian ryegrass *(Lolium multiflorum)* plants from microprojectile bombardment of embryogenic suspension cells. *Plant Cell Reports* 16:379-384.

Ye, X., X. Wu, H. Zhao, M. Frehner, J. Nosberger, I. Potrykus, and G. Spangenberg (2001). Altered fructan accumulation in transgenic *Lolium multiflorum* plants expressing a *Bacillus subtilis sacB* gene. *Plant Cell Reports* 20:205-212.

Yu, T.T., B.R. Huang, D.Z. Skinner, S. Muthukrishnan, and G.H. Liang (2000). Genetic transformation of creeping bentgrass. Annual Meeting Abstracts:154. Crop Science Society of America, Madison, WI.

Yu, T.T., D.Z. Skinner, G.H. Liang, H.N. Trick, B. Huang, and S. Muthukrishnan (2000). *Agrobacterium*-mediated transformation of creeping bentgrass using GFP as a reporter gene. *Hereditas* 133:229-233.

Zaghmout, O.M.F. and W.A. Torello (1989). Somatic embryogenesis and plant regeneration from suspension cultures of red fescue. *Crop Science* 29:815-817.

Zhang, G.-Y., S. Lu, S.-Y. Chen, T.A. Chen, W.A. Meyer, and C.R. Funk (2001). Transformation of triploid bermudagrass (*Cynodon dactylon* × *C. transvaalensis,* cv. TifEagele) with BADH gene for drought tolerance. In *Proceedings of the Tenth Annual Rutgers Turfgrass Symposium* (p. 21). New Brunswick, NJ: Rutgers University.

Zhao, Z.-Y., T. Cai, L. Tagliani, M. Miller, N. Wang, H. Pang, M. Rudert, S. Schroeder, D. Hondred, J. Seltzer, and D. Pierre (2000). *Agrobacterium*-mediated sorghum transformation. *Plant Molecular Biology* 44:789-798.

Zhong, H., M.G. Bolyard, C. Srinivasan, and M.B. Sticklen (1993). Transgenic plants of turfgrass (*Agrostis palustris* Hud.) from microinjectile bombardment of embryogenic callus. *Plant Cell Reports* 13:1-6.

Zhong, H., C. Srinivasan, and M.B. Sticklen (1991). Plant regeneration via somatic embryogenesis in creeping bentgrass (*Agrostis palustris* Huds). *Plant Cell Reports* 10:453-456.

Zhou, H., J.W. Arrowsmith, M.E. Fromm, C.M. Hironaka, M.L. Taylor, D. Rodrigues, M.E. Pajeau, S.M. Brown, C.G. Santino, and J.E. Fry (1995). Glyphosphate-tolerant CP4 and GOX genes as a selectable marker in wheat transformation. *Plant Cell Reports* 15:159-163.

Zoubenko, O., F. Uckun, Y. Hur, I. Chet, and N.E. Tumer (2003). Plant resistance to fungal infection induced by nontoxic pokewood antiviral protein mutants. *Nature Biotechnology* 15:992-996.

Chapter 14

Risks Associated with Genetically Engineered Crops

Paul St. Amand

INTRODUCTION

When the potential for genetically engineered plants (GEPs) to become a part of world agriculture developed into a reality, much concern was raised over the safety of such a practice. The U.S. Department of Agriculture (USDA) responded in 1990 by formally establishing the Biotechnology Risk Assessment Research Grants Program (BRARGP). The BRARGP was established to assist federal regulatory agencies in making science-based decisions on the safety of introducing any genetically modified organism into the environment. Competitive grants from the BRARGP were first awarded in 1992, and the program continues today with annual cycles of funding and review. Information generated as part of this program is freely available to the public in the form of popular and scientific journal articles and in the published proceedings of the annual meeting of the BRARGP research program review. These research results figure prominently in the decision-making process that has gone into the approval of GEPs for large-scale production.

In addition to the competitive grants program, in 2002 the Agricultural Research Service of the USDA initiated several long-term research projects designed to document the potential risks, as well as the benefits, of genetically engineered (GE) crops. More than ten new, multiyear projects have been initiated throughout the United States with diverse objectives, including developing ways to prevent the spread of engineered genes; confining the expression of engineered genes to specific, nonedible tissues; and assessing the effects of transgenics on nontarget species. These projects stress comparisons to conventional production systems and seek to obtain data that will facilitate realistic comparisons of the advantages and disadvantages of conventional agricultural systems compared to systems incorporating genetically modified (GM) crops. The data will be made available for public scrutiny and will provide a more complete foundation for science-based regulation of transgenic crops.

Genetically engineered plants have been field grown in more than ten countries since 1987 (James, 2001), when the USDA Animal and Plant Health Inspection Service (APHIS) first began regulating field testing of GEPs. Experience gained over the past 17 years, across 79 GEP species, and in many environments has not ended discussion of the risks associated with GEPs. Many of these risks are real and can be anticipated from the nature of the inserted genes or from our knowledge of the genetically engineered species. Other risks have been difficult to define or predict. As with the adoption and use of any new technology, unanticipated outcomes will also carry unknown levels of risk. The lack of knowledge of practical transgenic methods among the general population, coupled with fears of the unknown, is likely one of the greatest reasons that some groups are opposed to any form of genetic engineering. We do know, however, that genetic engineering creates DNA that is fundamentally the same as that found in nature (National Research Council, 2002). Two recent publications provide an assessment of the current status of risk assessment associated with transgenic plants (Letourneau, 2002; Kjellsson and Strandberg, 2001).

RISKS VERSUS BENEFITS

Risks can be defined as a set of mathematical models and probabilities of outcomes (National Research Council, 1983). However, no formal risk analysis will be presented here, nor will a complete risk versus benefit analysis be derived. Here we discuss risk simply as plausible or possible outcomes based on the knowledge at hand. In many cases, however, incomplete knowledge of the factors involved leads to an incomplete knowledge of the possible risks. A complete understanding of risk must also include the benefits that accompany those risks. Often, when one considers the potential for GEPs to represent a hazard, examples of plants that have been introduced with negative unintended results are described. Although several introduced plants have caused negative environmental impacts, such as water hyacinth, kudzu, and purple loosestrife in the United States, numerous other introduced plants have made tremendous positive impacts. For example, maize, wheat, and soybeans are not native to the United States, yet they form the foundation of U.S. agriculture today. Based on this experience, a newly introduced GEP should not be considered strictly as a potential hazard, but also as something having potentially huge benefits. In the strictest sense, any genetic change occurring via traditional breeding or by genetic engineering will have some measurable impact on the environment. The most important criterion for evaluating new GEPs should be, Will the benefits to society and the environment outweigh the risks?

LEGAL RISKS

Many possible legal issues surround GEPs and have already started to affect society on personal, regional, national, and worldwide levels. A complete discussion of the legal risks surrounding the use of GEPs is beyond the scope of this chapter. However, we now know that pollen dispersal distances can be several kilometers (Rieger et al., 2002; St. Amand et al., 2000) and small-scale studies carried out to predict pollen movement underestimate the possible distances (St. Amand et al., 2000). Hence, the risk of contamination of nontransgenic crops with transgenic pollen can be very great. It is apparent from recent legal rulings that growers of non-GEPs who do not report genetic contamination to the company owning the patent on a gene may be sued for patent infringement even if they gain no economic benefit from the unwanted gene flow into their fields (*Monsanto Canada Inc. vs. Percy Schmeiser,* 2001). Uncontrolled gene flow via pollen or seeds raises many legal issues. It is possible that growers may lose the right to save the seeds from their own landraces or varieties that they have developed, possibly after many years of selection, if those populations are contaminated with patented genes. Should growers be obliged to adjust their crop rotation, herbicide schedules, and field layout in order to protect their crops from contamination from neighboring GEPs? Should growers of non-GEPs bear the cost of dealing with gene flow that is unwanted and arguably forced upon them? Will companies or nearby growers reimburse non-GEP growers if gene flow forces the non-GEP grower to use a more costly herbicide or less efficient crop rotation plan? Contamination of fields with engineered genes may prevent products from being marketed as "GMO free" or "organic." Who will reimburse losses of growers targeting these premium-price markets? If a consumer or company claims to be injured by the use, production, or transportation of GEPs, can growers be held liable?

It is possible that the widespread use of transgenic crop varieties will lead to the loss of markets for farmers producing the transgenic products, regardless of the legality of those crops. For example, widespread public disapproval of transgenic crops in Europe essentially removes the European market as an outlet for transgenic crops grown in the United States, even though no legal bans have been in place in the United States. In this situation, a farmer attempting to grow nontransgenic crops may lose access to a market if his or her crop were to become contaminated with transgenic pollen, as happened with the *Monsanto vs. Schmeiser* case.

These and many other issues clearly constitute a set of new risks that consumers, growers, and companies must deal with according to local and national laws, many of which may not adequately address the unique circumstances surrounding the use of GEPs.

RISKS TO HUMANS

Perhaps the greatest concern over the release of transgenic plants into the environment is the potential for new sources of toxins or allergens in the food supply. Nordlee and colleagues (1996) found that transgenic soybeans expressing a storage protein from Brazil nuts *(Bertholletia excelsa)* had the same allergenic properties as the Brazil nut, demonstrating that an allergenic factor from one plant species can be transferred into another by genetic engineering. It has long been known that allergenic properties of related plant species tend to be similar (e.g., Martin et al., 1985), suggesting that sources of genes for plant improvement may be guided by knowledge of related species to ameliorate the possibility of creating a new allergenic plant type. Lehrer and Reese (1997, p. 122) suggested that "[t]he safety evaluation of transgenic foods is relatively easy when the allergenicity of the gene source is known."

Nonetheless, a great deal of concern has been expressed by the public over the presence of transgenic plant materials in food products. The finding of any transgenic plant material not approved for human consumption in food products will cause a great public outcry. By far, the most significant such incident involved the now-famous 'StarLink' corn contamination of food products. The 'StarLink' contamination resulted in the recall of more than 300 food products, the temporary closing of food manufacturing facilities, with concomitant job losses, and several claims of ill effects from consumers of products possibly contaminated with 'StarLink.' 'StarLink' had not been approved for human consumption because is had been engineered to express the Cry9c protein from a strain of *Bacillus thurengiensis* (Bt); Cry9c had been found to survive simulated human digestion much longer than other Bt-derived proteins that had been approved for human consumption. The U.S. Food and Drug Administration (FDA) commissioned a study by the Centers for Disease Control and Prevention (CDC) to determine the extent of involvement of the Cry9c protein in the manifestation of the reported ill effects. The final report from that study (Centers for Disease Control and Prevention, 2001, p. 3) concluded, "These findings do not provide any evidence that the reactions that the affected people experienced were associated with hypersensitivity to the Cry9c protein." Thus, there has not yet been a confirmed case of a person suffering any ill effects from consuming food derived from genetically modified plants.

Regardless of the lack of any tangible evidence of ill effects from the use of GM plants in foods, food and seed recalls continue to occur because of possible unintended inclusion of GM materials. In 2001, Reuters News Service (Hur, 2001) reported that in Japan, potato products were recalled due to possible inclusion of GM potatoes. The variety of potato in question, 'NewLeaf Plus', engineered to resist leafroll virus and Colorado potato beetle, had been approved for human consumption in the United States, but not in Japan. Seed of GM canola was recalled in 1997 from farmers in Canada because of contamination with transgenes other than the intended gene (Leite, 1997).

Other recalls can be expected for the next several years. However, the use of genetically modified plants in the production of foods continues to grow. Writing in the U.S. FDA publication, *FDA Consumer Magazine,* Raymond Formanek Jr. stated that "[a]n estimated two-thirds of the processed foods in U.S. supermarkets contain genetically engineered corn, soybeans or other crops" (Formanek, 2001, p. 11). With this kind of impact on the food industry, it is likely that GM plant products will be part of food production from now on. Accurate and thorough assessments of all risks associated with GM plants are vital to the continued diversification of GM plant technology, and many risk-assessment objectives and approaches are currently in use.

Because crops may be easily transformed and produce large amounts of proteins, oils, starches, or other compounds, they make excellent biofactories. Because of this, GEPs are currently used to produce nonfood products such as avidin in corn (Hood et al., 1997). It also is likely that GEP crops will soon produce enzymes, antigens, pharmaceuticals, and other nonfood products (Owen and Pen, 1996; Mor et al., 1998; Ohya, 1998; Perrin et al., 2000). Crop plants could even be engineered to reduce toxic pollutants, such as mercury or cadmium, in contaminated soils via phytoremediation (Heaton et al., 1998; Zhu et al., 1999). The use of such GEPs could have tremendous environmental benefits. However, as already demonstrated by the StarLink incident, complete segregation of nonedible and potentially harmful GE crop products from the food stream may be difficult or impossible, even with strict regulatory oversight. Also, gene flow via pollen from fertile nonfood GE crops will likely be impossible to contain completely (St. Amand et al., 2000; Ellstrand, 2001; Rieger et al., 2002). Extreme caution must be practiced if nonfood transgenes are engineered into crops.

Most current methods of plant transformation insert transgenic DNA into the recipient genome at random points, although recent advances in gene targeting in flowering plants have been reported (Hanin et al., 2001). The actual number of insertions cannot be controlled. These factors may cause insertional mutagenesis, which commonly occurs and is a useful tool in functional genomics research (Pereira, 2000), but it may also lead to unintended phenotypes in plants designed for commercial production. Physical rearrangement of the DNA around the transgene has been reported (Windels et al., 2001), but no instances of altered regulation of a gene in commercially produced transgenic crops have been reported.

There also is concern that selectable marker genes, such as antibiotic resistance genes commonly used as selection factors in the transformation process, may be expressed in the commercial crop to unnecessary levels. It recently was reported that an antibiotic resistance transgene in tobacco was transferred to and expressed by bacteria inhabiting the plant (Kay et al., 2002). Antibiotic-resistant bacteria developing on transgenic crops is not desirable, hence, removal of the selectable marker before commercialization should be encouraged. Various methods of removing the selectable marker gene from transgenic plants have been developed; the recent refinement of the cre-lox system (Corneille et al., 2001) is an example. Hare and Chua (2002) and Ozcan and colleagues (1993) have discussed the issue of removing the selectable marker and its possible impact on the public acceptance of transgenic crops.

RISKS TO THE ENVIRONMENT

Considerable discussion has ensued concerning the impact transgenic crops may have on surrounding plant and animal populations. Perhaps the two most widely discussed cases to date are those of possible contamination of wild populations of maize in Mexico with transgenic pollen, and the possible killing of monarch butterfly larvae by maize pollen expressing Bt toxin incidentally deposited on milkweed *(Asclepias syriaca)* plants. Quist and Chapela (2001) indicated they discovered evidence of transgenes in landraces of maize growing in remote regions of Mexico. These transgene fragments presumably had been transferred large distances from commercial plantings of GEP maize. The conclusions reached by Quist and Chapela now have been strongly criticized (Kaplinsky et al., 2002; Metz and Futterer, 2002; Mann, 2002).

Losey and colleagues (1999) and Jesse and Obrycki (2000) reported evidence that Bt maize pollen could cause significant death of monarch butterfly larvae feeding on milkweed plants in and around transgenic maize fields. This conclusion has been challenged (Gatehouse et al., 2002), which led to a spate of investigations and reports

(Hellmich et al., 2001; Oberhauser et al., 2001; Pleasants et al., 2001; Sears et al., 2000, 2001; Shelton and Shears, 2001; Stanley-Horn et al., 2001; Tschenn et al., 2001; Wraight et al., 2000; Zangerl et al., 2001). The general consensus of these reports indicates that the impact of Bt maize on the monarch butterfly is negligible. Nonetheless, the potential for the kind of impact reported by Jessel and Obrycki (2000) cannot be ignored and should be investigated for any new transgenic crops prior to their release.

Plant breeders and geneticists have been very successful at improving crop resistance to diseases and pests by incorporating naturally occurring resistance genes into newer varieties. These genes are typically found in the center of diversity for the crop in question. Centers of diversity are the geographic region in which the crop species exhibits maximum genetic diversity and are therefore extremely valuable. Many centers of diversity are at risk due to habitat destruction and possibly the reduction of genetic diversity caused by gene flow from cultivated varieties. Hybridization between cultivated rice *(Oryza sativa)* and Taiwanese wild rice (*Oryza rufipogon* ssp. *formosana*) resulted in the progressive loss of wild rice traits in native populations in Taiwan (Kiang et al., 1979). In less than 80 years, gene flow from cultivated rice, coupled with environmental factors, has driven the populations of Taiwanese wild rice to near extinction. This rate of introgression would be considered rapid on any time scale. It also has been shown that the genetic diversity decreased measurably over a ten-year period in one particular wild rice population (Akimoto et al., 1999) and that an allele common to cultivated rice was present in the wild rice population.

The introduction of even a single gene can cause large changes in a species. The colonizing ability of rose clover *(Trifolium hirtum)* was tremendously increased by a single allelic replacement (Martins and Jain, 1979). Also, the weediness of johnsongrass *(Sorghum halepense)* apparently is controlled by only a few genes (Paterson et al., 1995). Given the proper conditions, gene flow from GEPs, even single genes, could reduce genetic diversity in centers of diversity. However, gene flow issues are not unique to GEPs; domesticated crops have contributed to gene flow into centers since domestication. What may be unique to GEPs is the high rate of adoption of useful transgenes causing heavy and constant gene flow. If transgenes providing broad, horizontal resistance to fungi, bacteria, and pests are stacked and placed into one variety, then the fitness of the transgenes may be quite high. Increased fitness coupled with wide use and habitat erosion would likely cause population bottlenecking and decreased genetic diversity in crop centers. Current GE varieties are no more likely to reduce genetic diversity than are current non-GE varieties, but the future potential for such GE crop-mediated reduction is high. Testing new transgenes for fitness should be a priority before they are released, and additional care should be exercised for crops grown near their own center of diversity.

THE UNEXPECTED

There may be unexpected benefits to growing transgenic crop plants. The Green Revolution was made possible largely by the introduction of a very few genes (Hanson et al., 1982; Dalrymple, 1986). Even though Green Revolution varieties greatly increased crop yields in much of the developing word, these varieties also required intensive fertilizer and pesticide inputs. In the 1980s, many environmentalists began to oppose the use of high amounts of inputs because of environmental pollution. However, those environmentalists overlooked a major environmental benefit of the Green Revolution, a tremendous worldwide drop in deforestation. Norman Borlaug explains that "[w]ithout high-yield agriculture, either millions would have starved or increases in food output would have been realized through drastic expansion of acres under cultivation—losses of pristine land a hundred times greater than all losses to urban and suburban expansion" (Easterbrook, 1997, p. 78). It has been estimated that in India alone, natural wilderness equal to the area of California would have been lost to cultivation if not for the Green Revolution (Easterbrook, 1997). One or a few transgenes of the future may have similar, unexpected benefits.

REFERENCES

Akimoto, M., Y. Shimamoto, and H. Morishima (1999). The extinction of genetic resources of Asian wild rice, *Oryza rufipogon* Griff.: A case study in Thailand. *Genetic Resources and Crop Evolution* 46:419-425.

Centers for Disease Control and Prevention (2001). *Investigation of Human Health Effects Associated with Potential Exposure to Genetically Modified Corn.* A Report to the U.S. Food and Drug Administration from the Centers for Disease Control and Prevention. Available from the Centers for Disease Control and Prevention <http://www.cdc.gov>.

Corneille, S., K. Lutz, Z. Svab, and P. Maliga (2001). Efficient elimination of selectable marker genes from the plastid genome by the CRE-lox site-specific recombination system. *The Plant Journal* 27:171-178.

Dalrymple, D.G. (1986). *Development and Spread of High-Yielding Rice Varieties in Developing Countries.* Washington, DC: Agency for International Development.

Easterbrook, G. (1997). Forgotten benefactor of humanity. *The Atlantic Monthly* 279(January):74-82.

Ellstrand, N.C. (2001). When transgenes wander, should we worry? *Plant Physiology* 125:1543-1545.

Formanek, R. Jr. (2001). Proposed rules issued for bioengineered foods. *FDA Consumer Magazine* 35 (March-April):9-11.

Gatehouse, A.M., N. Ferry, R.J. Raemaekers (2002). The case of the monarch butterfly: A verdict is returned. *Trends in Genetics* 18:249-251.

Hanin, M., S. Volrath, A. Bogucki, M. Briker, E. Ward, and J. Paszkowski (2001). Gene targeting in *Arabidopsis. The Plant Journal* 28:671-677.

Hanson, H., N.E. Borlaug, and R.G. Anderson (1982). *Wheat in the Third World.* Boulder, CO: Westview Press.

Hare, P.D. and N.H. Chua (2002). Excision of selectable marker genes from transgenic plants. *Nature Biotechnology* 20:575-580.

Heaton, A.C.P., L.C. Rugh, N.J. Wang, and R.B. Meagher (1998). Phytoremediation of mercury- and methylmercury-polluted soils using genetically engineered plants. *Journal of Soil Contamination* 7:497-509.

Hellmich, R.L., B.D. Siegfried, M.K. Sears, D.E. Stanley-Horn, M.J. Daniels, H.R. Mattila, T. Spencer, K.G. Bidne, and L.C. Lewis (2001). Monarch larvae sensitivity to *Bacillus thuringiensis*-purified proteins and pollen. *Proceedings of the National Academy of Sciences of the United States of America* 98(21):11925-11930.

Hood, E.E., D.R. Witcher, S. Maddock, I. Meyer, C. Baszczynski, M. Bailey, P. Flynn, J. Register, L. Marshall, D. Bond, et al. (1997). Commercial production of avidin from transgenic maize: Characterization of transformant, production, processing, extraction, and purification. *Molecular Breeding* 3:291-306.

Hur, Jae (2001). Japan food recall revives StarLink bitech scare. Reuters, May 25.

James, C.A. (2001). *Global Review of Commercialized Transgenic Crops.* ISAAA Briefs No. 24. Ithaca, NY: International Service for the Acquisition of Agribiotech Applications (ISAAA).

Jesse, L.C.H. and J.J. Obrycki (2000). Field deposition of Bt transgenic corn pollen: Lethal effects on the monarch butterfly. *Oecologia* 125(2):241-248.

Kaplinsky, N., D. Braun, D. Lisch, A. Hay, S. Hake, and M. Freeling (2002). Biodiversity: Maize transgene results in Mexico are artefacts. *Nature* 416:601-602.

Kay, E., T.M. Vogel, F. Bertolla, R. Nalin, and P. Simonet (2002). In situ transfer of antibiotic resistance genes from transgenic (transplastomic) tobacco plants to bacteria. *Applied and Environmental Microbiology* 68:3345-3351.

Kiang, Y.T., J. Antonovics, and L. Wu (1979). The extinction of wild rice *(Oryza perennis-formosana)* in Taiwan. *Journal of Asian Ecology* 1:1-9.

Kjellsson, G. and M. Strandberg (2001). *Monitoring and Surveillance of Genetically Modified Higher Plants: Guidelines for Procedures and Analysis of Environmental Effects.* Boston, MA: Birkhäuser-Verlag.

Lehrer, S.B. and G. Reese (1997). Recombinant proteins in newly developed foods: Identification of allergenic activity. *International Archives of Allergy and Immunology* 113:122-124.

Leite, G.T. (1997). Monsanto statement regarding Limagrains recall of Roundup Ready canola seed varieties LG3315 and LG3295. Monsanto publication.

Letourneau, D.K. (Ed.) (2002). *Genetically Engineered Organisms: Assessing Environmental and Human Health Effects.* Boca Raton, FL: CRC Press.

Losey, J.E., L.S. Rayor, and M.E. Carter (1999). Transgenic pollen harms monarch larvae. *Nature* 399:214.

Mann, C.C. (2002). Mexican maize: Transgene data deemed unconvincing. *Science* 296:236-237.

Martin, B.G., L.E. Mansfield, and H.S. Nelson (1985). Cross-allergenicity among the grasses. *Annals of Allergy* 54:99-104.

Martins, P.S. and S.K. Jain (1979). Role of genetic variation in the colonizing ability of rose clover (*Trifolium hirtum* All.). *The American Naturalist* 114:591-595.

Metz, M. and J. Futterer (2002). Biodiversity: Suspect evidence of transgenic contamination. *Nature* 416:600-601.

Monsanto Canada Inc. vs. Percy Schmeiser (2001). <http://decisions.fct-cf.gc.ca/fct/2001/2001fct256.html>.

Mor, T.S., M.A. Gomes-Lim, and K.E. Palmer (1998). Perspective: Edible vaccines—A concept coming of age. *Trends in Microbiology* 6:449.

National Research Council (1983). *Risk Assessment in the Federal Government: Managing the Process.* Washington, DC: National Academy Press.

National Research Council (2002). *Environmental Effects of Transgenic Plants: The Scope and Adequacy of Regulation.* Washington, DC: National Academy Press.

Nordlee, J.A., S.L. Taylor, J.A. Townsend, L.A. Thomas, and R.K. Bush (1996). Identification of a Brazil-nut allergen in transgenic soybeans. *New England Journal of Medicine* 334:688-692.

Oberhauser, K.S., M.D. Prysby, H.R. Mattila, D.E. Stanley-Horn, M.K. Sears, G. Dively, E. Olson, J.M. Pleasants, W.K.F. Lam, and R.L. Hellmich (2001). Temporal and spatial overlap between monarch larvae and corn pollen. *Proceedings of the National Academy of Sciences of the United States of America* 98:11913-11918.

Ohya, K. (1998). Expression of mammalian genes encoding biologically active proteins or peptides with potential pharmaceutical applications in transgenic plants. *Japanese Journal of Veterinary Research* 46:136-137.

Owen, M.R.L. and J. Pen (1996). *Transgenic Plants: A Production System for Industrial and Pharmaceutical Proteins.* New York: J. Wiley.

Ozcan, S., S. Firek, and J. Draper (1993). Can elimination of the protein products of selectable marker genes in transgenic plants allay public anxieties? *Trends in Biotechnology* 11:219.

Paterson, A.H., K.F. Schertz, Y.R. Lin, S.C. Liu, and Y.L. Chang (1995). The weediness of wild plants: Molecular analysis of genes influencing dispersal and persistence of johnsongrass, *Sorghum halepense* (L.) Pers. *Proceedings of the National Academy of Sciences of the United States of America* 92:6127-6131.

Pereira, A. (2000). A transgenic perspective on plant functional genomics. *Transgenic Research* 9:245-260.

Perrin, Y., C. Vaquero, I. Gerrard, M. Sack, J. Drossard, E. Stoger, P. Christou, and R. Fischer (2000). Transgenic pea seeds as bioreactors for the production of a single-chain Fv fragment (scFV) antibody used in cancer diagnosis and therapy. *Molecular Breeding* 6:345-352.

Pleasants, J.M., R.L. Hellmich, G.P. Dively, M.K. Sears, D.E. Stanley-Horn, H.R. Mattila, J.E. Foster, P. Clark, and G.D. Jones (2001). Corn pollen deposition on milkweeds in and near cornfields. *Proceedings of the National Academy of Sciences of the United States of America* 98:11919-11924.

Quist, D. and I.H. Chapela (2001).Transgenic DNA introgressed into traditional maize landraces in Oaxaca, Mexico. *Nature* 414:541-543.

Reuters (2001). Japan food recall revives StarLink biotech scare. May 25.

Rieger, M.A., M. Lamond, C. Preston, S.B. Powles, and R.T. Roush (2002). Pollen-mediated movement of herbicide resistance between commercial canola fields. *Science* 296:2386-2388.

Sears, M.K., R.L. Hellmich, D.E. Stanley-Horn, K.S. Oberhauser, J.M. Pleasants, H.R. Mattila, B.D. Siegfried, and G.P. Dively (2001). Impact of Bt corn pollen on monarch butterfly populations: A risk assessment. *Proceedings of the National Academy of Sciences of the United States of America* 98:11937-11942.

Sears, M.K., D.E. Stanley-Horn, and H.R. Matilla (2000). *Preliminary Report on the Ecological Impact of BT Corn Pollen on the Monarch Butterfly in Ontario.* Montreal, Quebec: Canadian Food Inspection Agency Plant Health and Production Division Plant Biotechnology Office.

Shelton, A.M. and M.K. Sears (2001). The monarch butterfly controversy: Scientific interpretations of a phenomenon. [Erratum: Mar 2002, v. 29 (5), p. 679.] *The Plant Journal* 27:483-488.

St. Amand, P.C., D.Z. Skinner, and R.N. Peaden (2000). Risk of alfalfa transgene dissemination and scale-dependent effects. *Theoretical and Applied Genetics* 101:107-114.

Stanley-Horn, D.E., G.P. Dively, R.L. Hellmich, H.R. Mattila, M.K. Sears, R. Rose, L.C.H. Jesse, J.E. Losey, J.J. Obrycki, and L. Lewis (2001). Assessing the impact of Cry1Ab-expressing corn pollen on monarch butterfly larvae in field studies. *Proceedings of the National Academy of Sciences of the United States of America* 98:11931-11936.

Tschenn, J., J.E. Losey, L.H. Jesse, J.J. Obrycki, and R. Hufbauer (2001). Effects of corn plants and corn pollen on monarch butterfly (Lepidoptera: Danaidae) oviposition behavior. I 30:495-500.

Windels, P., I. Taverniers, E. Van Bockstaele, and M. De Loose (2001). Characterisation of the Roundup Ready soybean insert. *European Food Research and Technology* 213:107-112.

Wraight, C.L., A.R. Zangerl, M.J. Carroll, and M.R. Berenbaum (2000). Absence of toxicity of *Bacillus thuringiensis* pollen to black swallowtails under field conditions. *Proceedings of the National Academy of Sciences of the United States of America* 97:7700-7703.

Zangerl, A.R., D. McKenna, C.L. Wraight, M. Carroll, P. Ficarello, R. Warner, and M.R. Berenbaum (2001). Effects of exposure to event 176 *Bacillus thuringiensis* corn pollen on monarch and black swallowtail caterpillars under field conditions. *Proceedings of the National Academy of Sciences of the United States of America* 98:11908-11912.

Zhu, Y.L., E.A.H. Pilon-Smits, L. Jouanin, and N. Terry (1999). Overexpression of glutathione synthetase in Indian mustard enhances cadmium accumulation and tolerance. *Plant Physiology* 119:73-79.

Index